Lecture Notes in Mathematics

Edited by A. Dold and B. Eckmann

Subseries: Fondazione C.I.M.E., Firenze
Adviser: Roberto Conti

T0210846

1206

Probability and Analysis

Lectures given at the 1st 1985 Session of the Centro Internazionale Matematico Estivo (C.I.M.E.) held at Varenna (Como), Italy
May 31 – June 8, 1985

Edited by G. Letta and M. Pratelli

Springer-Verlag
Berlin Heidelberg New York London Paris Tokyo

Editors

Giorgio Letta
Maurizio Pratelli
Dipartimento di Matematica, Università di Pisa
Via Buonarroti 2, 56100 Pisa, Italy

Mathematics Subject Classification (1980): Primary: 28A15, 43A17, 46B20, 60G42, 60G46, 60J60; Secondary: 28B05, 53B21, 60G15, 47H10

ISBN 3-540-16787-0 Springer-Verlag Berlin Heidelberg New York
ISBN 0-387-16787-0 Springer-Verlag New York Berlin Heidelberg

Printing and binding: Druckhaus Beltz, Hemsbach/Bergstr.
2146/3140-543210

PREFACE

This volume collects the texts of the lectures presented at the C.I.M.E. School on PROBABILITY AND ANALYSIS, held at Varenna (Como), Italy, during the period May 31-June 8, 1985.

The purpose of this Session was not so much to offer a complete view on the subject, which is extremely wide and scattered, as to furnish some valid examples of the current areas of research interest in this direction.

We sincerely thank the four main speakers of these courses (J.M. Bismut, D.L. Burkholder, S.D. Chatterji and G. Pisier), who presented a very precise and accurate exposition of important results, which were largely obtained by themselves.

The present volume also contains the texts of two seminars by M. Sion and W. Schachermayer.

We thank the Scientific Committee of C.I.M.E., who kindly agreed to devote a Session of the C.I.M.E. Courses to this stimulating topic, as well as Professor A. Moro, Secretary of C.I.M.E., for his considerable and efficient efforts in the organization of the Session.

Pisa, May 1986.

G. Letta

M. Pratelli

TABLE OF CONTENTS

C.I.M.E. Session on Probability and Analysis

List of Participants

C. Asperti, Scuola Normale Superiore, Piazza dei Cavalieri, 56100 Pisa

J. Azéma, Laboratoire de Probabilités, Tour 56 (3e étage), 4 place Jussieu, 75230 Paris Cedex 05

P. Baldi, Dipartimento di Matematica, Università, Via Buonarroti 1, 56100 Pisa

J.-M. Bismut, Département de Mathématique, Université Paris Sud, Bâtiment 425, 91405 Orsay, France

D.L. Burkholder, Department of Mathematics, University of Illinois, Urbana, Illinois 61801

D. Candeloro, Dipartimento di Matematica, Università, Via Pascoli, 06100 Perugia

R. Carrese, Dipartimento di Matematica e Appl., Università, Via Mezzocannone 8, 80134 Napoli

E. Casini, Via Tesio 11, 20151 Milano

M. Chaleyat-Maurel, Laboratoire de Probabilités, Tour 56 (3e étage), 4 place Jussieu, 75230 Paris Cedex 05

S.D. Chatterji, Département de Mathématiques, Ecole Polytechnique Fédérale, Lausanne

F. Chersi, Istituto di Matematica, Università, Piazzale Europa 1, 34127 Trieste

M. Chiarolla, Dipartimento di Matematica, Università, Via G. Fortunato, 70125 Bari

M. Defant, Mathematisches Seminar der Universitat, Ohlshausenstr. 40-60, D-2300 Kiel 1, West Germany

U. Dinger, Department of Mathematics, Chalmers Univ. of Technology, S-41296 Goteborg

V. Esposito, Dipartimento di Matematica e Appl., Università, Via Mezzocannone 8, 80134 Napoli

F. Fagnola, Scuola Normale Superiore, Piazza dei Cavalieri, 56100 Pisa

T. Jeulin, Laboratoire de Probabilités, Tour 56, 4 place Jussieu, 75230 Paris Cedex 05

G. Kerkyacharian, Université de Nancy I, Faculté des Sciences, B.P. n.239, 54056 Vandoeuvre Les Nancy, France

F. Marchetti, Dipartimento di Matematica, Università di Roma "La Sapienza", Piazzale A. Moro 2, 00185 Roma

R. Morkvenas, Istituto di Matematica, Università, Via L.B.Alberti 4, 16132 Genova

A. Moro, Istituto Matematico Università, Viale Morgagni 67/A, 50134 Firenze

A.M. Picardello, Dipartimento di Matematica, Università di Roma "La Sapienza", Piazzale A. Moro, 2, 00185 Roma

G. Pisier, Université Paris 6, Equipe d'Analyse, Tour 4610, 4e étage, Pl. Jussieu, 75230 Paris Cedex 05, France

M. Sanz-Solé, Departament d'Estadistica, Facultad de Matemàtiques, Universitat de Barcelona, Gran Via 585, 08007 Barcelona, Spain

W. Schachermayer, Institut fur Mathematik, Universitat Wien, Strudlhofgasse 4, A-1090 Wien

C. Sempi, Via Monte S. Michele 10, 73100 Lecce

M. Sion, Department of Mathematics, University of British Columbia, Vancouver, Canada

H.-O. Tylli, Department of Mathematics, University of Helsinki, Hallituskatu 15, 00100 Helsinki 10, Finland

L. Tubaro, Dipartimento di Matematica, Università, 38050 Povo, Trento

C. Zanco, Dipartimento di Matematica, Università, Via C. Saldini 50, 20133 Milano

P.A. Zanzotto, Via S. Antonio 7, 56100 Pisa

F. Zimmermann, Mathematisches Seminar, CAU, Ohlshausenstr. 40-60, D-2300 Kiel 1

PROBABILITY and GEOMETRY

Jean-Michel BISMUT

Département de Mathématique
Université Paris Sud
Bâtiment 425
91405 ORSAY
France

The purpose of these notesis to give an introduction to a few applications of probability theory in a variety of situations which are connected with analysis or geometry.

In Section 1, we show how to use stochastic differential equations to obtain an explicit expression of the law of diffusion bridges by integrating "along the fiber" the infinite dimensional Wiener measure. The simplicity of the method - which is in principle based on an elementary application of the implicit function theorem in infinite dimensions - limits its applicability to elliptic diffusions and certain few hypoelliptic ones. However it at least offers a conceptual framework to attack more complicate hypoelliptic situations. We here mostly follow our work [17] which is based on the Malliavin calculus and uses large deviation techniques. The main outcome is a compact formula which may be used to calculate asymptotic expansions over bridges for any functional of the considered bridgc. For another approach to the same problem, we refer to Azencott [9], [10].

In Section 2, and following our work [19], we describe a probabilistic proof of the Index Theorem of Atiyah-Singer for Dirac operators. Our presentation of the Index Theory is extremely brief. The proof is based on the formula obtained in Section 1. This formula is in fact more precise than what is needed since it contains the whole asymptotic expansion of the heat kernel. Also P. Lévy's well known area formulas [57] play a key role in Index Theory (and also in fixed point theory). We also refer to the proofs of Alvarez-Gaume [1], Friedan-Windey [31], which are based on physical considerations, and to Getzler [35], [36] and Berline-Vergne [15] for analytic proofs. In particular the second proof of Getzler [36] gives the analytic counterpart of the probabilistic proof. For fixed point Theory, we refer to [5], [8], [14], [19], [21], [37].

In [4], Atiyah and Witten had exhibited a remarkable formal similarity between the Index formula for the Dirac operator on the spin complex and localization formulas in equivariant cohomology by Berline-Vergne [13], [14], Duistermaat-Heckman [28]. In Section 3, and following [25], we give a new proof of such formulas and show that the heat equation method, when interpreted probabilistically, is the corresponding infinite dimensional rigorous extension. Still following [20], [25] we also show that the functional integral constructs infinite dimensional

characteristic classes.
A precise understanding of the localization formulas of [13], [14], [28] and an idea of Quillen [49] led us in [24] to the heat equation proof of the Atiyah-Singer Index Theorem for families [8], which we also briefly discuss.

In Section 4, we introduce the Witten complex, and describe Witten's proof of the Morse inequalities [56]. Instead of spectral theory [56], we still use the results of Section 1, and follow our work [23].

References are only briefly indicated. Most of the time, we refer to the original work (including ours) for more detailed references. This is in particular true for Sections 1 and 2.

For the non probability oriented reader, the survey [18] might be useful for a guided tour to Brownian motion.

1. Short time asymptotics of diffusion bridges and heat kernels.

In this section, we will give a short survey of the results we obtained in [17] on the short time asymptotics of diffusion bridges.

Two ideas will be exploited :

a) The first idea is the fact that the law of a smooth diffusion $x.$ can be obtained as the image of the Brownian measure P by solving a stochastic differential equation.

b) The second idea is that if x_1 is the end point of the diffusion, the conditional law of $x.$ can be obtained by integration along the fiber $(x_1=y)$ of the measure P considered as a differential form on the infinite dimensional Wiener space.

In practice two instruments are used simultaneously :

a) The differential analysis on Wiener space, or Malliavin calculus, combined with the theory of stochastic flows. This permits us essentially to deal with Brownian motion almost as if it was living in a finite dimensional space. Also dependence of solutions on parameters can be considered in the usual C^∞ sense, as envisioned in earlier work by

Blagoveshchenskii-Freidlin [26].

b) The formulation by Azencott of large deviation theory [9]. Indeed Azencott established large deviation results as consequences of large deviation on the Wiener space by studying the stochastic differential equation map.

Localization is obtained using Azencott's approach, and the differential analysis is done using the Malliavin calculus. Of course the necessary estimates are obtained in [17] using the stochastic calculus (martingale inequalities, Garcia-Rodemich-Rumsey's Theorem [33], etc...) We refer to [17] for a much more systematic and precise exposition of the ideas contained in this section. In particular, we want to underline strong connections with earlier work of De Witt-Morette-Maheswari-Nelson [27] and also Elworthy-Truman [29].

The reader should keep in mind the following facts.
a) In the parametrix method for second order elliptic operators [32], one is interested in the asymptotic expansion of the heat equation semigroup $p_t(x,y)$. Here we want much more, i.e. understand the structure of the laws of the whole path.

b) In [17], a truly satisfactory answer has been given only in the elliptic case. This is so because in the hypoelliptic situations (Hörmander [39]) $w \to x_1(w)$ in non singular only almost surely (this is the essence of the Malliavin calculus [44]).

The non differentiability of Brownian motion plays here a key role. Now the considered mapping may well become singular on differentiable deterministic paths. In general for small time, the considered bridge measures concentrate on singular points. This makes the problem of understanding the asymptotic structure of the bridge very difficult and interesting.

c) Rough estimates on the heat equation semi-group are needed, like the estimate of Varadhan [52] for elliptic diffusions. Similar uniform estimates are needed for non elliptic diffusions (see Gaveau [34], Léandre [43]).

Other approches exist to attack the previously described problems.

a) In Molchanov [46], the asymptotic structure of the bridges in the elliptic case was described in Gaussian approximation . However stochastic differential equations were not used. Also Kifer [42] obtained small time expansions of certain semi-groups using a mixture of analytic and probabilistic techniques.

b) In [10], Azencott obtained such expansions in the elliptic case using the stochastic differential equation describing the bridge itself. As is well-known, such equations are singular and necessitate a careful analysis .

Also note that in Fefferman [30], Sanchez-Calle [51], much progress has been done in the understanding of the time behavior of the heat equation semi-group associated with $\sum_1^m X_i^2$ under Hörmander's assumptions [39], in relation with the geometry of the vector fields $X_1 \dots X_m$. This might also lead ultimately to a better understanding of the bridges themselves.

In the next sections we will apply the techniques of Section 1 only in comparatively simple situations where only a few terms of the small time expansions are needed. However the fact of having a global description of the law of the bridge will help us considerably, specially in proving the Index Theorem.

This section is organized as follows.

In a), a finite dimensional analogue is described. In b), certain stochastic differential equations are introduced. In c), the non singularity of a certain mapping is considered. In d), an integration along the fiber procedure in infinite dimensions is given, which leads to the asymptotic representation of certain bridges. In e), a proof of the results of d) is sketched. In f), asymptotic expansions are considered.

In the whole text, d will denote Stratonovitch differentials, and δ Itô differentials [18].

a) <u>A finite dimensional analogue</u>.

Let Φ be a C^∞ mapping from R^n into R^m . We endow R^n with the Gaussian measure $dP(w) = \frac{e^{-\frac{|w|^2}{2}} dw}{(\sqrt{2\pi})^n}$. Let $dP_\Phi(y)$ be the image measure by Φ . By definition for any $f \in C_\infty^b(R^m)$

(1.1) $$\int_{R^n} f[\Phi(w)]dP(w) = \int_{R^m} f(y)\,dP_\Phi(y) .$$

Now assume that $m \leq n$, and that Φ is a submersion, i.e. that the rank of $\Phi'(w)$ is m at every w .

Set

$$K_y = \Phi^{-1}\{y\} .$$

If K_y is non empty, it is a $n-m$ dimensional submanifold of R^n. Also $dP_\Phi(y)$ is of the form

(1.2) $$dP_\Phi(y) = m(y)\,dy$$

In the sense of classical differential geometry, $m(y)dy$ is the integral along the fiber K_y of $\frac{e^{-\frac{|w|^2}{2}} dw}{(\sqrt{2\pi})^n}$, which we write in the form :

(1.3) $$m(y)\,dy = \int_{K_y} \frac{e^{-\frac{|w|^2}{2}} dw}{(\sqrt{2\pi})^n}$$

In particular if under a C^∞ change of coordinates in R^n $w \to \psi(w) = x$, $\Phi'(x) = \Phi(\psi^{-1}(x))$ is a projection

(1.4) $$x = (y,y') \to \Phi'(x) = y$$

if $g(x)dx$ is the law dP evaluated in the new coordinates, then

(1.5) $$m(y) = \int_{R^{n-m}} g(y,y')\,dy'$$

However in general (1.4) is only valid locally. Another may of rewriting (1.3) is to use an explicit factorization of dx . Namely let $d\sigma^y(w)$ be the area element in K_y . Let $\Phi'^*(w) : R^m \to R^n$ be the adjoint of $\Phi'(w)$. Then (1.3) can be explicitly rewritten as

(1.6) $$m(y) = \int_{K_y} \frac{e^{-\frac{|w|^2}{2}}}{(\sqrt{2\pi})^n} \frac{d\sigma^y(w)}{[\det(\Phi'\Phi'^*)(w)]^{1/2}}$$

(1.3) - (1.6) contains in fact more information than $m(y)$. It is obvious that

$$(1.7) \qquad dP^y(w) = \frac{e^{-\frac{|w|^2}{2}} d\sigma^y(w)}{m(y)(\sqrt{2\pi})^n [\det \Phi'\Phi'^*(w)]^{1/2}}$$

is the law of w conditional on $\Phi(w)=y$. This shows that

$$(1.8) \qquad \int_{R^n} f[\Phi(w)]\, h(w)\, dP(w) = \int_{R^m} f(y)\, dy \left[\int_{K_y} \frac{h(w)\, e^{-\frac{|w|^2}{2}} d\sigma^y(w)}{(\sqrt{2\pi})^n \ \det [\Phi'\Phi'^*(w)]^{1/2}}\right]$$

We now fix $y \in R^m$.

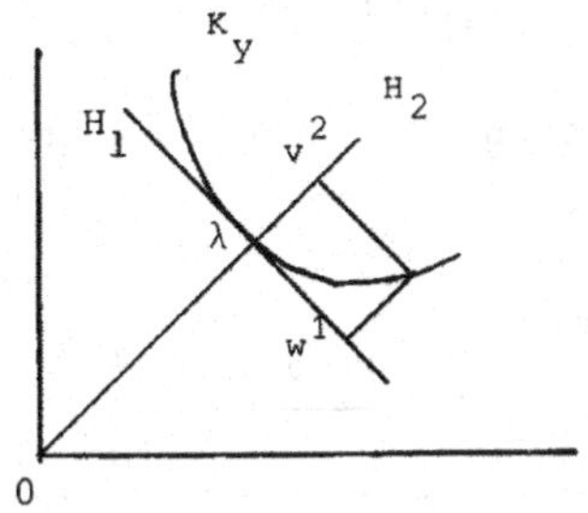

Figure 1

We assume that $\lambda \in R^n$ is the unique element of minimal Euclidean norm in K_y. Let V be a "small" neighborhood of λ in R^n.

Set

$$(1.9) \qquad H_1 = T_\lambda K_y \ ; \ H_2 = H_1^\perp$$

Clearly $\lambda \in H_2$.

As shown in Figure 1, we can parametrize locally K_y by H_1. For $w^1 \in H_1$ small enough, there is a unique v^2 (with v^2 small enough) such that

$$(1.10) \qquad (\lambda + w^1, v^2) \in K_y .$$

v^2 depends smoothly on w^1.

We now find that in the local coordinates w^1

$$(1.11)\qquad m(y)\,dP^{y}(w) = \frac{e^{-\frac{|\lambda+v^2(w^1)|^2}{2}}}{(\sqrt{2\pi})^m \det[\frac{\partial\Phi}{\partial v^2}(\lambda+w^1+v^2(w^1))]}\;\frac{e^{-\frac{|w^1|^2}{2}}\,dw^1}{(\sqrt{2\pi})^{n-m}}$$

Also

$$\det[\frac{\partial\Phi}{\partial v^2}(\lambda+w^1+v^2)] = \frac{\det[\Phi'(\lambda+w^1+v^2)\,\Phi'^*(\lambda)]}{[\det(\Phi'\Phi'^*(\lambda)]^{1/2}}$$

Set

$$(1.12)\qquad dP_1(w^1) = \frac{e^{-\frac{|w^1|^2}{2}}\,dw^1}{(\sqrt{2\pi})^{n-m}}$$

$dP_1(w^1)$ is the Gaussian measure on H_1. We find that

$$(1.13)\qquad \int_V h(w)\,m(y)\,dP^{y}(w) = \int_{|w^1|\leq\varepsilon} \frac{h[\lambda+w^1+v^2(w^1)]\,e^{-\frac{|\lambda+v^2(w^1)|^2}{2}}\,[\det(\Phi'\Phi'^*)(\lambda)]^{1/2}\,dP_1(w^1)}{(\sqrt{2\pi})^m \det[\Phi'(\lambda+w^1+v^2(w^1))\,\Phi'^*(\lambda)]}$$

Now replace $\Phi(w)$ by

$$\Phi_t(w) = \Phi(\sqrt{t}w)\ .$$

$m_t(y)$, $dP^{y}_t(w)$ will denote the corresponding objects. Then for a bounded smooth h (eventually depending on t)

$$(1.14)\qquad \int_V h(w)\,m_t(y)\,dP^{y}_t(w) = \int_{|w^1|\leq\varepsilon/\sqrt{t}} \frac{h(\lambda+\sqrt{t}w^1+v^2(\sqrt{t}w^1))}{(\sqrt{2\pi t})^m}\;\frac{e^{-\frac{|\lambda+v^2(\sqrt{t}w^1)|^2}{2t}}\,[\det\Phi'\Phi'^*(\lambda)]^{1/2}\,dP_1(w^1)}{[\det\Phi'(\lambda+\sqrt{t}w^1+v^2(\sqrt{t}w^1))\,\Phi'^*(\lambda)]}$$

Now (1.14) is interesting for the following reasons :

- Assume that as $t\downarrow 0$, $m_t(y)\,dP^{y}_t(w)$ localizes on λ , in the sense that for any $k\in\mathbb{N}$

(1.15) $$m_t(y)P_t^y({}^cV) = e^{-\frac{|\lambda|^2}{2t}}\, o(t^k)$$

(we may eventually replace $o(t^k)$ by $e^{-\chi/t}$ with $\chi>0$). The r.h.s. of (1.14) is then an accurate evaluation of

$$\int h(w)\, m_t(y)\, dP_t^y(w)$$

up to $e^{-\frac{|\lambda|^2}{2t}}\, o(t^k)$.

• If (1.15) is verified, we have a very precise evaluation of $m_t(y)$, but also a precise understanding of P_t^y as $t\downarrow\downarrow 0$. In particular, we may Taylor expand $m_t(y)$ in the form

(1.16) $$m_t(y) = \frac{e^{-\frac{|\lambda|^2}{2t}}}{(\sqrt{2\pi t})^n}[a_o(y)+a_1(y)t+\ldots]$$

and more generally, we may expand

(1.17) $$\int h_t(w)\ m_t(y)\, dP_t^y(w)\,.$$

However all the information is contained in the compact formula (1.14).

• In a neighborhood of λ , all the P_t^y are described by means of the unique probability space $(H_1\ , dP_1(w^1))$. This will be of utmost importance in an infinite dimensional context.

In general (1.15) is an easy estimate only if K_y is compact. Otherwise some work has to be done to prove (1.15).

b) Differential equations and stochastic differential equations.

The basic idea which we will develop is that in certain infinite dimensional situations, (1.14) still makes sense.

Set

(1.18) $$H = L_2([0,1]\ ;\ R^k)\ .$$

Let $X_1\ \ldots, X_k$ be a family of vector fields on R^m whose components belong to $C_\infty^b(R^m)$.

Fix $x_o \in R^m$. For $h \in H$, consider the differential equation

(1.19)
$$dx^h = \sum_1^k X_i(x^h) h^i ds$$
$$x^h(0) = x_o$$

(from now on, the summation signs will be omitted).

Set

(1.20)
$$\Phi(h) = x_1^h$$

We are in a situation analogue to what is done in Section 1, except that R^n is replaced by H.

Let P be the Wiener measure on $C([0,1]\ ;\ R^k)$. $w_t = (w_t^1 \ldots w_t^k)$ denotes the corresponding Brownian motion.

For $t > 0$, consider the stochastic differential equation

(1.21)
$$dx = X_i(x) . \sqrt{t} dw^i$$
$$x(0) = x_o$$

where the differential dw is the Stratonovitch differential of w [18]. Set

(1.22)
$$\Phi(\sqrt{t} dw) = x_1$$

The notations (1.20)-(1.22) are compatible.
Indeed :

a) P should be thought of as the Gaussian cylindrical measure on H. Of course a.s. $\frac{dw}{ds} \notin H$, i.e.

$$P(H) = 0 .$$

b) As is well-known, by a result of Wong-Zakai, Stroock-Varadhan [50], if w is replaced by its piecewise linear interpolation w^n on dyadic time intervals $[\frac{k}{2^n}, \frac{k+1}{2^n}[$, as $n \to +\infty$ $x^{\sqrt{t}\dot{w}^n}$ converges to the solution x of (1.21) in probability.

In particular

$$\Phi(\sqrt{t}\dot{w}^n) \to \Phi(\sqrt{t}dw) \text{ in probability .} \tag{1.23}$$

More generally, we will take for granted all the results on stochastic flows [16], [17], [18], which quarantee that a.s., differentiation with respect to parameters is possible on stochastic differential equations.

c) Non singularity of Φ .

Let us now try to reproduce what has been done in a). We will study the correspondence

$$h \to \Phi(h)$$

If x^h is the solution of (1.19), set

$$x_t^h = \varphi_t^h(x_o) \quad . \tag{1.24}$$

Let φ_t^{h*} be the derivative $\frac{\partial\varphi^h}{\partial x}t$. It is easy to verify that $\varphi_t^{h*}(x_o)$ is invertible. By proceeding as in [17, Chapter 1], we find that for $v \in H$

$$\Phi'(h)v = \varphi_1^{h*} \int_o^1 (\varphi_s^{h*-1}X_i)(x_o)v^i \, ds \quad .$$

$(\Phi'\Phi'^*)(h)$ is the matrix

$$p \to C^h p = \varphi_1^{h*} \int_o^1 (\varphi_s^{h*-1} X_i)(x_o) < \varphi_1^{h*} \varphi_s^{h*-1} X_i(x_o), p > ds . \tag{1.25}$$

The non singularity of Φ is equivalent to the invertibility of C^h.

Modulo the irrelevant invertible φ_1^{h*} , C^h is exactly the Malliavin covariance matrix of the problem [17, Chapter 1].

As is well-known by the Malliavin calculus [44], if $X_1,\dots,X_m$ verify the assumptions of Hörmander [39] (i.e. if $X_1 \dots X_m$ and their Lie brackets span R^m at each $x \in R^m$), then C^{dw} is a.s. invertible. This fact, and estimates on the L_p norm of $\|(C^{dw})^{-1}\|$ are the key steps in the probabilistic proof of Hörmander's theorem. The key point is that w is not differentiable.

However recall that if we go back to the picture of Figure 1, and think in terms of large deviations, as $t \downarrow \downarrow 0$, in principle $dP_t^y(w)$ concentrates on a $\lambda \in H$, on which in general, even under Hörmander's assumption, C^λ is not invertible.

For a given $y \in R^m$, if K_y is defined by

$$K_y = \{h \in H \; ; \; \Phi(h) = y\}. \tag{1.26}$$

K_y is generally a singular subset of H.

Conditions under which C^h is invertible are given in [17, Chapter 1] They include

a) The elliptic case : This is the case where $X_1, \dots X_m$ span R^n everywhere. In this case Φ is obviously a submersion.

b) Certain few hypoelliptic situations which are modelled on the three dimensional Heisenberg group (X_1, X_2 being the first two generators of its Lie algebra).

Except these two cases, there is little hope that the method outlined in a) could work.

d) Integration along the fiber in infinite dimensions.

We now concentrate on the elliptic case.

For more general situations see [17] and Léandre [43]. Remember that our purpose is to understand fully the small time asymptotics of certain diffusion bridges (which live on the sets K_y for $y \in R^m$).

To simplify, we do the assumption that $k=m$. We then assume that $X_1(x_0) \dots X_m(x_0)$ span R^m.

Also, we will only interest ourselves in the disintegration or the diffusion measure on K_{x_0}, i.e. on the law of diffusion loops which start and end at $x_0 \in R^m$. For the general case see [17].

Here

$$K_{x_0} = \{h \in H \; ; \; \Phi(h) = x_0\}.$$

Obviously the element of minimal norm in K_{x_o} is $\lambda=0$.

We now use the picture of Figure 1.

Using (1.24), it is clear that H_1 (which coincides with TK_{x_o}) is exactly

(1.27) $$H_1 = \{v\in H \ ; \int_o^1 v\,ds = 0\} .$$

Its orthogonal H_2 is given by

(1.28) $$H_2 = \{v\in H \ ; \ v = \text{cst}\}.$$

The cylindrical gaussian measure P_1 on H_1 is the Brownian bridge measure. Under $dP_1(w^1)$, $w^1(0\leq s\leq 1)$ will be a Brownian bridge in R^m, with $w_o^1 = w_1^1 = 0$. $\frac{dw^1}{ds}$ should be thought as living in H_1 (although it doesn't !).

Our next problem is to make sense of Figure 1.

Namely we should make sense of the following equation

(1.29) $$dx = X_i(x)[\sqrt{t}\ dw^{1,i} + v^{2,i}\ ds]$$

$$x(o) = x_o .$$

where in (1.29), for a given w^1 , for t small enough, v^2 should be chosen uniquely so that

(1.30) $$x_1 = x_o$$

But then obviously v^2 will anticipate on x, w^1 and so we get out of the usual probabilistic folklore ! Also, as far as estimates are concerned, it is well-known that if w^1 is described as a semi-martingale, its drift is singular at $t=1$.

So let us explain how to solve these difficulties.

1). Take a usual Brownian motion w' . Set

$$w_s^1 = w_s' - sw'_1 .$$

w^1 is then a Brownian bridge.

2). Replace equation (1.29) by

$$(1.31) \qquad dx = X_i(x)[\sqrt{t}\, dw'^i + (v^{2,i} - \sqrt{t} w'^{1,i})\, ds]$$

$$x(o) = x_o$$

We will have to make sense of (1.31) and satisfy (1.30).

We now use stochastic flows [16, Chapter I - III]. Consider the equation

$$(1.32) \qquad dy = X_i(y)\sqrt{t}\, dw'^i$$

$$y(o) = y_o ,$$

and its associated stochastic flow of C^∞ diffeomorphisms $\psi_s(\sqrt{t}\, dw')$ so that in (1.32), $y_s = \psi_s(\sqrt{t}\, dw', y_o)$.

Consider the <u>differential</u> equation

$$(1.33) \qquad dz = (\psi_s^{*-1} X_i)(z)(v^{2,i} - \sqrt{t}\, w'^{1,i})\, ds$$

$$z(o) = x_o .$$

Of course (1.33) is solvable as a differential equation, if the flow ψ does not grow too much at infinity. This is the case if $X_1, \ldots X_m$ have compact support.

So in the sequel, we assume that $X_1, \ldots X_m$ have compact support. This is no restriction since we are essentially interested in what happens in a neighborhood of x_o .

Since w' is now an usual Brownian motion, estimations on (1.31) - (1.33) are fairly standard [16, Chapter I-III].

So as a <u>definition</u> of a solution of (1.31), we set

$$(1.34) \qquad x_s = \psi_s(\sqrt{t}\, dw', z_s) .$$

Again this definition is sound from the point of view of the approximation of stochastic flows with the flows of differential equations, as shown by the approximation result of [16,Theorems I.1.2 and I.2.1].

A trivial application of the ordinary implicit function theorem (with w' as a parameter) shows that indeed for t small enough, a unique v^2 of small norm exists such that (1.30) is verified. The reader may wander wether v^2 depends on w' or only on w^1. Approximation or the theory of enlargement of filtrations [40] shows that in fact v^2 is a function of w^1. We will write $v^2(\sqrt{t}\, dw^1)$.

In order to make sense formally of formula (1.14) in this situation, we must show how to define the determinants appearing in (1.14).

This is very easy. As shown by (1.33), $x.$ defined by (1.31) is C^∞ in v^2. For t small enough, the analogue of the determinant in (1.14) does not vanish and its inverse is uniformly bounded.

It is not entirely trivial to find how to replace the now meaningless $|w^1|\leq\varepsilon/\sqrt{t}$. We will instead introduce a function $G(t,dw^1)$ which has the following properties [17, Chapter 4].

- $0 \leq G \leq 1$
- If $G \neq 0$, $v^2(\sqrt{t}\, dw^1)$ is well-defined.
- For any $k \in N$

  $P[G(t,dw^1) \neq 1] = o(t^k)$

The explicit definition of G is directly connected with the applicability of the implicit function theorem.

We now state in this situation the main result of [17, Section 4].

Theorem 1.1. For any bounded function h on $\mathcal{C}([0,1]\,;\,R^m)$ the following equality holds

(1.35)
$$m_t(x_o) \int_{\mathcal{C}[0,1];R^m)} h(w)G(t,dw^1)\,dP_t^{x_o}(w)$$

$$= \int_{\mathcal{C}[0,1];R^m)} h(\sqrt{t}w^1 + sv^2)\frac{G(t,dw^1)}{(\sqrt{2\pi t})^m}\exp -\frac{|v^2|^2}{2t}$$

$$\frac{dP_1(w^1)}{\det \frac{\partial\Phi}{\partial v^2}(\sqrt{t}w^1 + v^2(\sqrt{t}dw^1))} .$$

Also for any $k \in N$, and uniformly on compact subsets of R^m

(1.36)
$$m_t(x_o)\; P_t^{x_o}(G(t,dw^1) \neq 1) = o(t^k) \;.$$

Remark 1. (1.35) tells us in particular that in the region $G(t,dw^1)=1$, the description of the measure $m_t(x_o)dP_t^{x_o}(w)$ which we gave using the implicit function theorem is exact.

Recall that in the probabilistic litterature [40], a Brownian bridge is usually described in terms of a u-process (in the sense of Doob). Namely let $p_t(x_o,y)$ be the heat kernel associated with the operator $e^{\frac{t\sum_i^n X_i^2}{2}}$. Then under $P_t^{x_o}$, x. is described by the stochastic differential equation

(1.37)
$$dx = X_i(x)\left(\sqrt{t}dw^i + \frac{tX_i p_{t(1-s)}(x,x_o)\; ds}{p_{t(1-s)}(x,x_o)}\right)$$

$$x(0) = x_o \;.$$

However note that the drift is singular at $s=1$ [17, Section 2]. The description of the bridge given in (1.35) eliminates the singularity. Also we now will use the bridge to obtain informations in particular on p_t . On the contrary (1.37) uses p_t to construct the bridge. However note that Azencott [10] was able to use (1.37) to also obtain asymptotic expansions.

e) A sketched proof.

We now briefly sketch the proof of Theorem 1.1.

The basic idea is to split the Brownian measure P on H_1, H_2.

Clearly under the measure

$$dP(w^1)\ \frac{\exp -\frac{|v^2|^2}{2t} dv^2}{(\sqrt{2\pi t})^n}$$

if w is given by

$$w_s = w_s^1 + \frac{v^2 s}{\sqrt{t}} \tag{1.38}$$

w is a standard Brownian motion. We now consider the stochastic differential equation .

$$\begin{aligned} dx &= X_i(x)(\sqrt{t}\,dw^{1,i} + v^{2,i}ds) \,. \\ x(0) &= x_o \end{aligned} \tag{1.39}$$

which we solve as indicated in (1.29) - (1.34).
Approximation theory or enlargement of filtrations show that if w is given by (1.38) and x by (1.39) then

$$x_1 = \Phi(\sqrt{t}dw) \,.$$

Now

$$\int g(x_1)h(x)\,dP(w) = \int g(x_1)h(x)\,dP_1(w^1)\frac{e^{-\frac{|v^2|^2}{2t}}dv^2}{(\sqrt{2\pi t})^m} \tag{1.40}$$

The idea is now to change variables in (1.39), i.e. for a given w^1, to make the change of variables $v^2 \to x_1$. This makes appear the determinant $\frac{\partial\Phi}{\partial v^2}$. By showing that a smooth disintegration of (1.40) is possible we obtain (1.35).

The most difficult point is obviously (1.36), which is the analogue of (1.15). In [17], to prove (1.36), we use large deviation results on diffusions defined by stochastic differential equations by

- using large deviation results on Brownian motion
- establishing simple "continuity" properties of stochastic differential equations.

Also we use the uniform estimates on the heat kernel for elliptic diffusions.

$$\exp\{-\frac{d^2(x_o,y_o)-\chi}{2t}\} \leqslant p_t(x_o,y_o) \leqslant \frac{c}{t^N}\exp\{-\frac{d^2(x_o,y_o)}{2t}\} \tag{1.41}$$

of Molchanov [46], Azencott and al. [12].

To establish large deviation results for bridges, we divide [0,1] in the intervals [0,1/2] and [1/2,1] and use standard large deviation theory for free diffusions as well as (1.41) and Varadhan's technique [52], [53].

A key step is to prove that under $P_t^{x_o}$, the conditions of the implicit function theorem are verified up to $o(t^k)$. By proceeding as previously indicated, this is equivalent to establishing a large deviation result on standard stochastic flows.

This is done in [17] by an indirect technique which uses (1.37) and large deviation estimates on

$$\frac{\text{grad } p_t}{p_t} .$$

More recently, Kusuoka [59] has shown us how to establish large deviation results for stochastic flows, which are as good as the corresponding results for usual stochastic differential equations.

The idea of Kusuoka is as follows :

a) On a single trajectory of a flow $\psi_s(\sqrt{t}dw,x_o)$, standard large deviations give bounds like $e^{-\chi/t}(\chi>0)$.

b) Take $\eta>0$ arbitrarily small. If we now choose $e^{\eta/t}$ points x_o on a uniform grid in a bounded region (with $\eta<\chi$), the bound will become

$e^{-(\chi - \eta)/t}$.

c) Any point in a bounded region is at a distance at most $e^{-\frac{\eta}{m}/t}$ of a point in the grid. Standard estimates on the module of continuity of a flow [16, Chapter I] show that the remaining probability which should be estimated is like $C_\delta \, e^{-\delta/t}$ (with δ arbitrary large).

f) A remark on the asymptotic expansions.

In [17, Chapter 4], we show that (1.35) can be expanded in powers of t , thus obtaining the Minakshishundaram-Pleijel expansion of $m_t(x_o) = p_t(x_o, x_o)$ when $h = 1$.

This is done by expanding $\frac{|v^2|^2}{2t}$, $\frac{1}{\det \frac{\partial \phi}{\partial v^2}}$ in the variable $\sqrt{t}$ using iterated stochastic integrals.

However note that classically, in finite dimensions (see Hörmander [58]), expansions of Laplace type integrals are obtained using a certain second order differential operator associated with the Laplace integral.

Here the situation is infinite dimensional. So the corresponding differential operator is infinite dimensional.

2. The Atiyah-Singer Index Theorem : a probabilistic proof.

In this section, we give a probabilistic proof of the Atiyah-Singer Index Theorem for Dirac operators, based on the asymptotic representation of Section 1. We here follow our paper [19].

In a), we give a few definition . In b), we introduce the Dirac operator D. In c), the heat equation method is briefly described. In d), the heat equation semi-group associated with D^2 is constructed probabilistically. In e), an asymptotic representation of the supertrace of the heat kernel is given. In f), the asymptotics of the supertrace is explicitly found in terms of a Brownian bridge. In g), the local Index formula is calculated using the well-known area formula of P. Lévy [57]. This formula is proved again by means of an infinite determinant. This will be of utmost importance in the next section in connection with Atiyah [4].

This section includes neither motivation nor applications to the Index Theorem. We refer to Atiyah-Bott [5], Atiyah-Singer [8], Atiyah [3], in which the backround material is developed. The heat equation method was introduced in Mc Kean-Singer [45], Patodi [47], Gilkey [37], Atiyah-Bott-Patodi [7]. Physicist's proofs of the Index Theorem based on supersymmetry were given by Alvarez-Gaume [1] Friedan-Windey [31], Getzler gave a rigorous proof in [35] using pseudodifferential operators techniques. Berline-Vergne [15] gave another proof using heat equation in the bundle of frames.

The second proof of Getzler [36] is more directly related to our proof in [19]. We refer to [36] for more details.

a) <u>A few definitions</u>.

$n = 2\ell$ is an integer. R^n is the Euclidean space endowed with an orthogonal oriented base $e_1, \dots e_n$.

Recall that the Clifford algebra $c(R^n)$ is the algebra generated over R by 1, $e_1, \dots e_n$ and the commutation relations.

(2.1) $$e_i e_j + e_j e_i = -2\delta_{ij} .$$

$c(R^n)$ is spanned by the products $e_{i_1} \dots e_{i_p}$ (with $i_1 < i_2 \dots < i_p$).

As a representation of $SO(n)$, $c(R^n)$ is isomorphic (as a vector space) to the exterior algebra $\Lambda(R^n)$. $c(R^n)$ has a natural grading (which corresponds to the grading of $\Lambda(R^n)$). In particular $c(R^n)$ is Z_2 graded, and so has even and odd elements.

$c(R^n) \otimes_R C$ identifies to $\operatorname{End} S$, where S is a 2^ℓ dimensional Hermitian space of spinors. Set

(2.2) $$\tau = i^\ell e_1 \dots e_n .$$

Then

(2.3) $$\tau^2 = 1$$

Set

$$S_+ = \{s\in S \ ; \ \tau s = s\} \ ; \ S_- = \{s\in S \ ; \ \tau s = -s\}.$$

Then

$$S = S_+ \oplus S_- \ .$$

S_+, S_- are orthogonal subspaces of S, of dimension $2^{\ell-1}$. The elements of $c^{even}(R^n)$ (resp $c^{odd}(R^n)$) commute (resp. anticommute) with τ.

Definition 2.1. If $A\in c(R^n)$, the supertrace $Tr_s A$ is defined by

(2.4) $$Tr_s A = Tr[\tau A] \ .$$

If $A\in c(R^n)$, as an element of $End(S_+\oplus S_-)$ we may write

(2.5) $$A = \begin{bmatrix} E & F \\ G & H \end{bmatrix}$$

and so

(2.6) $$Tr_s A = Tr \ E - Tr \ H \ .$$

It may be easily shown (see Atiyah-Bott [5, p. 484]) that if $i_1 < \ldots < i_p$, $p < n$

(2.7) $$Tr_s \ e_{i_1} \ldots e_{i_p} = 0$$

and that

(2.8) $$Tr_s \ e_1 \ldots e_n = (-2i)^{\ell}$$

Spin(n) denotes the double cover of $SO(n)$. Since for $n \geq 3$, $\pi_1(SO(n)) = Z_2$, for $n \geq 3$, Spin(n) is the double cover of $SO(n)$.

Spin(n) can be embedded in $c^{even}(R^n)$. We partly describe this embedding.

Let $\mathfrak{a}$ be the set of (n,n) real antisymmetric matrices. $\mathfrak{a}$ is the Lie algebra of $SO(n)$ and Spin(n).

If $A = (a_i^j) \in \mathfrak{a}$, we identify A with the element of $c^{even}(R^n)$

(2.9) $$k(A) = \frac{1}{4} a_i^j e_i e_j .$$

The key facts are that k is a Lie algebra homomorphism and that if $f \in R^n$ is considered as an element of $c(R^n)$

(2.10) $$[k(A), f] = Af .$$

In the r.h.s. of (2.10), Af is still considered as an element of $c(R^n)$.

Spin(n) acts unitarily and irreducibly on S_+ and S_-.

Set σ be the projection $\mathrm{Spin}(n) \to SO(n)$.

If $x \in \mathrm{Spin}(n)$, for $f \in R^n$, $xfx^{-1} \in R^n$ and also

(2.11) $$\sigma(x) f = xfx^{-1}$$

If $A \in \mathfrak{a}$, we also identify A with the 2-form $X, Y \to \langle Y, AY \rangle$, i.e. with

(2.12) $$\frac{1}{2} a_j^i \, dx^i \wedge dx^j .$$

$A^{\wedge k}$ denotes the kth power of A in the exterior algebra $\Lambda(R^n)$.

Definition 2.2. The Pfaffian Pf(A) of $A \in \mathfrak{a}$ is defined by the relation

(2.13) $$\frac{A^{\wedge \ell}}{\ell !} = \mathrm{Pf}A \, dx^1 \wedge \dots \wedge dx^n .$$

Note the relation

$$\det A = (\mathrm{Pf}A)^2$$

The easiest way to see the deep relation of the Clifford algebra with the heat kernel is the following [19, Theorem 1.5]

Theorem 2.3. Let $t \to x_t$ be a continuous mapping from R into Spin(n), which is C_1 at $t = 0$, such that x_o is the identity.

If $A \in \mathfrak{a}$ is defined by

(2.14) $$A = \frac{dx}{dt}\, t = o$$

then

(2.15) $$\lim_{t \downarrow \downarrow 0} \frac{Tr_s[x_t]}{t^{n/2}} = i^{\ell} PfA .$$

Proof : For t small enough, we may write

(2.16) $$x_t = \exp[t\,A + o(t)]$$

where $o(t)$ is of length 2 in the Clifford algebra. Now

(2.17) $$\exp(tA+o(t)) = I + \frac{tA+o(t)}{1!} + \ldots \frac{(tA+o(t))^{n/2}}{(n/2)!} + o(t^{n/2}) .$$

Using (2.7)-(2.9), it is clear that the first $n/2-1$ elements have length $\leq n-2$, i.e. give a 0 contribution to Tr_s.

So we find that the l.h.s. of (2.15) is given by

(2.18) $$Tr_s \frac{A^{n/2}}{(n/2)!}$$

However in (2.18) in the Clifford product $A^{n/2}$, terms containing two identical factors like $e_1\, e_2\, e_1 \ldots$ give a 0 contribution since their length is $< n$. In the r.h.s. of (2.18) we can replace Clifford multiplication by exterior multiplication. Using (2.8), (2.18), (2.15) follows. □

b) The Dirac operator.

M is a connected compact Riemannian manifold of dimension $n=2\ell$. Let N be the $SO(n)$ bundle of oriented orthonormal frames in TM.

We assume that M is spin, i.e. N lifts as a spin bundle N', with

$$N' \xrightarrow{\sigma} N \xrightarrow{\pi} M$$

so that σ induces the covering projection $Spin(n) \to SO(n)$.

The topological obstruction to the spin structure is the second Stiefel- Whitney class $w_2(M)$.

Let $F, F_{\pm}$ be the vector bundles

$$F = N' \times_{Spin(n)} S \tag{2.19}$$

$$F_{\pm} = N' \times_{Spin(n)} S_{\pm}$$

F , $F_{\pm}$ are unitary bundles over M . TM acts by Clifford multiplication over F and exchanges F_+ and F_- .

The Levi-Civita connection on N lifts naturally to N' . So $F, F_{\pm}$ are endowed with a unitary connection. Let ∇ be the covariant differentiation operator.

R denotes the curvature tensor of TM . Using (2.9), we can identify R with the curvature tensor of $F_{\pm}$.

ξ denotes a unitary bundle over M . We endow ξ with an unitary connection. We still note ∇ the covariant differentiation operator on ξ .

TM acts on $F \otimes \xi$ by Clifford multiplication on F, i.e. $e \in TM$ is identified with $e \otimes 1$.

$\Gamma(F_{\pm} \otimes \xi)$ denotes the C^∞ sections of $F_{\pm} \otimes \xi$.

There is a natural L_2 scalar product on $\Gamma(F_{\pm} \otimes \xi)$ given by

$$< h,h' > = \int_M < h,h' >(x)dx$$

Definition 2.4. If $e_1, \ldots e_n$ is an orthonormal base of TM , the Dirac operator D acting on the C^∞ sections of $F \otimes \xi$ is given by

$$D = \sum_1^n e_i \nabla_{e_i} . \tag{2.20}$$

D interchanges $\Gamma(F_+ \otimes \xi)$ and $\Gamma(F_- \otimes \xi)$. D is formally self-adjoint on $\Gamma(F \otimes \xi)$. Its principal symbol is $\sqrt{-1}\xi$, where $\xi \in T^*M$ (identified with TM) is a Clifford multiplication operator. Since

$$\xi^2 = -|\xi|^2$$

D is an elliptic operator

Let $D_\pm$ be the restriction of D to $\Gamma(F_\pm \otimes \xi)$.

By standard results in index theory (see Treves [62, Chapter 3]) $\text{Ker } D_+$, $\text{Ker } D_-$ are finite dimensional.

Definition 2.5. The index of D_+ is defined by

$$\text{Ind } D_+ = \dim \text{Ker } D_+ - \dim \text{Ker } D_- \quad . \tag{2.21}$$

This is the definition of the analytical index of D_+ . The purpose of the Atiyah-Singer Index Theorem is to prove the equality of the analytical index with a topological index determined by the principal symbol of D .

c) The heat equation method.

If $(e_1(x)\ldots e_l(x))$ is a locally defined smooth section of N , let us recall that the horizontal Laplacian Δ^H acting on $\Gamma(F\otimes\xi)$ is given by

$$\Delta^H = \sum_1^n [(\nabla_{e_i(x)})^2 - \nabla_{\nabla_{e_i(x)} e_i(x)}]$$

Also let K be the scalar curvature of M .

We first recall Lichnerowicz's formula [60].

Theorem 2.6. If $e_1,\ldots e_n$ is an orthonormal base of TM , D^2 is given by

$$D^2 = -\Delta^H + \frac{K}{4} + \frac{1}{2}\sum e_i e_j \otimes L(e_i,e_j) \quad . \tag{2.22}$$

Proof : See [60] and [19, Theorem 1.9].

Remark 1. The remarkable cancellations which explain Lichnerowicz's formula should not be thought of as "accidental". The proof of the Index Theorem has already started. Indeed as we shall see, in Quillen's formalism [49], D should be thought of as a generalized connection, and D^2 as its curvature.

By standard elliptic theory, for $t > 0$, $e^{-\frac{tD^2}{2}}$ is given by a smooth kernel $P_t(x,y)$. For $(x,y) \in M \times M$, $P_t(x,y)$ maps linearly $(F_\pm \otimes \xi)_y$ into $(F_\pm \otimes \xi)_x$.

We can still define the supertrace of $H \in \mathrm{End}(F \otimes \xi)$ by an obvious extension of (2.4), i.e. by setting

$$\mathrm{Tr}_s H = \mathrm{Tr}[(\tau \otimes 1)H] .$$

In particular $\mathrm{Tr}_s P_t(x,x)$ is well-defined.

The first step in the heat equation method is the following (Mc Kean-Singer [45], Atiyah-Bott-Patodi [7]).

Theorem 2.7. For any $t > 0$

$$\mathrm{Ind}\, D_+ = \int_M \mathrm{Tr}_s[P_t(x,x)]dx \tag{2.23}$$

Proof : The operators D_-D_+ and D_+D_- have a discrete spectrum. For $\lambda > 0$, let $H_\pm^\lambda$ be the corresponding finite dimensional eigenspace in $\Gamma(F_\pm \otimes \xi)$. Since $\lambda > 0$, it is easy to find that D_+ is an isomorphism from H_+^λ on H_-^λ. The r.h.s. is obviously given by

$$\mathrm{Ind}\, D_+ + \sum_{\lambda > 0} e^{-\lambda t}(\dim H_+^\lambda - \dim H_-^\lambda) . \tag{2.24}$$

Since $\dim H_+^\lambda = \dim H_-^\lambda$, (2.23) is proved. □

d) A probabilistic construction of the heat equation semi-group.

Let $X_1^*, \ldots X_m^*$ be the standard horizontal vector fields on N. If ω is the connection form on N, X_i^* is defined by

$$\omega(X_i^*) = 0 \tag{2.25}$$

$$u^{-1}\pi_* X_i^* = e_i$$

(here $e_1, \ldots e_n$ is the canonical base of R^n).

Take $x_o \in M$, $u_o \in N_{x_o}$. For $t > 0$, consider the stochastic differential equation

$$(2.26) \qquad du^t = \sum_1^h X_i^*(u^t).\sqrt{t}\, dw^i$$

$$u(o) = u_o$$

Then, as is well known by the construction of Malliavin [61], Eells-Elworthy [63], if $x_s^t = \pi u_s^t$, $x_{s/t}$ is a standard Brownian motion on M starting at x_o. The curve in $T_{x_o}M$ $\beta_s \to u_o w_s$ is the development of x_s in $T_{x_o}M$, and is a standard Euclidean Brownian motion in $T_{x_o}M$. In fact note that the generator of the diffusion (2.26) is the operator $t\mathcal{L}$, where $\mathcal{L}$ is given by

$$\mathcal{L} = \frac{1}{2}\sum_1^n X_i^{*2} \quad .$$

It is well-known that $\mathcal{L}$ projects equivariantly on M as the Laplacian $\frac{\Delta^H}{2}$.

Incidently note that $\mathcal{L}$ is not elliptic on N. However since we will consider the mapping $w \to x_1$, the submersion property used in section 1 is preserved. So what will be done is an adaptation of what we did in Section 1. For more details see [17, Chapter 4].

In particular $u_s u_o^{-1}$ is the parallel transport operator from $T_{x_o}M$ on $T_{x_s}M$. This parallel transport makes sense because using approximations [16, Chapter 1], it is a limit of standard parallel transport operators along piecewise C^∞ curves.

More generally, let $\tau_s^{o,t}$ be the parallel transportation operator from fibers over x_o into fibers over x_s^t, and set

$$\tau_o^{s,t} = [\tau_s^{o,t}]^{-1}.$$

To construct the semi-group $e^{-\frac{tD^2}{2}}$, we use a matrix version of the Feynman-Kac formula.

$e_1, \ldots e_n$ is an orthonormal oriented base of $T_{x_o}M$.

$\tau_o^{s,t} L_{x_s^t}(\tau_s^{o,t} e_i, \tau_s^{o,t} e_j)$ denotes the element of End ξ_{x_o}

$\tau_o^{s,t} L_{x_s^t}(\tau_s^{o,t} e_i, \tau_s^{o,t} e_j)\tau_s^{o,t}$.

Definition 2.8. U_s^t denotes the process in $\mathrm{End}(F\otimes\xi)_{x_o}$ defined by the differential equation

(2.27)
$$dU_s^t = U_s^t[-\frac{1}{2}t \sum_{i<j} e_i e_j \otimes \tau_o^{s,t} L_{x_s^t}(\tau_s^{o,t} e_i, \tau_s^{o,t} e_j)]ds$$
$$U_o^t = I$$

We now have

Theorem 2.9. If $h\in\Gamma(F\otimes\xi)$, the following identity holds.

(2.28)
$$e^{-\frac{tD^2}{2}} h(x_o) = E[\exp\{-t \int_o^1 \frac{K(x_s^t)ds}{8}\}U_1^t \tau_o^{1,t} h(x_1^t)]$$

Proof: Itô's formula (see [16,IX, Theorems 1.2 - 1.3]) shows that

(2.29)
$$\tau_o^{s,t} h(x_s^t) = h(x_o)+\int_o^s \frac{t}{2} \tau_o^{v,t} \Delta^H h(x_v)dv$$
$$+ \int_o^s \tau_o^{v,t} \nabla_{\tau_v^{o,t} e_i} h(x_v^t).\delta w^i .$$

Using (2.29), it is then easy to apply Itô's formula [18] to the process

(2.30)
$$\exp\{-t \int_o^s \frac{K(x_v^t)dv}{8}\} U_s^t \tau_o^{s,t} h(x_s^t)$$

Taking expectations (which makes disappear the terms containing δw), we get (2.28). □

e) An asymptotic representation of $\mathrm{Tr}_s P_t(x_o, x_o)$.

We now disintegrate (2.28) the way we did in Section 1.

H_1, H_2 are still given by (1.27), (1.28).

P_1 is still defined as after (1.28). (1.29) is replaced by

(2.31)
$$du^t = X_i^*(u^t)(\sqrt{t}\, dw^{1,i} + v^{2,i}ds)$$
$$u^t(0) = u_o$$

Set

(2.32) $$x_s^t = \pi\, u_s^t$$

v^2 is adjusted so that

$$x_1^t = x_o$$

Using (1.35), and (2.28), we see that for any $k \in N$

(2.33) $$Tr_s[P_t(x_o\,,x_o)] = \int \frac{1}{(\sqrt{2\pi t})^n}\, Tr_s[U_1^t\, \tau_o^{1,t}]\exp\{-t\int_o^1 \frac{K(x_s^t)ds}{8}\}$$

$$\frac{\exp - \dfrac{|v^2(\sqrt{t}dw^1|^2}{2t}}{\det \dfrac{\partial\Phi}{\partial v^2}(\sqrt{t}dw^1+v^2)}\, G(t,dw^1)\,dP_1(w^1) + o(t^k) \;,$$

where $o(t^k)$ is uniform on M .

We are now left with the task of studying the asymptotics as $t \downarrow\downarrow 0$ of (2.33).

f) <u>The asymptotics of</u> $Tr_s P_t(x_o\,,x_o)$.

In (2.33), as $t \downarrow\downarrow 0$

- $G(t,dw^1) \to 1$ boundedly
- $\exp\{-t\int_o^1 \frac{K(x_s^t)ds}{8}\} \to 1$ boundedly .

Also Figure 1 shows that at $t = 0$

(2.34) $$v^2 = \frac{\partial v^2}{\partial\sqrt{t}} = 0$$

so that

(2.35) $$\exp - \frac{|v^2|^2}{2t} \to 1 \qquad \text{boundedly .}$$

We now show the critical fact that

(2.36) $$\frac{Tr_s\, U_1^t\, \tau_o^{1,t}}{(\sqrt{2\pi t})^n}$$

has a limit.

Indeed if $\tau_o^{1,t}$ is considered as acting on $T_{x_o}M$, the very definition of the curvature R [19, Section 4] shows that

$$\tau_o^{1,t} = I - \frac{t}{2}\int_o^1 R_{x_o}(u_o\, dw^1, u_o w^1) + o(t) .$$

If $L = 0$, in view of Theorem 2.3, the sequence of events should be obvious. Indeed, if $\tau_o^{1,t}$ is now considered as acting on $F_{\pm,x_o}$, we have

$$\tau_o^{1,t} = \exp\{ -\frac{t}{2}\int_o^1 R_{x_o}(u_o\, dw^1, u_o w^1) + o(t)\}$$

where $o(t)$ is taken in $\mathfrak{a}$ (identified with the antisymmetric elements of End $T_{x_o}M$).

We expand $\tau_o^{1,t}$ as in (2.17), i.e., if $\tau_o^{1,t}$ is now considered as acting on $(F\otimes\xi)_{x_o}$

$$(2.37) \qquad \tau_o^{1,t} = [1 + (-\frac{t}{2}\int_o^1 R_{x_o}(u_o\, dw^1, u_o w^1) + o(t)) + \dots + \frac{(-\frac{t}{2}\int_o^1 R_{x_o}(u_o\, dw^1, u_o w^1) + o(t))^{n/2} + o(t^{n/2})}{(n/2)!}] \otimes [I + O(t)]$$

Similarly, we expand U_1^t in the form

$$(2.38) \qquad U_1^t = I + \int_o^1 -\frac{t}{2}\sum_{i<j} e_i\, e_j \otimes \tau_o^{s,t} L_{x_s^t}(\tau_s^{o,t} e_i, \tau_s^{o,t} e_j)\, ds$$
$$+ \int_{0\leq s\leq s'\leq 1} [-\frac{t}{2}\sum_{i<j} e_i\, e_j \otimes \tau_o^{s,t} L_{x_s^t}(\tau_s^{o,t} e_i, \tau_s^{o,t} e_j)]$$
$$[-\frac{t}{2}\sum_{k<\ell} e_k\, e_\ell \otimes \tau_o^{s',t} L_{x_{s'}^t}(\tau_{s'}^{o,t} e_k, \tau_{s'}^{o,t} e_\ell)]\, ds\, ds'$$
$$+ \dots + o(t^{n/2}) .$$

Now if we calculate $\mathrm{Tr}_s[U_s^t\, \tau_o^{1,t}]$ and use (2.7), (2.8), proceeding as in (2.16) - (2.18), we see that the limit of (2.36) is simply obtained by cancelling the factor t and replacing the Clifford products $e_i\, e_j$

by the Grassmann products $(-2i)dx^i \wedge dx^j$.

Let η be the Riemannian orientation form of M. $\mathrm{Tr}_s[P_t(x_o,x_o)]\eta(x_o)$ is now a differential form of degree n over M . If $\omega,\omega' \in \overset{n}{\underset{o}{\oplus}}\Lambda(M)$ we write

$$\omega \overset{n}{=} \omega'$$

if ω and ω' have the same component in the maximal degree n .

We finally obtain [19, Theorem 3.15].

<u>Theorem 2.10</u>. The following relation holds

$$(2.39) \qquad \lim_{t\downarrow\downarrow 0} \mathrm{Tr}_s[P_t(x_o,x_o)]\eta(x_o) \overset{n}{=} \int \exp^{\wedge}\{-\frac{i}{4\pi}\int_o^1 R_{x_o}(u_o\, dw^1, u_o w^1)\}$$

$$dP_1(w^1) \wedge \mathrm{Tr}\exp[-\frac{L_{x_o}}{2i\pi}] \quad .$$

<u>Remark 2</u>. In [19] we proceeded as follows. Let γ be a Brownian motion valued in a independent of w^1 . Consider the stochastic differential equations

$$(2.40) \qquad dV_s^{1,t} = V_s^{1,t}(-\frac{t}{2}\sum_{i<j} e_i e_j d\gamma_i^j)$$

$$V_o^{1,t} = I_{F_{x_o}}$$

$$dV_s^{2,t} = V_s^{2,t}[\sum_{i<j} \tau_o^{s,t} L_{x_s^t}(\tau_s^{o,t} e_i, \tau_s^{o,t} e_j)\delta\gamma_i^j]$$

$$V_o^2 = I_{\xi_{x_o}}$$

Using the stochastic calculus, it is elementary to prove that

$$(2.41) \qquad U_1^t = E^{\gamma}[V_1^{1,t} \otimes V_1^{2,t}]\exp[\frac{n(n-1)t^2}{16}]$$

We can then directly use Theorem 2.3 to calculate

$$(2.42) \qquad \lim_{t\downarrow\downarrow 0} \mathrm{Tr}_s[V_1^{1,t}\tau_o^{1,t}] \otimes \mathrm{Tr}[V_1^{2,t}\ \tau_o^{1,t}].$$

In fact if V^2 is the solution of

$$(2.43) \qquad dV_s^2 = V_s^2[L_{x_o}(e_i, e_j)\delta\gamma_i^j]$$
$$V_o^2 = I$$

(2.42) is given by

$$(2.44) \qquad (-i)^{n/2}\, \mathrm{Pf}\Big[\int_o^1 \frac{R(u_o\, dw^1, u_o w^1)}{4\pi} + \frac{u_o\, \gamma_1\, u_o^{-1}}{2\pi}\Big] \mathrm{Tr}\, V_1^2$$

Using (2.13), we can replace $\mathrm{Pf}[\ldots]$ by $e^{\wedge[\ldots]}$ and proceed as before to obtain (2.39).

This avoids the explicit expansions (2.37) - (2.38). In Section 3, we will see the interpretation of (2.41).

For a related proof, see Azencott [11].

g) <u>The Index Theorem of Atiyah-Singer</u>

In (2.39), $\mathrm{Tr}\exp(\frac{-L}{2i\pi})$ is a representative of the Chern character $\mathrm{ch}\ \xi$ of ξ .

We now evaluate the first term in the r.h.s. of (2.39).

<u>Definition 2.11</u>. $\hat{A}$ denotes the ad $O(n)$ invariant analytic function on a such that if $C \in a$ has diagonal entries

$\begin{bmatrix} o & x_i \\ -x_i & o \end{bmatrix}$, then

$$(2.45) \qquad \hat{A}(C) = \prod_1^{\ell} \frac{x_i/2}{\mathrm{sh}\,\frac{x_i}{2}}$$

In the sequel $\hat{A}(\frac{R}{2\pi})$ should be understood as the analytic function $\hat{A}$ evaluated on $\frac{R}{2\pi}$ where all the products are replaced by exterior products.

We now have

<u>Theorem 2.12</u>. The following identity holds :

$$(2.46) \qquad \int \exp{}^{\wedge}\Big\{\frac{-i}{4\pi}\int_o^1 R(u_o\, dw^1, u_o w^1)\Big\}\, dP_1(w^1) = \hat{A}\Big(\frac{R}{2\pi}\Big) .$$

Proof : Since R is the Levi-Civita curvature tensor, it is known that that

(2.47) $$\int_0^1 \langle X, \int_0^1 R(u_o\, dw^1, u_o w^1) Y\rangle = -\int_0^1 \langle R(X,Y) u_o\, dw^1, u_o w^1\rangle .$$

We should now evaluate

(2.48) $$\int \exp^\wedge\{\frac{i}{4\pi} \int_0^1 \langle R u_o\, dw^1, u_o w^1\rangle\} .$$

However now R can be replaced by any antisymmetric $C \in \mathfrak{a}$. Also since Brownian bridge is rotation invariant, we can as well assume that C has diagonal entries

$\begin{bmatrix} 0 & x_i \\ -x_i & 0 \end{bmatrix}$. Using a well known formula of P. Lévy [57], we then know that (2.48) is given by

$$\frac{\prod_1^\ell x_i/4\pi}{\operatorname{sh} x_i/4\pi}$$

The proof is finished. □

We finally obtain the Index Theorem of Atiyah-Singer [8],[7].

Theorem 2.13. Ind D_+ is given by

(2.49) $$\operatorname{Ind} D_+ = \int_M \hat{A}(\frac{R}{2\pi}) \operatorname{ch} \xi .$$

Remark 3. In [57], Yor shows that

(2.50) $$E \exp -i y [\int_0^1 w^{1,1} dw^{1,2} - w^{1,2} dw^{1,1}] =$$

$$(E[\exp -\frac{y^2}{2} \int_0^1 |w^{1,1}|^2 ds])^2 = \frac{y}{\operatorname{sh} y}$$

the final equality being classical (for the simple proof by Williams, see Yor [57]).

Also (2.50) can be directly calculated using the stochastic calculus on the Brownian bridge as in [17, Theorem 4.17].

We now explain another derivation of (2.50). Consider the symmetric kernel $K(s,t)$ on $[0,1]^2$ given by

$$K^{1,1} = K^{2,2} = 0 \tag{2.51}$$

$$K^{1,2}(s,t) = \begin{cases} y & s \leq t \\ -y & s > t \end{cases}$$

$$K^{2,1}(s,t) = K^{1,2}(t,s) .$$

We now use the notations H_1, H_2 in (1.27), (1.28) with $m = 2$.

If $f = (f^1, f^2) \in H_1$, then

$$\int_{[0,1]^2} < K(s,t) f_s, f_t > ds\,dt = 2y[\int_{0 \leq s \leq t \leq 1} (f_s^1 f_t^2 - f_s^2 f_t^1)\, ds\, dt]$$

In particular

$$\int_{0 \leq s \leq t \leq 1} < K(s,t)\, \delta w_s^1, \delta w_t^1 > = y\int_0^1 (w_s^{1,1} dw_s^{1,2} - w_s^{1,2} dw_s^{1,1}) \tag{2.52}$$

In the r.h.s. of (2.52) d can be replaced by δ.

Let $\overline{P}_1$ be the projection operator from H on H_1.

$\overline{P}_1 K$ is a Hilbert-Schmidt operator on H_1. Let $\{\lambda_k\}$ be its eigenvalues, $\{\varphi_k\}_{k \in N}$ a corresponding base of orthonormal eigenvectors in H_1. Since

$$\overline{P}_1 K = \Sigma\, \lambda_k\, \varphi_k(s) \otimes \varphi_k(t) \tag{2.53}$$

we find that

$$\int_{0 \leq s \leq t \leq 1} < K(s,t)\, \delta w_s^1, \delta w_t^1 > = \Sigma \frac{\lambda_k}{2} [[\int_0^1 < \varphi_k(s), \delta w_s^1 >]^2 - 1] . \tag{2.54}$$

Since $\{\int_0^1 < \varphi_k(s), \delta w_s^1 >\}$ are mutually independent Gaussian variables,

we finally find

$$E[\exp - iy\int_0^1 w^{1,1} dw^{1,2} - w^{1,2} dw^{1,1}] = \frac{1}{\{\det_2[I + i K]\}^{1/2}} \tag{2.55}$$

where

(2.56) $$\det_2(I + iK) = \Pi[1 + i\lambda_k]e^{-i\lambda_k}$$

The λ_k can be easily explicitly calculated and are given by

(2.57) $$\lambda_k = \frac{y}{k\pi}, k \in Z_* ,$$

and each eigenspace is of dimension 2 .

So we find that

(2.58) $$[\det_2(I + iK)]^{1/2} = \prod_{k \in Z_{+*}} [1 + \frac{y^2}{k^2\pi^2}]$$

Of course this proves again the well khown result

(2.59) $$\frac{y}{\text{shy}} = \frac{1}{\prod_{k \in Z_{+*}} [1 + \frac{y^2}{k^2\pi^2}]}$$

As we shall see, in the formalism of Atiyah-Witten [4], replacing $A(\frac{R}{2\pi})$ by the corresponding infinite product has a very interesting geometric interpretation.

3. Localization formulas in equivariant cohomology and the Index Theorem.

In [4] Atiyah and Witten made a very interesting remark on the path integral representation of the Index of the Dirac operator for the Spin complex. They showed that if one applies formally in infinite dimensions a localization formula of Berline-Vergne [13], [14], Duistermaat-Heckman [28], the path integral should be equal on a priori grounds to the integral of a certain characteristic class over M , which turns out to be exactly equal to $\hat{A}(R/2\pi)$. The formal application of the localization formula gives then the right answer for the Index.

In [5], Atiyah and Bott discussed various aspects of localization formulas, which also follow from purely algebraic arguments, and suggest that a proof of such formulas should be given in infinite dimensions.

In [20], we extended the Atiyah-Witten formalism to the case of a twisted spin complex. We also shew that the Lefschetz fixed point formulas of Atiyah-Bott [5], Atiyah-Singer [8] could be given such a formal interpretation. This formalism also helped us to find a heat equation proof of the infinitesimal Lefschetz formulas [21], [22].

Now the proofs of Berline-Vergne [13], Duistermaat-Heckman [28] are difficult to extend in infinite dimensions, since they make an explicit use of Stokes formula on small balls. As is well known, it is difficult to integrate on balls in infinite dimensions.

In [25], we took the opposite direction. Since the heat equation, when interpreted probabilistically, produced the required localization in infinite dimensions, could it be that such a proof could now be extended in finite dimensions so as to give another proof of the localization formulas of [13] - [14] - [28]? We gave in [25] such a proof, which reproduces in detail the various technical steps of the proof of the Index Theorem given in Section 2. Later on [24], [25], inspired by the finite dimensional baby proof, we could give two heat equation proofs of the Index Theorem for families.

In this section, we want to introduce the reader to some of these problems. In a), we give the main assumptions and notations. In b), we give a brief sketch of our proof [25] of the localization formulas of Berline-Vergne [13], [14], Duistermaat-Heckman [28]. In c), we summarize the remarks of Atiyah and Witten [4]. In d), we show that the proof of the Index Theorem in Section 2, and the proof of the localization formula in b), are strictly parallel. In e), following [20], we show how to extend the formalism of Atiyah and Witten [4] to twisted spin complexes. In f), we briefly summarize various applications and extensions. In particular, we briefly explain the superconnection formalism of Quillen [49], and our heat equation proofs of the Index Theorem for families [24].

a) Assumptions and notations.

M is a C^∞ connected compact oriented Riemannian manifold of dimension n.

X denotes a Killing vector field on M, i.e. X generates a group of isometries of M.

∇ denotes the covariant differentiation operator for the Levi-Civita connection of TM . R is the curvature tensor of TM . Set

$$J_X = \nabla . X \tag{3.1}$$

Then J_X is an antisymmetric (1,1) tensor. Also since ∇ is X invariant, we know [13] that

$$\nabla_Y J_X + R(X,Y) = 0 . \tag{3.2}$$

$\Lambda(M)$ denotes the set of C^∞ sections of the exterior algebra of T^*M . An element μ of $\Lambda(M)$ can be written as

$$\mu = \mu_o + \ldots + \mu_n \tag{3.3}$$

where μ_j is a j-form on M . If S is a submanifold of M of dimension j , by definition

$$\int_S \mu = \int_S \mu_j . \tag{3.4}$$

Set

$$F = \{X = 0\} \tag{3.5}$$

It is well-known that F is a finite union of totally geodesic connected submanifolds of M . Let N be the normal bundle of F in M. Since F is totally geodesic, the Levi-Civita connection of TM produces an Euclidean connection on N . R^N denotes the restriction of R to N .

Since N is orthogonal to TF which is stable by J_X , N is stable by J_X . Also by (3.2) J_X is parallel on F . J_X is non degenerate on N , which has even dimension. The 2-form $\langle Y, J_X Z \rangle$ being non degenerate on N , N is naturally oriented. The orientation of N makes $\mathrm{Pf}\, J_X > 0$. TF is then also oriented, so that $TF \oplus N$ is oriented like TM .

The exterior differentiation operator d increases the degree of μ_i by 1 , the interior multiplication operator i_X decreases it by 1 .

Following Berline-Vergne [13], Witten [56], we now set the following definition.

Definition 3.1. $\mu \in \Lambda(M)$ will be said to be X equivariantly closed (X e.c.) if

$$(d + i_X)\mu = 0 . \tag{3.6}$$

(3.6) shows that if $\mu = \mu_o + \ldots + \mu_n$, then

$$d\mu_j + i_X \mu_{j+2} = 0 \tag{3.7}$$

b) Localization formulas in equivariant cohomology.

We now will give a new proof of the localization formulas of Berline-Vergne [13], [14], Duistermaat-Heckman [28] (also see Atiyah-Bott [6]).

It will be more detailed than necessary since our purpose is to compare it with the proof of the Index Theorem.

In what follows we use the notation $\dfrac{1}{\mathrm{Pf}[\frac{J_X + R^N}{2\pi}]}$ in the following sense.

Recall that on N $\mathrm{Pf}(J_X) > 0$. We can then expend analytically $\mathrm{Pf}[\frac{J_X + R^N}{2\pi}]^{-1}$ in the variable R^N, replacing products by exterior products. We then get a finite sum of differential forms. $\dfrac{1}{\mathrm{Pf}[\frac{J_X + R^N}{2\pi}]}$ is an even element in $\Lambda(F)$.

We now have the formula of Berline-Vergne [13]-[14], Duistermaat-Heckman [28].

Theorem 3.2. If μ is X e.c., then

$$\int_M \mu = \int_F \frac{\mu}{\mathrm{Pf}[\frac{J_X + R^N}{2\pi}]} \tag{3.8}$$

Proof : This proof is taken from [25]. Let X' be the 1-form dual to X.

We first claim that for any s

(3.9) $$\int_M \mu = \int_M \exp - [s(d + i_X)X']\mu$$

Indeed (3.9) holds at 0 . Moreover.

(3.10) $$\frac{\partial}{\partial s} \int_M \exp\{- s(d + i_X)X'\} \quad \mu = - \int (d + i_X)X' \ \exp\{- s(d + i_X)X'\}\mu$$

Now if L_X is the Lie derivative operator associated with X

(3.11) $$(d + i_X)^2 = L_X .$$

Since X is Killing, $L_X X' = 0$. It follows that

(3.12) $$(d + i_X)[\exp\{- s(d + i_X)X'\} \mu] = 0 .$$

So the r.h.s. of (3.10) is equal to

(3.13) $$-\int (d + i_X)[X' \ \exp\{- s(d + i_X)X'\} \mu]$$

Now if $v \in \Lambda(M), \int_M dv = 0$. Also since $i_X v$ has degree $\leqslant n-1$ $, \int_M i_X v = 0$.

(3.13) is then clearly 0 and (3.9) holds.

We get for any $t > 0$

(3.14) $$\int_M \mu = \int_M \exp\{-\frac{dX' + |X|^2}{2t}\} \mu .$$

As $t \downarrow\downarrow 0$, (3.14) clearly localizes on F . By making the change of coordinates $y = \sqrt{t}y'$ in N , we get that as $t \downarrow\downarrow 0$, if V is a small neighborhood of F , (3.14) is very close to

(3.15) $$t^{\frac{\dim N}{2}} \int_N \exp\{ - \frac{dX'(x,\sqrt{t},y) + |X|^2(x,\sqrt{t}y)}{2t}\} \ \mu(x,\sqrt{t}y)$$

Now

(3.16) $$\frac{|X|^2(x,\sqrt{t}y)}{2t} \simeq \frac{|J_X y|^2}{2} + o(\sqrt{t})$$

Using (3.2) and a non entirely trivial argument on differential forms [25], we find that if $R^{TF}(J_X y,y)$ is the restriction of the antisymmetric matrix $R(J_X y,y)$ to TF , (3.15) converges to

$$(3.17) \qquad \int_F [\int_N \exp\{J_X - \frac{R^{TF}(J_X y,y)}{2} - \frac{|J_X y|^2}{2}\}]\mu$$

Now the symmetries of the Levi-Civita curvature tensor R show that if $Y,Z \in TF$

$$(3.18) \qquad < Y, R^{TF}(J_X y,y)Z > = < R^N(Y,Z)y, J_X y > .$$

(3.17) is then equal to

$$(3.19) \qquad \int_F [\int_N \exp\left\{J_X - \frac{< R^N(.,.)y, J_X y>}{2} - \frac{|J_X y|^2}{2}\right\}]\mu$$

where R^N is considered as a 2 form on F with values in End N. The gaussian integral (3.19) is readily evaluated to be equal to

$\mathrm{Pf}[\frac{J_X + R^N}{2\pi}]^{-1}$. The proof is finished. □

c) A remark of Atiyah and Witten.

We go back to the assumptions of Section 2.

We first consider the case of the spin complex i.e. ξ is the trivial line bundle (with $L = 0$) .

Take $\alpha \in SO(n)$, $\beta \in Spin(n)$ such that $\sigma(\beta) = \alpha$.
Let $\theta_1, \dots \theta_\ell$ be the angles associated with α. Then it follows from (2.7), (2.8) and [5,p. 484], [19, Proposition 1.2] that

$$(3.20) \qquad i^{-\ell}\, \mathrm{Tr}_s \beta = \pm \prod_1^n 2 \sin \frac{\theta_j}{2} .$$

We now follow Atiyah [4]. Let $\overline{M}^\infty$ be the space of smooth loops in M $s \in S_1 = R/Z \to x_s$. $T_x \overline{M}^\infty$ is identified with the space of smooth periodic vector fields. If $X, Y \in T_x \overline{M}^\infty$, we define the scalar product

$$(3.21) \qquad < X,Y > = \int_0^1 < X_s, Y_s > ds .$$

$\overline{M}^\infty$ has a Riemannian structure.

In the sequel, we will do as if the Brownian measure was carried by smooth paths. Note that although Brownian motion is not smooth, all the standard operations like parallel transportation are well defined and are limits of the corresponding operations on approximating smooth paths [16, Chapters 1 - 3].

S_1 acts isometrically on $\overline{M}^{\infty}$ by the mapping k_s defined by

$$k_s x = x_{s+} .$$

k_s acts isometrically on $\overline{M}^{\infty}$. The generating Killing vector field $X(x)$ is given by

$$X(x)_s = \frac{dx_s}{ds} \tag{3.22}$$

The associated one form X' is given by

$$X'(Y) = \int_0^1 \langle Y, dx \rangle \tag{3.23}$$

dX' is the 2-form given by

$$dX'(Y,Z) = 2 \int_0^1 \langle \frac{DY}{Ds}, Z \rangle \, ds . \tag{3.24}$$

In (3.24), $\frac{DY}{Ds}$ is the covariant differential of Y along x.

In particular

$$\frac{(d + i_X)X'}{2} = \frac{1}{2} \int_0^1 \left|\frac{dx}{ds}\right|^2 ds + \frac{dX'}{2} \tag{3.25}$$

Now Atiyah considers in [4] the eigenvalue problem on $Y \in T_x \overline{M}^{\infty}$

$$\frac{DY}{Ds} = \lambda Y . \tag{3.26}$$

Recall that $Y_o = Y_1$. (3.26) can be put in the equivalent form

$$\frac{dZ}{ds} = \lambda Z \tag{3.27}$$

$$\tau_1^o Z_1 = Z_o .$$

If $\theta_1, \dots \theta_\ell$ are the angles of τ_1^o, it is trivial to verify that λ takes the values

$$\pm 2i\pi m \pm i\theta_j \tag{3.28}$$

The Pfaffian of $-\frac{dX'}{2}$ is formally given by

$$(3.29)\qquad Pf[-\frac{dX'}{2}] = \prod_{j=1}^{\ell} \theta_j \prod_{M=1}^{+\infty} [4\pi^2 m^2 - \theta_j^2]$$

By dividing formally by the infinite normalizing constant $\prod_{m=1}^{+\infty} (4\pi^2 m^2)^{\ell}$, we get

$$(3.30)\qquad \frac{Pf[-\frac{dX'}{2}]}{(\prod_1^{\infty} 4\pi^2 m^2)^{\ell}} = \prod_{j=1}^{\ell} \theta_j \prod_{m=1}^{+\infty} [1 - \frac{\theta_j^2}{4\pi^2 m^2}]$$

$$= \prod_{j=1}^{\ell} 2 \sin \frac{\theta_j}{2} .$$

The idea of Atiyah and Witten is to use the equality of (3.20) and (3.30) to rewrite (2.28) in a different way. Namely if $Q_t^{x_o}$ is the law of the scaled Brownian bridge starting and ending at x_o, if $p_t(x,y)$ is the heat kernel on M, they write formally

$$(3.31)\qquad p_t(x_o,x_o)\,dx_o\,dQ_t^{x_o}(x) = \frac{1}{(\sqrt{2\pi t})^d} \exp\{-\int_o^1 \frac{|\dot{x}|^2}{2t} ds\} dD(x)$$

In (3.31), d is the dimension of $\overline{M}^{\infty}$ $(d=+\infty)$ and $dD(x)$ is the "volume element" of $\overline{M}^{\infty}$.

Using (2.23), (3.20), (3.30), (3.31) and neglecting $\exp\{-\frac{t}{8}\int_o^1 K(x_s)\,ds\}$, we can write the Index of D_+ as

$$(3.32)\qquad \text{Ind } D_+ = \frac{(\prod_1^{+\infty} m^2)^{\ell}\, i^{\ell}}{(2\pi)^{\ell}} \int \exp\{-\frac{\int_o^1 |\frac{dx}{ds}|^2 ds}{2t}\} Pf[-\frac{dX}{2}] dD(x) .$$

Now if $d' < +\infty$, if A is a (d',d') antisymmetric matrix

$$(3.33)\qquad \frac{A^{\wedge \frac{d'}{2}}}{(d'/2)!} = PfA\, dx^1 \wedge \ldots \wedge dx^{d'} = e^A_{d'}$$

We now use (3.33) formally, and so we obtain

$$(3.34)\qquad \text{Ind } D_+ = \frac{(\prod_1^{+\infty} m^2)^{\ell}\, i^{\ell}}{(2\pi)^{\ell}} \int \exp - \frac{(d + i_X)X'}{2t} .$$

Since $L_X X' = 0$, we have

$$(d + i_X) \exp - \frac{(d + i_X)X'}{2t} = 0 .$$

Now Atiyah and Witten [4] apply formula (3.8) in a formal way. They note that

$$M = (X = 0)$$

and so we should get

$$\text{(3.35)} \qquad \operatorname{Ind} D_+ = \frac{(\prod_1^{+\infty} m^2)^{\ell}\, i^{\ell}}{(2\pi)^{\ell}} \int_M \frac{1}{\operatorname{Pf}\left[\frac{J_X + R^N}{2\pi}\right]}$$

Now at $x \in M, N_x$ is the set of f taking values in T_xM such that $\int_0^1 f ds = 0$. Also $J_X = \frac{d}{ds}$. The eigenvalues of J_X are $\pm 2i\pi m$. Proceeding as in Atiyah [4], we find that if A is a (n,n) antisymmetric matrix with diagonal entries $\begin{bmatrix} 0 & x_i \\ -x_i & 0 \end{bmatrix}$, then

$$\text{(3.36)} \qquad \operatorname{Pf} \frac{J_X + A}{2\pi} = \prod_{\substack{m=1 \\ 1 \leq i \leq \ell}}^{+\infty} \left[m^2 - \frac{x_i^2}{4\pi^2}\right]$$

so that

$$\text{(3.37)} \qquad \frac{(\prod_1^{+\infty} m^2)^{\ell}}{\prod_{m=1}^{+\infty}\left[m^2 - \frac{x_i^2}{4\pi^2}\right]} = \frac{1}{\prod_{\substack{m=1 \\ 1 \leq i \leq \ell}}^{+\infty}\left[1 - \frac{x_i^2}{4\pi^2 m^2}\right]} = \frac{\prod_1^{\ell} \frac{x_i}{2}}{\sin \frac{x_i}{2}}$$

and so using (3.35), we should get

$$\text{(3.38)} \qquad \operatorname{Ind} D_+ = \int_M \hat{A}\left(\frac{R}{2\pi}\right)$$

(3.38) is of course the correct answer.

In [4], Atiyah suggested that these formal considerations should be proved. The proofs of Berline-Vergne [13], [14], Duistermaat-Heckman [28] were in fact impossible to extend in infinite dimensions.

d) The heat equation probabilistic proof as a rigorous proof of localization.

Our idea was that the heat equation proof of the Index Theorem should be taken as a model proof of localization in infinite dimensions, which could be extended in finite dimensions.

This is the reason why we gave the new proof of Theorem 3.2.

Remarkably enough, even the intermediary steps of the proof of Theorem 2.10 and Theorem 3.2 are closely related. In particular the intermediary formulas (2.39) and (3.17), the key steps (2.47) and (3.18) can be easily compared.

Note that the normalizations which had to be explicitly done in (3.34) are automatically performed by Brownian motion.

This point of view is fully developed in our papers [20] and [25].

e) Extension of the formalism of Atiyah and Witten to twisted spin complexes.

We now briefly explain how in [20], we extended the formalism of Atiyah - Witten [4] to twisted spin complexes.

Namely we used formula (2.41). We find from (2.41) that

$$\mathrm{Tr}_s\, U_1^t \tau_o^{1,t} = E^\gamma[\mathrm{Tr}_s[V_1^{1,t}\tau_o^{1,t}]\mathrm{Tr}[V_1^{2,t}\tau_o^{1,t}]]\exp\left(\frac{n(n-1)t^2}{16}\right) \tag{3.39}$$

Now doing as if γ was differentiable and proceeding as in (3.30), we find that if r^γ is the (random) 2-form on $T_x\overline{M}^\infty$

$$r^\gamma(Y,Z) = \int_0^1 <\tau_o^s Y_s, e_i><\tau_o^s Z_s, e_j> d\gamma_j^i$$

then

$$\frac{i^\ell\, \mathrm{Pf}[-\frac{dX'}{2} - tr^\gamma]}{(\prod_1^{+\infty} 4\pi^2 m^2)^\ell} = \pm\, \mathrm{Tr}_s[V_1^{1,t}\tau_o^{1,t}] \tag{3.40}$$

Forgetting about $\exp[\frac{n(n-1)t^2}{16}]$, we obtain formally that if $\sigma = \frac{dX'}{2}$

(3.41) $$\text{Ind } D_+ = \frac{(\prod_1^{+\infty} m^2)^{\ell}\, i^{\ell}}{(2\pi)^{\ell}} \int \exp\{\frac{-\int_0^1 |\frac{dx}{ds}|^2}{2t}\} \frac{Pf[-(\sigma+tr^{\gamma})]}{t^d}$$

$$Tr[V_1^{2,t}\, \tau_0^{1,t}]dP(x)\; dP'(\gamma).$$

Proceeding as in (3.33), we get

(3.42) $$\text{Ind } D_+ = \frac{(\prod_1^{+\infty} m^2)^{\ell}\, i^{\ell}}{(2\pi)^{\ell}} \int_{\overline{M}^{\infty}} \exp \frac{-(d+i_X)X'}{2t} [\int \exp(-r^{\gamma}) Tr[V_1^{2,t}\, \tau_0^{1,t}]$$

$$dP'(\gamma)]$$

Now the final term in (3.42) can be explicitly computed using the stochastic calculus. We briefly summarize the results in [20].

<u>Definition 3.3</u>. H_s denotes the solution of the differential equation

(3.43) $$dH_s = H_s \tau_0^s L_{x_s}\; ds\;.$$

$$H_0 = I\;.$$

(3.43) should be interpreted as follows.

$\tau_0^s L_{x_s}$ is a 2-form on $T_x\overline{M}^{\infty}$ with values in $\text{End}(\xi_{x_0})$ defined by

$$(Y,Z)\in T_x\overline{M}^{\infty} \to \tau_0^s L_{x_s}(Y_s,Z_s)\tau_s^0$$

H sould then be expanded as a series

$$H_s = I + \int_0^s \tau_0^v L_{x_v} \tau_v^0\, dv + \int_{0\le v_1\le v_2\le s} \tau_0^{v_1} L_{x_{v_1}} \wedge \tau_0^{v_2} L_{x_{v_2}}\, dv_1\, dv_2 + \ldots$$

In [20], we find that if $\beta = Tr[H_1\, \tau_0^1]$, then

(3.44) $$\text{Ind } D_+ = \frac{(\prod_1^{\infty} m^2)^{\ell}\, i^{\ell}}{(2\pi)^{\ell}} \int \exp \frac{-(d+i_X)X'}{2t}\; \beta$$

Now we proved in [20] that β is X equivariantly closed i.e.

(3.45) $$(d+i_X)\beta = 0\;.$$

β appears as an infinite dimensional characteristic class associated with a bundle whose structure group is a Kac-Moody group.

f) Applications and extensions.

It should now be clear that there is indeed a theory of differential forms on the loop space $\overline{L}^{\infty}(M)$, and that Brownian motion produces as such all the required normalizations, so that this theory makes sense. The very comparison of the proof of Theorems 2.10 and 3.2 shows that this is not just a formal remark, but that computations reflect this in detail.

The comparison of (2.23)-(2.28) with (3.43) is instructive. Brownian motion selects by itself the form of maximal degree (which is here infinite) which is to be integrated. It does so by using spinors (which are finite dimensional objects), instead of differential forms. However by doing so, Brownian motion hides the deep algebraic structure which reveals itself in the non rigorous formula (3.44).

Such observations can be pushed much beyond what is done here.

First of all, the Lefschetz fixed point formulas of Atiyah-Bott [5], Atiyah-Singer [8] have been proved probabilistically in Bismut [19] Another proof has been given in Berline-Vergne [14]. It was shown in [20] that the equivariant cohomology formalism also applies in this case. It G is the group of isometries of M , G acts obviously on $\overline{M}^{\infty}$, and the action of G commutes with the action of S_1 . S_1 is then replaced by $S_1 X G$. Also the infinitesimal Lefschetz formulas were proved in [21] - [22] using heat equation. The proof was inspired by a formal representation of the type (3.44) .

We want to mention the recent proof by us [24] of the Index Theorem for families of Dirac operators. Let M be a compact connected manifold, f a submersion from M onto the compact connected manifold B Let G_y be the fiber $G_y = f^{-1}(\{y\})$. For each $y \in B, D_y$ is a Dirac operator operating along the fiber G_y . The Index Theorem of Atiyah-Singer for families calculates $\text{Ker } D_+ - \text{Ker } D_-$ as an element of $K(B)$. In [24], we gave two heat equation proofs of the computation of

$$\text{ch}(\text{Ker } D_+ - \text{Ker } D_-) \ . \tag{3.46}$$

The proofs in [24] are based on two facts :

1). The extension to infinite dimensions of the formalism of Quillen [49]. In [49], Quillen had considered the case of a Z_2 graded vector bundle $E = E_0 \oplus E_1$ endowed with an "odd" linear mapping u exchanging E_0 and E_1. He had produced non trivial representatives of $ch(E_0 - E_1)$ using explicitly the linear mapping u. If v is the restriction of u to E_0, clearly in the sense of K theory

$$\text{Ker } v - \text{Coker } v = E_0 - E_1 .$$

In particular in Quillen's formalism, ordinary connections ∇ on $E_0 \oplus E_1$ (which respect the splitting) and odd linear mappings like u are on the same footing. $\nabla + u$ is called a superconnection by Quillen [49]. If $R = (\nabla + u)^2$, Quillen shows that $Tr_s[e^{-R/2i\pi}]$ still represents $ch(E_0 - E_1) = ch(\text{Ker } v - \text{Coker } v)$.

Now for $y \in B$, let H_+^∞, H_-^∞ be the set of C^∞ sections over the fiber G_y of the considered fiber bundles. The whole purpose of Index Theory for families is to make sense of the equality

$$\text{Ker } D_+ - \text{Coker } D_+ = H_+^\infty - H_-^\infty . \tag{3.47}$$

However, contrary to what happens in finite dimensions, the r.h.s. can be well defined only if the operator D_+ is given. Two unequivalent elliptic operators of course produce different $H_+^\infty - H_-^\infty$. So in infinite dimensions, (3.47) is tautological.

In [24], we could extend Quillen's formalism to the infinite dimensional H_+^∞, H_-^∞. Using D instead of u we proved that indeed $ch(\text{Ker } D_+ - \text{Ker } D_-)$ was given by the infinite dimensional analogue of Quillen's formula.

2). Using formal formulas like (3.44), we could find the right objects to be considered so as to calculate $ch(\text{Ker } D_+ - \text{Coker } D_+)$ explicitly.

This makes the formula (2.23) for Ind D_+.

$$\text{Ind } D_+ = Tr_s \, e^{-\frac{D^2}{2}} \tag{3.48}$$

as crying to be considered as a formula for the Chern character of

Ker D_+- Coker D_+ (this is irrelevant if B is a point !). Also D^2 should now be viewed as a curvature.

For the full proofs of these facts, we refer to [24]. The proof has been explained in detail in [25].

4. The Witten complex and the Morse inequalities.

In this section, we briefly discuss the proof by Witten [56] of the Morse inequalities for a Morse function having isolated critical points. In [56], Witten has shown how to construct other complexes than the complex of the Rham which have the same cohomology as the de Rham complex. In [56] Witten introduced a small parameter $t \downarrow\downarrow 0$. By studying the lower part of the spectrum of a modified Laplacian, he proved the Morse inequalities.

In this section, we will show how to use instead the small time behavior of the heat kernel. Equality of the Index with a heat equation trace is now replaced by an inequality. The proof then becomes closely related to the heat equation proof of the Lefschetz fixed point formulas when the fixed points are isolated (see [15], [19] for the general case). This point of view has been developed by us in [23].

In [56], Witten also suggested that instanton considerations could prove analytically that a geometrically constructed finite dimensional complex associated with h gives the real (and even the integer) cohomology of M, as shown by Smale [64]. This has been shown analytically by Helffer and Sjöstrand [38] using remarkably efficient techniques to study tunnelling problems (which are problems where energy wells interact). The analysis of Helffer-Sjöstrand goes much beyond what is done here. However note that probabilistic work of Ventcell-Freidlin [54] - [55], was also applied in Ventcell [65] to eigenvalue problems and may have a direct bearing on this problem. Ventcell-Freidlin techniques were used in tunnelling problems by Jona Lasinio-Martinelli-Scoppola in [41].

a) Assumptions and notations.

M is a connected compact Riemannian manifold of dimension n.

∇ denotes the covariant differentiation operator for the Levi-Civita connection.

$\Lambda^p(M)$ denotes the set of p-forms on TM. $\Lambda(M)$ is the exterior algebra

$$\Lambda(M) = \bigoplus_{o}^{n} \Lambda^p(M).$$

The operator d acts on C^∞ sections of $\Lambda(M)$. Let δ be its adjoint for the standard Riemannian structure on differential forms. The Laplacian $\Box$ is defined by

$$\Box = (d+\delta)^2 = d\delta + \delta d .$$

Hodge's theory shows that if $\Box_p$ is the restriction of $\Box$ to sections of $\Lambda^p(M)$, $\mathrm{Ker}\,\Box_p$ is isomorphic to the pth real cohomology group $H^p(M;R)$.

B_p $(o \leq p \leq n)$ are the Betti numbers of M, i.e. $B_p = \dim H^p(M;R)$.

h is a C^∞ function on M with values in R.

If A is a (n,n) tensor, A acts as a derivation on $\Lambda(M)$, so that if $\alpha \in \Lambda^1(M)$, $X \in TM$

$$A\alpha(X) = -\alpha(AX) \tag{4.1}$$

Following Witten [56], we now define the operators

$$d^h = e^{-h}\, d e^h \tag{4.2}$$

$$\delta^h = e^h\, \delta\, e^{-h}$$

$$\Box^h = d^h \delta^h + \delta^h d^h .$$

As pointed out by Witten [56], the cohomology of the operator d^h is the same as the cohomology of d. Hodge's theory still holds, so that if $\Box_p^h$ is the restriction of $\Box^h$ to the section $\Lambda^p(M)$

$$B_p = \dim[\mathrm{Ker}\,\Box_p^h]. \tag{4.3}$$

Clearly

(4.4) $$d^h = d + dh\wedge$$

$$\delta^h = \delta + i_{dh} .$$

In (4.4), $dh \in \Lambda^1(M)$ is identified with an element of TM by the metric, and so i_{dh} is well-defined.

An elementary computation shows that

(4.5) $$\Box^h = \Box + |dh|^2 - \Delta h - 2\nabla . dh$$

b) The basic inequality.

For $\alpha > o$, $t > o$, let $P_t^\alpha(x,y)$ be the C^∞ kernel associated with the operator $e^{-\frac{\alpha t \Box^{h/t}}{2}}$.

$J_p(\alpha,t,x)$ denotes the trace of $P_t^\alpha(x,x)$ acting on $\Lambda_x^p(M)$.

$K_p(\alpha,t)$ is defined by

$$K_p(\alpha,t) = \int_M J_p(\alpha,t,x)dx$$

Definition 4.1. If $(a_p)_{o \leq p \leq n}$, $(b_p)_{o \leq p \leq n}$ are two sequences of real numbers, we write

(4.6) $$(a_p) \geqslant (b_p)$$

if for any $q (o \leq q \leq n)$

(4.7) $$a_q - a_{q-1} + \ldots \geqslant b_q - b_{q-1} + \ldots$$

with equality in (4.7) for $q = n$.

We now have

Theorem 4.2. For any $\alpha > 0$, $t > 0$

(4.8) $$(K_p(\alpha,t)) \geqslant (B_p)$$

Proof : For $\lambda > 0$, let F_λ be the eigenspace of $\alpha t \Box^{h/t}$. F_λ

splits into

$$F_\lambda = \overset{n}{\underset{o}{\oplus}} F_\lambda^p \tag{4.9}$$

where F_λ^p are the corresponding forms of degree p.
Also the sequence

$$0 \to F_\lambda^o \xrightarrow[d^{h/t}]{} F_\lambda^1 \xrightarrow[d^{h/t}]{} \cdots \xrightarrow[d^{h/t}]{} F_\lambda^n \to 0 \tag{4.10}$$

is exact. Indeed if $\omega \in F_\lambda^p$, $\alpha t \square^{h/t} \omega = \lambda\omega$, if $d^{h/t}\omega = 0$ then

$$\omega = \frac{\alpha t d^{h/t}}{\lambda} \delta^{h/t} \omega$$

and so $d^{h/t}$ is exact on F_λ.

Trivially, this implies that

$$(\dim F_\lambda^p) \geqslant 0 \tag{4.11}$$

Now

$$K_p(\alpha,t) - K_{p-1}(\alpha,t) + \ldots = \operatorname{Tr} e^{-\frac{\alpha t \square_p^{h/t}}{2}} - \operatorname{Tr} e^{-\frac{\alpha t \square_{p-1}^{h/t}}{2}} + \ldots \tag{4.12}$$

$$= B_p - B_{p-1} + \ldots + \sum_{\lambda > 0} e^{-\lambda t}[\dim F_\lambda^p - \dim F_\lambda^{p-1} \ldots].$$

Using (4.11), (4.8) is obvious. □

Remark 1. The inequality (4.8) is the analogue of the equality (2.23) in Index Theory.

c) The Morse inequalities : the non degenerate case.

h is now assumed to be a Morse function, i.e. h has a finite number of critical point $x_1, \ldots x_\ell$ at which d^2h is non degenerate. Recall that we use the convention (4.1).

We now claim [23, Theorem 1.4].

<u>Theorem 4.3</u>. As $t\downarrow\downarrow 0$, $K_p(\alpha,t)$ has a limit $K_p(\alpha)$ given by

(4.13) $$K_p(\alpha)=\sum_{i=1}^{\ell}\frac{\mathrm{Tr}_p\exp(\alpha\nabla.dh(x_i))}{|\det(I-\exp(-\alpha\nabla.dh(x_i)))|}$$

<u>Proof</u> : Weitzenböck's formula shows that

$$\square=-\Delta^H+L$$

where L is a 0 order matrix valued operator.

We now prove (4.13) with $\alpha=1$. With a general α the proof is identical.

$u^t_\cdot$, $x^t_\cdot$ are taken as in (2.31), (2.32). Let U^s_t be the process of linear operators acting on $\Lambda_{x_o}(M)$ given by

(4.14) $$dU^t_s=U^t_s[\tau^s_o\,\nabla dh(x^t_s)-\tfrac{1}{2}t\,\tau^s_o\,L(x^t_s)]ds$$

$$U^t_o=I$$

Then for any $x_o\in M$, for any $k\in N$

(4.15) $$J_p(1,t,x_o)=\int\frac{1}{(\sqrt{2\pi t})^n}\exp\{-\int_o^1\frac{|dh|^2(x^t_s)ds}{2t}+\int_o^1\frac{\Delta h(x^t_s)ds}{2}\}$$

$$\exp\{-\frac{|v^2(\sqrt{t}dw^1,x_o)|^2}{2t}\}\mathrm{Tr}_p[U^\alpha_1\tau^{1,t}_o]\frac{G(t,dw^1)dP_1(w^1)}{\det[C'(\sqrt{t}dw^1,x_o)]}+o(t^k).$$

Here $v^2(\sqrt{t}dw^1,x_o)$ is explicitly depending on x_o.

Also $\det[C'(\sqrt{t}dw^1,x_o)]$ is the determinant appearing in (1.35).

As $t\downarrow\downarrow 0$, since $|dh|\neq 0$ out of the x_i, and since a Brownian bridge with parameter t escapes with small probability far from the starting point x_o, we find that if V is a small neighborhood of the $\{x_i\}$, as $t\downarrow\downarrow 0$

(4.16) $$\int_M J_p(1,t,x_o)dx_o\sim\int_V J_p(1,t,x_o)dx_o\,.$$

Taking geodesic coordinates around each x_i and doing the change of variables $X = \sqrt{t}X'$ in $T_{x_i}M$, we finally find that if $u_1 \dots u_\ell$ are orthogormal frames at $x_1 \dots x_\ell$

(4.17) $$\lim_{t \downarrow \downarrow 0} \int J_p(1,t,x_o)dx_o = \sum_{i=1}^{\ell} \int \exp\{\frac{1}{2}\Delta h(x_i) -$$

$$\frac{1}{2}\int_o^1 |\nabla_{c_i+u_i w_s^1} dh(x_i)|^2 ds\} \mathrm{Tr}_p[\exp \nabla.dh(x_i)] \frac{dP_1(w^1)dc_i}{(\sqrt{2\pi})^n}$$

Now if b is a one dimensional Bridge starting and ending at x (at time 1), by [66, p.206]

(4.18) $$E \exp\{-\frac{\beta^2}{2}\int_o^1 |b|^2 ds\} = \sqrt{\frac{\beta}{sh\beta}} \exp[-\beta x^2 \; th\frac{\beta}{2}] .$$

Putting $\nabla.dh(x_i)$ in diagonal form and using (4.18), we get (4.13) □

Remark 2. (4.18) is directly related to the harmonic oscillator. In [56] Witten instead studies directly the lower part of the spectrum of $\square^{h/t}$.

Also note that using the Morse Lemma, we could as well assume that on a neighborhood of each x_i, h is quadratic, and the metric is flat, so that on this neighborhood, $\square^h$ is exactly the harmonic oscillator.

Let us recall that x_i is of index p if the number of negative eigenvalues of $\nabla\, dh(x_i)$ is exactly p.

Let M_p be the number of x_i of index p.

We now have [25, Theorem 1.5]

Theorem 4.4. The following relations holds.

(4.19) $$\lim_{\alpha \to +\infty} K_p(\alpha) = M_p$$

Proof : The argument is now directly related to Witten [56].

Let $\mu_1, \dots \mu_n$ be the eigenvalues of $\nabla.dh(x_i)$. Using the convention (4.1), we know that the eigenvalues of $\nabla.dh(x_i)$ acting on

$\Lambda^p_{x_i}(M)$ are all the sums

$$-\sum_{k=1}^{p} \mu_{j_k}$$

where $j_1 < j_2 < \ldots < j_p$. The eigenvalues of $\exp(\alpha\nabla . dh(x_i))$ are then given by $\exp\{-\alpha \sum_{k=1}^{p} \mu_{j_k}\}$. If the index of x_i is p , the corresponding term in (4.13) tends to 1 as $\alpha \to +\infty$. Otherwise it tends to 0 .

Using Theorems 4.2 and 4.4, we now find the Morse inequalities.

Theorem 4.5. $(M_p) \geqslant (B_p)$.

Remark 3. In [56], Witten suggested that his argument extended to the case where the critical points of h form a submanifold of M , in order to obtain the degenerate Morse inequalities of Bott. In [23], we found out that in the case of degenerate critical points, the metric of M has to be directly related in a non trivial way to h . We were able to prove the degenerate Morse inequalities using the existence of the Thom class of the normal bundle of the critical point set.
The proof is not a trivial extension of what has been done before.

REFERENCES

[1] ALVAREZ-GAUME L. : Supersymmetry and the Atiyah-Singer Index Theorem. Commun. Math. Physics 90(1983), 161-173.

[2] ATIYAH M.F. : K-Theory. Benjamin 1967.

[3] ATIYAH M.F. : Classical groups and classical operators on manifolds. In "Differential operators manifolds". CIME, Cremonese, Rome 1975.

[4] ATIYAH M.F. : Circular symmetry and stationary phase approximation. . In "Proceedings of the Conference in Honor of L. Schwartz." Astérisque 131 (1985), 43-59

[5] ATIYAH M.F., BOTT R. : A Lefschetz fixed point formula for elliptic complexes. I, Ann. of Math. 86(1967), 394-407 ;
II, 88(1968), 451-491.

[6] ATIYAH M.F., BOTT R. : The moment map and equivariant cohomology. Topology 23,1-28(1984).

[7] ATIYAH M.F., BOTT R., PATODI V.K. : On the heat equation and the Index Theorem. Invent. Math. 19,279-330(1973).

[8] ATIYAH M.F., SINGER I.M. : The Index of elliptic operators. I, Ann. of Math. 87(1968), 485-530 ; III, Ann. of Math. 87(1968), 546-604 ; IV, Ann. of Math. 93(1971), 119-138.

[9] AZENCOTT R. : Grandes déviations et applications. Cours de probabilité de Saint-Flour. Lecture Notes in Math. n° 774. Berlin Springer 1978.

[10] AZENCOTT R. : Densités des diffusions en temps petit : développements en temps petit. In Séminaire de Probabilités n° XVIII, p. 402-498. Lecture Notes Math. n° 1059, Berlin, Springer 1984.

[11] AZENCOTT R. : Une approche probabiliste du Théorème d'Atiyah-Singer, d'après J.M. Bismut. Séminaire Bourbaki 1984-1985. Astérisque 133-134 (1986), 7-18

[12] AZENCOTT R., BALDI P., BELLAICHE A., BELLAICHE C., BOUGEROL P., CHALEYAT-MAUREL M., ELIE L., GRANARA J. : Géodésiques et diffusions en temps petit. Astérisque 84-85. Paris 1981.

[13] BERLINE N., VERGNE M. : Zéros d'un champ de vecteurs et classes caractéristiques équivariantes. Duke Math. J.50, 539-549(1983).

[14] BERLINE N., VERGNE M. : The equivariant index and Kirillov's character formula. Am. J. Math. 107 (1985), 1159-1190

[15] BERLINE N., VERGNE M. : A computation of the equivariant index of the Dirac operator. Bull. Soc. Math. Fr. 113 (1985), 305-345

[16] BISMUT J.M. : Mécanique aléatoire. Lecture Notes in Math. n° 866. Berlin : Springer 1981.

[17] BISMUT J.M. : Large deviations and the Malliavin calculus. Progress in Math. n° 45. Birkhäuser 1984.

[18] BISMUT J.M. : Transformations différentiables du mouvement Brownien. Proceedings of the Conference in honor of L. Schwartz. Astérisque 131 Paris 1985. 61-87

[19] BISMUT J.M. : The Atiyah-Singer Theorems : a probabilistic approach. I, The Index Theorem. J. Funct. Anal. 57(1984). 56-99. II, The Lefschetz fixed-point formulas.J. Funct. Anal. 57(1984), 329-348.

[20] BISMUT J.M. : Index Theorem and Equivariant Cohomology on the loop space. Comm. in Math. Physics. 98,213-237(1985).

[21] BISMUT J.M. : The infinitesimal Lefschetz formulas : A heat equation proof. J. Funct. Anal. 62 (1985), 435-457

[22] BISMUT J.M. : Formules de Lefschetz délocalisées. Séminaire Bony-Sjöstrand-Meyer 1984 - 1985. Exp. 7. 13pp. École Polytechn. Palaiseau (1985)

[23] BISMUT J.M. The Witten complex and the degenerate Morse inequalities. To appear in J. Diff. Geom. (May 1986)

[24] BISMUT J.M. : The Atiyah-Singer Index Theorem for families of Dirac operators : two heat equation proofs. Invent. Math. 83 (1986), 91-151

[25] BISMUT J.M. Localization formulas, superconnections and the Index Theorem for families. Comm. Math. Phys. 103 (1986), 127-166

[26] BLAGOVESHCHENSKII Y.N., FREIDLIN M.I. : Certain properties of processes depending on parameters. Sov. Math. Dokl. 2(1961), 633-636.

[27] DE WITT-MORETTE C., MAHESHWARI A., NELSON B. : Path Integration in non relativistic quantum mechanics. Physics Reports 50(1979), 255-372.

[28] DUISTERMAAT J.J., HECKMAN G.J. : On the variation of the cohomology of the symplectic form of the reduced phase space. Invent. Math. 69(1982), 259-268 ; Addendum 72(1983),153-158.

[29] ELWORTHY K.D., TRUMAN A. : Classical mechanics, the diffusion heat equation and the Schrödinger equation on a Riemannian manifold. J. Math. Phys. 22, 2144-2166(1981).

[30] FEFFERMAN C. : The uncertainty principle. Bull AMS 9(1983), 129-206.

[31] FRIEDAN D., WINDEY H. : Supersymmetric derivation of the Atiyah-Singer Index and the Chiral anomaly. Nuclear Physics B. 235, (1984)395-416.

[32] FRIEDMAN A. : Partial differential equation of parabolic type. Englewood-Cliffs. Prentice Hall 1964.

[33] GARSIA A.M., RODEMICH E., RUMSEY H. : A real variable lemma and the continuity of paths of some Gaussian processes. Indiana Univ. Math. J. 20(1970), 565-578.

[34] GAVEAU B. : Principe de moindre action, propagation de la chaleur et estimées sous-elliptiques sur certains groupes nilpotents. Acta Math. 139(1977), 96-153.

[35] GETZLER E. : Pseudodifferential operators on supermanifolds and the Atiyah-Singer Index Theorem. Commmun. Math. Phys. 92(1983), 163-178.

[36] GETZLER E. : A Short proof of the Atiyah-Singer Index Theorem. Topology 25 (1986), 111-117

[37] GILKEY P. : Curvature and the eigenvalues of the Laplacian. Advan. in Math. 10(1973), 344-382.

[38] HELFFER B., SJÖSTRAND J. : Sur le complexe de Witten. To appear.

[39] HÖRMANDER L. : Hypoelliptic second order differential equations Acta Math. 119(1967), 147-171.

[40] JEULIN T. : Semi-martingales et grossissement d'une filtration. Lecture Notes in Math. n° 833. Berlin Springer 1980.

[41] JONA-LASINIO G., MARTINELLI F. SCOPPOLA E. : New approach to the semiclassical limit of Quantum Mechanics. Commun. Math. Phys. 80, 223-254(1981).

[42] KIFER Y. : On the asymptotics of the transition density of processes with small diffusion. Theory Prob. and Appl. 21,512-522(1976).

[43] LEANDRE R. : To appear .

[44] MALLIAVIN P. : Stochastic calculus of variations and hypoelliptic operators. In "Proceedings of the Conference on Stochastic differential equations of Kyoto", 1976 (K. Itô ed.) pp 155-263. Wiley, New-York 1978.

[45] Mc KEAN H., SINGER I.M. : Curvature and the eigenvalues of the Laplacian. J. of Diff. Geom. 1(1967),43-69.

[46] MOLCHANOV S.A. : Diffusion processes and Riemannian geometry. Russian Math. Surveys 30(1975),1-53.

[47] PATODI V.K. : Curvature and the eigenforms of the Laplace operator. J. of Diff. Geom. 5(1971),251-283.

[48] PATODI V.K. : An analytic proof of the Riemann-Roch-Hirzehruch Theorem for Kaehler manifolds. J. of Diff. Geom. 5(1971),251-283.

[49] QUILLEN D. : Superconnections and the Chern character. Topology 24 (1985), 89-95

[50] STROOCK D., VARADHAN S.R.S. : On the support of diffusion processes. 6th Berkeley Symposium on probability and statistics. Vol. III p. 333-359. Berkeley : Univ. of Calif. Press. 1972.

[51] SANCHEZ-CALLE A. : Fundamental solutions and Geometry of the sum of square of vectors fields. Invent. Math. 78(1984), 143-160.

[52] VARADHAN S.R.S. : Diffusion processes on a small time interval. Commun. Pure and Appl. Math. 20(1967),659-685.

[53] VARADHAN S.R.S. : Asymptotic probabilitiesanddifferential equations Commun. Pure and Appl. Math. XIX,(1966),261-286.

[54] VENTCELL A.D.,FREIDLIN M.I. : On small random perturbations of dynamical systems. Russian Math. Surveys 25(1970), p.1-55.

[55] VENTCELL A.D., FREIDLIN M.I. : Random perturbation of dynamical systems. Grundl. Math. Wiss. 260. Berlin : Springer 1984.

[56] WITTEN E. : Supersymmetry and Morse theory. J. of Diff. Geom. 17(1982),661-692.

[57] YOR M. : Remarques sur une formule de P. Lévy. In Séminaire de Probabilités n° XIV pp.343-346. Lecture Notes in Math. n°784. Berlin : Springer 1980.

[58] HÖRMANDER L. : The analysis of Linear Partial differential operators. Grundl. Math. Wiss. n°256. Berlin : Springer 1983.

[59] KUSUOKA S. To appear.

[60] LICHNEROWICZ A. : Spineurs harmoniques. CRAS. Math. 257,7.9(1963).

[61] MALLIAVIN P. : Géométrie différentielle Stochastique. Presses de l'Université de Montréal : Montréal(1978).

[62] TREVES F. : Introduction to pseudodifferential operators and Fourier Integral operators. Vol. 1 New-York : Plenum Press 1980.

[63] EELLS J., ELWORTHY K.D. : Wiener integration on certain manifolds of maps. In "Some problems in nonlinear analysis". CIME IV, pp. 69-94. Cremonese, Rome 1971.

[64] SMALE S. : Differentiable dynamical systems. Bull Am. Math. Soc. 7 (1967),747-817.

[65] VENTCELL A.D. : On the asymptotics of eigenvalues of matrices. Sov. Math.Dokl. 13(1972),65-68.

[66] ITÔ K., Mc KEAN H. : Diffusion processes and their sample paths. Grundl. Math. Wiss 125 Berlin : Springer 1974.

MARTINGALES AND FOURIER ANALYSIS IN BANACH SPACES

Donald L. Burkholder

Department of Mathematics
University of Illinois
Urbana, Illinois 61801

The power of martingale theory in the study of the Fourier analysis of scalar-valued functions is now widely appreciated. Our aim here is to describe some new martingale methods and their application to the Fourier analysis of functions having values in a Banach space.

One of the themes of this work is that the new methods developed for B-valued martingales can be used to obtain new information even in the real case, for example, the best constants in some basic inequalities. These sharp inequalities for real-valued martingales lead, in turn, to new information about some of the classical Banach spaces, for example, the unconditional constant of any monotone basis of $L^p(0,1)$ where $1 < p < \infty$.

TABLE OF CONTENTS

1. HISTORICAL PERSPECTIVE

Let us begin with a question that excited the interest of more and more mathematicians during the early part of this century (see Zygmund [50], the recent discussions of Gårding [31] and Cartwright [24], and the original papers cited in these three works): How does the size of a periodic function control the size of its conjugate? Consider, for example, the trigonometric polynomial

$$f(\theta) = \frac{a_0}{2} + \sum_{k=1}^{N} (a_k \cos k\theta + b_k \sin k\theta) \tag{1.1}$$

where the coefficients $a_0,\dots,a_N$, $b_1,\dots,b_N$ are real numbers and the size of f is measured by its norm $\|f\|_p$ in $L^p(0,2\pi)$:

$$\|f\|_p^p = \int_0^{2\pi} |f(\theta)|^p \, d\theta . \tag{1.2}$$

The polynomial conjugate to f is given by

$$g(\theta) = \sum_{k=1}^{N} (a_k \sin k\theta - b_k \cos k\theta) . \tag{1.3}$$

The orthogonality of the trigonometric functions implies that $\|g\|_2 \leq \|f\|_2$ and, for many years, the question was whether or not this inequality could be extended in some sense to L^p. In fact, as Marcel Riesz eventually proved ([42], [43]),

$$\|g\|_p \leq c_p \|f\|_p , \quad 1 < p < \infty , \tag{1.4}$$

where the constant depends only on p and not on the coefficients or the positive integer N. This work, with his related work on the interpolation of operators, has had a profound influence on 20th century analysis.

We shall also be interested in Riesz's inequality for the nonperiodic case, an inequality that can be derived from (1.4); see [43] and [50]. Let $f \in L^p(\mathbb{R})$ and $\varepsilon > 0$. If $H_\varepsilon f$ denotes the truncated Hilbert transform of f, that is,

$$H_\varepsilon f(x) = \frac{1}{\pi} \int_{|y| > \varepsilon} \frac{f(x-y)}{y} \, dy ,$$

then, for $1 \leq p < \infty$, the limit

(1.5) $$Hf(x) = \lim_{\epsilon \to 0} H_\epsilon f(x)$$

exists almost everywhere and, for $1 < p < \infty$,

(1.6) $$\|Hf\|_p \leq c_p \|f\|_p \,.$$

Actually, the constant in (1.6) is the same as the constant in (1.4). Recall that $f + iHf$ is the boundary function (the nontangential limit a.e.) of a function analytic in the upper half-plane.

Riesz's results have been greatly extended. A decisive step was taken by Calderón and Zygmund [22] who showed that for a large class of kernels $K: \mathbb{R}^n \setminus \{0\} \to \mathbb{C}$, the limit

(1.7) $$Tf(x) = \lim_{\epsilon \to 0} \int_{|y| > \epsilon} f(x-y)K(y)\,dy$$

exists a.e. if $f \in L^p(\mathbb{R}^n)$ for some p satisfying $1 \leq p < \infty$, and

(1.8) $$\|Tf\|_p \leq c_p \|f\|_p$$

if $1 < p < \infty$. Here, of course, the constant is not necessarily the same as before.

For many kernels, the Calderón-Zygmund inequality (1.8) can be derived from the Riesz inequality (1.6) with the use of the method of rotation introduced in [23]. Calderón and Zygmund show that this is the case, for example, if the kernel K satisfies

(1.9) $$K(x) = \Omega(x)/|x|^n$$

on $\mathbb{R}^n \setminus \{0\}$ where Ω is integrable with respect to surface measure σ on the unit sphere S of $\mathbb{R}^n$,

(1.10) $$\Omega(\alpha x) = \Omega(x) \quad \text{if } \alpha > 0 \,,$$

(1.11) $$\int_S \Omega \, d\sigma = 0 \,,$$

and

$$(1.12) \qquad \int_S |\Omega| \log^+ |\Omega| d\sigma < \infty .$$

If Ω is an odd function, that is, if $\Omega(-x) = -\Omega(x)$, then (1.11) necessarily holds and (1.12) can be eliminated.

Now let B be a real or complex Banach space with norm $|\cdot|$ and let $a_0,\ldots,a_N$, $b_1,\ldots,b_N$ be elements of B. Then (1.1) and (1.3) define elements f and g of the Lebesgue-Bochner space $L^p_B(0,2\pi)$ and the norm of f is given by (1.2) as before. Within a decade or two of Riesz's discovery of (1.4), its validity in the B-valued case began to be considered. For example, some work of Bochner and Taylor [5] implies that (1.4) does not hold if B is the Lebesgue sequence space ℓ^1. Even the inequality

$$(1.13) \qquad \|g\|_2 \leq c\|f\|_2$$

does not hold if $B = \ell^1$ or ℓ^∞. For what Banach spaces B does (1.4) hold? Or (1.13)? These turn out to be equivalent questions.

Similarly, during the period just after Calderón and Zygmund had written their early papers on singular integral operators, the question arose: What happens if the assumption that $f \in L^p(\mathbb{R}^n)$ is replaced by the assumption, a natural one from several different points of view, that $f \in L^p_B(\mathbb{R}^n)$. For some of the early work related to this question, see Schwartz [46], Benedek, Calderón, and Panzone [2], Stein [47], and Vági [48].

Before examining this question further, let us consider the martingale analogue of the Marcel Riesz inequality. Suppose that $d = (d_1, d_2, \ldots)$ is a sequence of integrable functions on the Lebesgue unit interval or any other probability space with the property that d_{n+1} is orthogonal to every bounded continuous function of $d_1,\ldots,d_n$ for all positive integers n. That is, suppose that d is a martingale difference sequence. Let $\varepsilon = (\varepsilon_1, \varepsilon_2, \ldots)$ be a sequence of numbers in $\{-1,1\}$. Write

$$(1.14) \qquad f_n = \sum_{k=1}^{n} d_k ,$$

$$(1.15) \qquad g_n = \sum_{k=1}^{n} \epsilon_k d_k .$$

Then [11], for $1 < p < \infty$,

$$(1.16) \qquad \|g_n\|_p \leq c_p \|f_n\|_p$$

where the constant depends only on p and not on d, ϵ, or n. This inequality leads to the square-function inequality and other inequalities for martingales and stochastic integrals; see, in particular, [11] and [16].

Not long after the author discovered (1.16), the question arose as to the validity of (1.16) for martingale difference sequences in the Lebesgue-Bochner space $L_B^p(0,1)$. Here the martingale condition is simply that the integral of the product of the strongly integrable B-valued function d_{n+1} and any bounded, continuous, scalar-valued function of $d_1,\ldots,d_n$ is equal to 0, the origin of B. (See Diestel and Uhl [26] for further discussion.) Maurey [37] and Pisier [41] examined this question and showed, for example, that if (1.16) is to hold for B, then B has to be superreflexive but that there do exist superreflexive spaces, hence uniformly convex spaces (see Enflo [29]), that do not satisfy the above martingale inequality. Also, see the work of Aldous [1].

But here, as with the Riesz inequality, the question remained: For what class of Banach spaces does the inequality (1.16) hold? In particular, how can this class be described geometrically?

2. TWO PROBLEMS

Let us now state precisely two of the problems that will be of principal interest to us here. Fix a number p satisfying $1 < p < \infty$. For a real or complex Banach space B with norm $|\cdot|$, let $\alpha_p(B)$ be the least $\alpha \in [0,+\infty]$ with the property that if N is a positive integer and $a_0,\ldots,a_N$, $b_1,\ldots,b_N$ belong to B, then f and g defined by (1.1) and (1.3) satisfy

$$\|g\|_p \leq \alpha \|f\|_p .$$

It follows as in the classical case [43] that if $\alpha_p(B)$ is finite, then, for $f \in L^p_B(\mathbb{R})$, the limit (1.5) exists a.e. and

(2.1) $$\|Hf\|_p \leq \alpha_p(B)\|f\|_p .$$

Conversely, if the Hilbert transform exists and is bounded in $L^p_B(\mathbb{R})$ then $\alpha_p(B)$ is finite and is, indeed, the norm of H.

For this reason we shall often write $B \in HT$ if $\alpha_p(B)$ is finite. Furthermore, if $B \in HT$, the method of rotation becomes available as in the classical case [23] so that, for a large class of singular integral operators T, the limit (1.7) exists and satisfies (1.8) for all $f \in L^p_B(\mathbb{R}^n)$.

One of the problems mentioned in Section 1 can now be phrased as follows.

PROBLEM 2.1. <u>Characterize geometrically those Banach spaces</u> B <u>satisfying</u> $B \in HT$.

The class HT does not depend on p for $1 < p < \infty$. This follows from the work of Schwartz [46] and Benedek, Calderón, and Panzone [2] but will be completely obvious from the geometrical characterization that we present in the next section.

Let us now define the analogue of $\alpha_p(B)$ for martingales and their $\pm$ 1-transforms. Let $\beta_p(B)$ be the least $\beta \in [0,+\infty]$ such that if $(d_1,d_2,\dots)$ is a B-valued martingale difference sequence, $(\epsilon_1,\epsilon_2,\dots)$ is a numerical sequence in $\{-1,1\}$, and n is a positive integer, then

(2.2) $$\left\|\sum_{k=1}^{n} \epsilon_k d_k\right\|_p \leq \beta\left\|\sum_{k=1}^{n} d_k\right\|_p .$$

We shall write $B \in UMD$ (B has the unconditionality property for martingale differences) if $\beta_p(B)$ is finite.

If $B \in UMD$, then many of the significant results for real-valued martingale transforms carry over to the B-valued case; see [13] and [14].

PROBLEM 2.2. <u>Characterize geometrically those Banach spaces</u> B <u>satisfying</u> $B \in$ UMD.

The class UMD does not depend on p for $1 < p < \infty$. This was noticed by Maurey and Pisier [37] but, again, is obvious from the geometrical characterization that we give in the next section.

3. THE CONDITION OF ζ-CONVEXITY

A Banach space B is ζ-*convex* if there is a biconvex function $\zeta: B \times B \to \mathbb{R}$ such that $\zeta(0,0) > 0$ and

$$\zeta(x,y) \leq |x + y| \quad \text{if} \quad |x| = |y| = 1 . \tag{3.1}$$

Biconvexity means that both $\zeta(\cdot,y)$ and $\zeta(x,\cdot)$ are convex on B for all x and y in B.

It will be clear later (see Lemma 3.1) that it is enough to have ζ defined and biconvex on the product of the closed unit ball of B with itself rather than on the whole of $B \times B$. But before examining the condition of ζ-convexity, a condition introduced in [13] in a slightly different but equivalent form, we shall state several theorems in which it plays a role.

THEOREM 3.1. $B \in$ UMD $\Longleftrightarrow$ B <u>is</u> ζ-<u>convex</u>.

This comes from [13] but our proof here will be a little different.

THEOREM 3.2. $B \in$ UMD $\Longleftrightarrow B \in$ HT.

The implication $B \in$ UMD $\Rightarrow B \in$ HT is due to McConnell and the author; see [15] where the result was announced and the main step proven. The truth of the converse was there suggested and Bourgain later succeeded in giving a proof. See his paper [6].

Combining the two theorems, we have the immediate corollary:

THEOREM 3.3. $B \in$ HT $\Longleftrightarrow$ B <u>is</u> ζ-<u>convex</u>.

So if B is ζ-convex, a large part of the Calderón-Zygmund theory of singular integral operators carries over. Moreover, it is not possible to carry over the major results if B is not ζ-convex.

Let us now examine the condition of ζ-convexity beginning with the simple case $B = \mathbb{R}$. Here one must find a biconvex function on $\mathbb{R}^2$ such that $\zeta(1,1) \leq 2$, $\zeta(-1,1) \leq 0$, $\zeta(-1,-1) \leq 2$, $\zeta(1,-1) \leq 0$, but $\zeta(0,0) > 0$. A little thought shows that a good choice is

$$\zeta(x,y) = 1 + xy .$$

Now let H be a real or complex Hilbert space with inner product $\langle x,y \rangle$ and let (x,y) denote the real part of $\langle x,y \rangle$. The above example suggests that we should consider

$$\zeta(x,y) = 1 + (x,y) . \tag{3.2}$$

This function is biconvex on $H \times H$, $\zeta(0,0) = 1$, and

$$\begin{aligned}[\zeta(x,y)]^2 &\leq 1 + 2(x,y) + |x|^2|y|^2 \\ &= |x + y|^2 + (1 - |x|^2)(1 - |y|^2) .\end{aligned}$$

So if $|x| = 1$ or $|y| = 1$, then $\zeta(x,y) \leq |x + y|$. Thus, the function defined by (3.2) satisfies the requirements of the condition of ζ-convexity. However, in a sense to be made clear later, it is not the optimal such function.

If B is nondegenerate, as we shall always assume, and ζ is a biconvex function on B satisfying (3.1), then

$$\zeta(0,0) \leq 1 . \tag{3.3}$$

To see this let $x \in B$ with $|x| = 1$. Then

$$\zeta(0,0) \leq [\zeta(x,0) + \zeta(-x,0)]/2 \leq |x| = 1 .$$

For Hilbert spaces, the upper bound 1 can be attained as we have seen. A natural question is: Does this characterize Hilbert spaces?

THEOREM 3.4. _Suppose that_ B _is a Banach space. If there is a biconvex function_ ζ _on_ $B \times B$ _such that_ (3.1) _is satisfied and_

$$\zeta(0,0) = 1 ,$$

then B _is a Hilbert space._

See [14] for a proof of this and related results.

We shall also consider biconvex functions u that satisfy a stronger condition than (3.1), namely that

$$u(x,y) \leq |x + y| \quad \text{if} \quad |x| \vee |y| \geq 1 . \tag{3.4}$$

LEMMA 3.1. *Suppose that* ζ *is a biconvex function on*

$$\{(x,y) \in B \times B: |x| \vee |y| \leq 1\} \tag{3.5}$$

satisfying (3.1). *Then there is a biconvex function* u *on* $B \times B$ *satisfying* (3.4) *and*

$$\zeta(0,0) \leq u(0,0) . \tag{3.6}$$

PROOF. First of all, (3.1) implies that

$$\zeta(x,y) \leq |x + y| \quad \text{if} \quad |x| \vee |y| = 1 . \tag{3.7}$$

To see this assume as we can that $|x| < |y| = 1$, so $x + y \neq 0$, and choose $\alpha \in (0,1)$ to satisfy $|z| = 1$ where

$$z = \alpha^{-1}(x + y) - y .$$

(Note that the norm of this expression is larger than 1 for α near 0 but is smaller than 1 for α near 1.) Then $x = \alpha z + (1-\alpha)(-y)$ and, by (3.1),

$$\begin{aligned}\zeta(x,y) &\leq \alpha\zeta(z,y) + (1-\alpha)\zeta(-y,y)\\ &\leq \alpha|z + y| = |x + y| ,\end{aligned}$$

which is (3.7). Now let

$$\begin{aligned} u(x,y) &= \zeta(x,y) \vee |x + y| \quad \text{if} \quad |x| \vee |y| < 1 ,\\ &= |x + y| \quad \text{if} \quad |x| \vee |y| \geq 1 . \end{aligned} \tag{3.8}$$

Then u is biconvex on $B \times B$. For example, if $|y| \geq 1$, then $u(x,y) = |x + y|$ and $u(\cdot,y)$ is convex on B. If $|y| < 1$, then $u(\cdot,y)$ is locally convex on the complement of the unit sphere of B. So suppose that $|y| < 1 = |x|$ and

$x = \alpha_1 x_1 + \alpha_2 x_2$ where the $\alpha_i \in (0,1)$ and $\alpha_1 + \alpha_2 = 1$. Then

$$\begin{aligned} u(x,y) &= |x + y| \\ &\leq \alpha_1 |x_1 + y| + \alpha_2 |x_2 + y| \\ &\leq \alpha_1 u(x_1,y) + \alpha_2 u(x_2,y) \ . \end{aligned}$$

This establishes the convexity of $u(\cdot,y)$. It is clear from the definition of u that both (3.4) and (3.6) are satisfied.

It is sometimes useful to have u symmetric. If u is not already symmetric, one can replace $u(x,y)$ by $u(x,y) \vee u(y,x)$.

LEMMA 3.2. *If u is a biconvex function on $B \times B$ satisfying (3.4), then for all $x,y,x',y' \in B$,*

(3.9) $$|u(x,y) - u(x',y')| \leq |x - x'| + |y - y'| \ .$$

PROOF. Suppose that $x \neq x'$. Then $|x + \lambda(x' - x)| > 1$ for all large λ and, by (3.4) and the convexity of $u(\cdot,y)$,

$$\begin{aligned} u(x',y) - u(x,y) &\leq \lambda^{-1}[u(x + \lambda(x' - x),y) - u(x,y)] \\ &\leq \lambda^{-1}[|x| + |y| + \lambda|x' - x| - u(x,y)] \ . \end{aligned}$$

The last expression converges to $|x' - x|$ as $\lambda \to \infty$. The inequality (3.9) is now obvious.

REMARK 3.1. The norm of a Hilbert space is determined by an inner product. The analogue for a ζ-convex space B is that the norm $|\cdot|$ of B is determined up to equivalence by any biconvex function $\zeta\colon B \times B \to \mathbb{R}$ satisfying (3.1) and $\zeta(0,0) > 0$. Specifically, it is possible to define a new norm $\|\cdot\|$ on B using ζ but not $|\cdot|$ so that

(3.10) $$\zeta(0,0)\|x\| \leq |x| \leq \|x\| \ .$$

To show this, we can and do assume that

(3.11) $$\zeta(x,y) = \zeta(-x,-y)$$

for if ζ does not already satisfy this condition, we can replace $\zeta(x,y)$ by $\zeta(x,y) \vee \zeta(-x,-y)$. Let V_0 be the set of all x in B satisfying $\zeta(\alpha x,-\alpha x) > 0$ for all scalars α such that $|\alpha| \leq 1$ and let V be the smallest convex set containing V_0. Then $\alpha V = V$ if $|\alpha| = 1$, and

$$\zeta(0,0)U \subset V \subset U \tag{3.12}$$

where $U = \{x \in B: |x| < 1\}$, as we shall show, so

$$\|x\| = \inf\{\lambda > 0: x \in \lambda V\}$$

defines a norm on B satisfying (3.10).

To prove (3.12), we define u by (3.8) and let V_0^u be the set of all x in B satisfying $u(\alpha x,-\alpha x) > 0$ for all scalars α such that $|\alpha| \leq 1$. By (3.1) and (3.8), both V_0 and V_0^u are subsets of U: If x belongs to either V_0 or V_0^u, then $|\alpha x| \neq 1$ for all scalars α with $|\alpha| \leq 1$ so x must belong to U. But if $x \in U$, then $u(\alpha x,-\alpha x) = \zeta(\alpha x,-\alpha x) \vee 0$ for all such α so

$$V_0 = V_0^u .$$

The right-hand side of (3.12) follows from $V_0 \subset U$. The left-hand side will follow from $\zeta(0,0)U \subset V_0$. We shall use the fact that u satisfies (3.11) also so that

$$u(0,0) \leq [u(x,0) + u(-x,0)]/2 = u(x,0) .$$

Let $x \in \zeta(0,0)U$ so that $|x| < \zeta(0,0)$. Let $|\alpha| \leq 1$. Then, by Lemmas 3.1 and 3.2,

$$\begin{aligned}\zeta(0,0) &= u(0,0) \leq u(\alpha x,0)\\ &\leq u(\alpha x,-\alpha x) + |\alpha x|\\ &< u(\alpha x,-\alpha x) + \zeta(0,0) ,\end{aligned}$$

which implies that $x \in V_0^u = V_0$.

Let ζ_B denote the greatest biconvex function u on $B \times B$ satisfying (3.4). It is convenient to say that ζ_B is the <u>optimal</u> <u>ζ-function on</u> $B \times B$. As we shall see later, it has a simple martingale interpretation.

THEOREM 3.5. *Suppose that* H *is a real or complex Hilbert space and* (x,y) *is the real part of the inner product of* x *and* y. *Then the optimal* ζ-*function on* $H \times H$ *is given by*

$$\zeta_H(x,y) = (1 + 2(x,y) + |x|^2|y|^2)^{\frac{1}{2}} \quad \text{if } |x| \vee |y| < 1 , \tag{3.13}$$
$$= |x + y| \quad \text{if } |x| \vee |y| \geq 1 .$$

PROOF. Let ζ be defined on $H \times H$ by

$$\zeta(x,y) = [1 + 2(x,y) + |x|^2|y|^2]^{\frac{1}{2}}$$
$$= [|x + y|^2 + (1 - |x|^2)(1 - |y|^2)]^{\frac{1}{2}} .$$

Then $\zeta(x,y) > |x + y|$ on the set where $|x| \vee |y| < 1$ and $\zeta(x,y) = |x + y|$ if $|x| \vee |y| = 1$. Also ζ is biconvex on $H \times H$: If $a + bt + ct^2$ is a nonnegative polynomial with real coefficients, then the mapping $t \to (a + bt + ct^2)^{\frac{1}{2}}$ is convex on $\mathbb{R}$. Let u be the right-hand side of (3.13), that is,

$$u(x,y) = (1 + 2(x,y) + |x|^2|y|^2)^{\frac{1}{2}} \quad \text{if } |x| \vee |y| < 1 ,$$
$$= |x + y| \quad \text{if } |x| \vee |y| \geq 1 .$$

The function u satisfies (3.4) and, by Lemma 3.1 and its proof, u is biconvex on $H \times H$. Hence

$$u \leq \zeta_H .$$

It is clear that $\zeta_H(0,0) \leq 1 = u(0,0)$ and $\zeta_H(x,y) \leq |x + y| = u(x,y)$ if $|x| \vee |y| \geq 1$. So suppose that $0 < |x| \vee |y| < 1$ and, without loss of generality, that $y \neq 0$. Let

$$z = x|y|^2 + y .$$

Then, for $t \geq - 1/|y|^2$,

$$\zeta(x + tz,y) = \zeta(x,y)(1 + t|y|^2) . \tag{3.14}$$

Our assumption about x and y implies that

$$|x + tz| < 1$$

for all t in an interval (t_0, t_1) with equality holding at t_0 and t_1 where

$$- 1/|y|^2 < t_0 < 0 < t_1 \ .$$

Let $\alpha = - t_0/(t_1 - t_0)$. Then $\alpha t_1 + (1-\alpha) t_0 = 0$ and, by (3.14),

$$\begin{aligned} \zeta_H(x,y) &\le \alpha \zeta_H(x + t_1 z, y) + (1-\alpha) \zeta_H(x + t_0 z, y) \\ &\le \alpha |x + t_1 z + y| + (1-\alpha) |x + t_0 z + y| \\ &= \alpha \zeta(x + t_1 z, y) + (1-\alpha) \zeta(x + t_0 z, y) \\ &= \zeta(x,y) [\alpha(1 + t_1 |y|^2) + (1-\alpha)(1 + t_0 |y|^2)] \\ &= \zeta(x,y) = u(x,y) \ . \end{aligned}$$

This completes the proof that $\zeta_H = u$.

REMARK 3.2. Let $d(E,F)$ denote the Banach-Mazur distance between two isomorphic Banach spaces E and F: $d(E,F) = \inf \alpha\beta$ where α is the norm of T^{-1} and β is the norm of T and the infimum is with respect to all bounded linear transformations T from E onto F. Then

$$\zeta_E(0,0)/d(E,F) \le \zeta_F(0,0) \le \zeta_E(0,0) d(E,F) \ . \tag{3.15}$$

To prove the right-hand side, fix T and let u be the biconvex function on $E \times E$ defined by

$$\alpha\beta u(x,y) = \zeta_F(\alpha Tx, \alpha Ty) \ .$$

Note that $|x| = |T^{-1}Tx| \le \alpha |Tx|$, so $|x| \vee |y| \ge 1$ implies that $|\alpha Tx| \vee |\alpha Ty| \ge 1$ and

$$\alpha\beta u(x,y) \le |\alpha Tx + \alpha Ty| \le \alpha\beta |x + y| \ .$$

Thus $u(x,y) \le |x + y|$ if $|x| \vee |y| \ge 1$. Since ζ_E is maximal,

$$\zeta_F(0,0) = \alpha\beta u(0,0) \le \alpha\beta \zeta_E(0,0) \ .$$

This implies the right-hand side of (3.15). The left-hand side follows at once.

Note two consequences of this remark: (i) If B is a ζ-convex Banach space, then any space isomorphic to B is also ζ-convex. (ii) If B is a finite-dimensional space with dimension n, then

$$\zeta_B(0,0) \geq 1/n^{\frac{1}{2}} . \tag{3.16}$$

In addition to (3.15), this rests on $\zeta_{\ell^2_n}(0,0) = 1$ (see Theorem 3.5) and the estimate of Fritz John: $d(B,\ell^2_n) \leq n^{\frac{1}{2}}$. Of course, (i) also follows from the equivalent UMD property.

Let $1 < p < \infty$. Term-by-term integration gives $\beta_p(\ell^p) = \beta_p(\mathbb{R})$. Therefore, $\ell^p \in$ UMD and hence ℓ^p is ζ-convex. Similarly, $L^p(0,1)$ is ζ-convex. Many of the classical reflexive Banach spaces are UMD or HT and hence are ζ-convex. For example, Gutiérrez [33] shows that if $1 < p < \infty$, then the trace class C_p, the class of compact operators A on a Hilbert space H satisfying

$$\|A\| = [tr(A^*A)^{p/2}]^{1/p} < \infty ,$$

belongs to HT. Hence, by Theorem 3.3, C_p is ζ-convex. Also, see Bourgain [9].

4. ZIGZAG MARTINGALES

Let $Z = (Z_1, Z_2, \ldots)$ be a martingale with values in $B \times B$. Write $Z_n = (X_n, Y_n)$ where both X_n and Y_n have their values in B. Then Z is a <u>zigzag martingale</u> if, for every positive integer n, either

$$X_{n+1} - X_n \equiv 0 \quad \text{or} \quad Y_{n+1} - Y_n \equiv 0 . \tag{4.1}$$

For example, if f is a B-valued martingale with difference sequence d and g is its transform by a sequence ε of numbers in $\{-1,1\}$, then

$$X_n = f_n + g_n = \sum_{k=1}^{n} (1 + \varepsilon_k) d_k$$

and

$$Y_n = f_n - g_n = \sum_{k=1}^{n} (1 - \varepsilon_k) d_k$$

define a zigzag martingale Z.

If Z converges almost everywhere, let Z_∞ denote its almost everywhere limit. A martingale Z is <u>simple</u> if every term Z_n is a simple function and, for some positive integer n,

$$Z_n = Z_{n+1} = \cdots = Z_\infty .$$

A set $S \subset B \times B$ is <u>biconvex</u> if

$$\{x: (x,y) \in S\} \text{ and } \{y: (x,y) \in S\}$$

are convex for all y and x in B.

The following lemma is fundamental.

LEMMA 4.1. *Suppose that Z is a simple zigzag martingale with values in a biconvex set $S \subset B \times B$ and $u: S \to \mathbb{R}$ is a biconvex function. Then*

$$\text{(4.2)} \qquad Eu(Z_1) \le Eu(Z_2) \le \cdots \le Eu(Z_\infty) .$$

If $u: S \to \mathbb{R}$ is biconcave, then the inequality signs are reversed.

PROOF. By the zigzag property, either $Z_{n+1} = (X_{n+1}, Y_n)$ or $Z_{n+1} = (X_n, Y_{n+1})$. Suppose the first alternative holds, the other case being similar. Let

$$\text{(4.3)} \qquad A = \{Z_1 = z_1, \ldots, Z_n = z_n\}$$

where $z_n = (x_n, y_n)$ and $P(A) > 0$. Then, since Z is a martingale,

$$\int_A (X_{n+1} - x_n)\,dP = \int_A (X_{n+1} - X_n)\,dP$$
$$= 0$$

so that

$$x_n = \int_A X_{n+1}\,dP/P(A) .$$

Therefore, by Jensen's inequality applied to the convex function $u(\cdot, y_n)$,

$$u(x_n, y_n) \le \int_A u(X_{n+1}, y_n)\,dP/P(A)$$

and this is equivalent to the inequality

$$\int_A u(Z_n)\,dP \leq \int_A u(Z_{n+1})\,dP \;. \tag{4.4}$$

Summing both sides of (4.4) over all sets of the form (4.3), we obtain

$$Eu(Z_n) \leq Eu(Z_{n+1}) \;.$$

This gives (4.2).

REMARK 4.1. By (4.4), the sequence $(u(Z_1), u(Z_2), \ldots)$ is a submartingale and this carries over to nonsimple Z if u is continuous and the $u(Z_n)$ are integrable. The monotonicity of the sequence of expectations is the most important property for us here and for this it is enough to assume that Z is a very weak simple zigzag martingale; see Section 13 of [16].

5. A BOUNDARY VALUE PROBLEM

Consider the problem of finding the greatest biconvex function $u: B \times B \to \mathbb{R}$ such that

$$u(x,y) \leq |x + y| \quad \text{if} \quad |x - y| \geq 2 \;. \tag{5.1}$$

We can describe the greatest such function using zigzag martingales. If $(x,y) \in B \times B$, let $\underset{\sim}{Z}(x,y)$ denote the set of all simple zigzag martingales Z on the Lebesgue unit interval that have values in $B \times B$, start at (x,y) but finish in the set $\{|x - y| \geq 2\}$: $Z_1 \equiv (x,y)$ but

$$|X_\infty - Y_\infty| \geq 2. \tag{5.2}$$

Clearly, $\underset{\sim}{Z}(x,y)$ is nonempty.

LEMMA 5.1. *Define* $L: B \times B \to \mathbb{R}$ *by*

$$L(x,y) = \inf\{E|X_\infty + Y_\infty| : Z \in \underset{\sim}{Z}(x,y)\} \;.$$

Then L *is the greatest biconvex function* u *on* $B \times B$ *satisfying* (5.1).

PROOF. Let u be a biconvex function on $B \times B$ satisfying (5.1) and suppose that $Z \in \underset{\sim}{Z}(x,y)$. Then, by (5.1) and (5.2), we have that $u(Z_\infty) \leq |X_\infty + Y_\infty|$. Therefore, by Lemma 4.1,

$$u(x,y) = Eu(Z_1) \leq \cdots \leq Eu(Z_\infty) \leq E|X_\infty + Y_\infty| .$$

So, by the definition of L, we have that

$$u(x,y) \leq L(x,y) .$$

If $|x - y| \geq 2$, then, by taking $Z_n \equiv (x,y)$ for all n, we see that $L(x,y) \leq |x + y|$ so L satisfies (5.1).

To see that $L(\cdot,y)$ is convex, choose $x_1 \in B$, $x_2 \in B$, and $0 < \alpha < 1$. Let $x = \alpha x_1 + (1-\alpha)x_2$. Suppose that $Z^i \in \underset{\sim}{Z}(x_i,y)$, $i = 1,2$. We may assume that Z^1 and Z^2 move horizontally and vertically together, in fact, that

$$X^i_{2n} - X^i_{2n-1} \equiv 0 \quad \text{and} \quad Y^i_{2n+1} - Y^i_{2n} \equiv 0 .$$

Let Z be Z^1 and Z^2 spliced together in the following way: $Z_1 \equiv (x,y)$,

$$\begin{aligned} Z_{n+1}(s) &= Z^1_n(s/\alpha) && \text{if } s \in [0,\alpha) , \\ &= Z^2_n((s-\alpha)/(1-\alpha)) && \text{if } s \in [\alpha,1) . \end{aligned}$$

It is easy to check that Z is in $\underset{\sim}{Z}(x,y)$ and that

$$\begin{aligned} (5.3) \qquad L(x,y) &\leq E|X_\infty + Y_\infty| \\ &= \alpha E|X^1_\infty + Y^1_\infty| + (1-\alpha)E|X^2_\infty + Y^2_\infty| . \end{aligned}$$

If M_1 and M_2 are real numbers satisfying $L(x_1,y) < M_1$ and $L(x_2,y) < M_2$, then Z^1 and Z^2 can be chosen so that the right-hand side of (5.3) is less than $\alpha M_1 + (1-\alpha)M_2$. Thus,

$$(5.4) \qquad L(x,y) \leq \alpha L(x_1,y) + (1-\alpha)L(x_2,y) .$$

Similarly, $L(x,\cdot)$ is convex and the lemma is proved.

REMARK 5.1. Let L^0 be the function defined on $B \times B$ by

$$L^0(x,y) = \inf\{E|X_\infty + Y_\infty| : Z \in \underset{\sim}{Z}^0(x,y)\}$$

where

$$\underset{\sim}{Z}^0(x,y) = \{Z \in \underset{\sim}{Z}(x,y) : Z \text{ is dyadic}\} .$$

(A martingale on the Lebesgue unit interval is dyadic if, for all $n \geq 1$, its n-th term and the norm of the (n+1)-st term of its difference sequence are both constant on the interval $[(k-1)/2^{n-1}, k/2^{n-1})$ for all $k = 1,2,\ldots,2^{n-1}$.) Since $\underset{\sim}{Z}^0(x,y)$ is a subset of $\underset{\sim}{Z}(x,y)$,

$$L(x,y) \leq L^0(x,y) .$$

But the reverse inequality is also true. The above splicing argument with $\alpha = 1/2$ shows that L^0 is midpoint biconvex. In fact, L^0 is the greatest midpoint biconvex function u on $B \times B$ satisfying (5.1). But L^0 is locally bounded from above: If $|x - y| \leq 2$ and $|z| = 4$, then

$$\begin{aligned} L^0(x,y) &\leq [L^0(x + z,y) + L^0(x - z,y)]/2 \\ &\leq |x + y| + 4 \end{aligned}$$

while, if $|x - y| \geq 2$, then $L^0(x,y) \leq |x + y|$. Thus, every function $\varphi: \mathbb{R} \to \mathbb{R}$ of the form

$$\varphi(\alpha) = L^0(\alpha x_1 + (1-\alpha)x_2,y)$$

is not only midpoint convex but is also locally bounded from above, hence is convex. Accordingly, L^0 is a biconvex function satisfying (5.1) and, since L is the greatest such function, $L^0 \leq L$. Therefore,

$$L^0 = L . \tag{5.5}$$

REMARK 5.2. Recall that ζ_B denotes the greatest biconvex function u on $B \times B$ satisfying (3.4). We claim that

$$\zeta_B = L . \tag{5.6}$$

Since (3.4) is more restrictive than (5.1), it is clear that $\zeta_B \leq L$. To prove the reverse inequality, we need to show only that L satisfies (3.4). Let $|x| \vee |y| \geq 1$. If $|x - y| \geq 2$ also holds, then $L(x,y) \leq |x + y|$. If $|x - y| < 2$ and $|y| \geq 1$, as we can assume, then $x + y \neq 0$ and, for all large λ,

$$\begin{aligned} L(x,y) &\leq (1 - \lambda^{-1})L(-y,y) + \lambda^{-1}L(-y + \lambda(x + y),y) \\ &\leq |x + y| . \end{aligned}$$

REMARK 5.3. It is often possible to restrict attention to those martingales in $\underset{\sim}{Z}(x,y)$ or $\underset{\sim}{Z}^0(x,y)$ with small jumps. For example, if $\delta > 0$, let $\underset{\sim}{Z}_\delta(x,y)$ denote those Z in $\underset{\sim}{Z}(x,y)$ satisfying

$$|X_{n+1} - X_n| + |Y_{n+1} - Y_n| \leq \delta , \quad n \geq 1 .$$

Then

$$L_\delta = L \tag{5.7}$$

where $L_\delta(x,y) = \inf\{E|X_\infty + Y_\infty| : Z \in \underset{\sim}{Z}_\delta(x,y)\}$. Similarly, $L_\delta^0 = L^0$ in the dyadic case. It is clear that $L \leq L_\delta$ since $\underset{\sim}{Z}_\delta \subset \underset{\sim}{Z}$. But the converse is also clear from Lemma 5.1. The biconvexity of L_δ follows from the local biconvexity of L_δ: If in the proof of the analogue of (5.4), we choose x_1 and x_2 to satisfy $|x_1 - x_2| \leq \delta$, then $Z \in \underset{\sim}{Z}_\delta(x,y)$.

6. B ∈ UMD ⇒ B IS ζ-CONVEX

To prove this half of Theorem 3.1, we shall show that if $1 < p < \infty$ and $\beta_p(B)$ is finite, then $\zeta_B(0,0) > 0$. In fact,

$$\zeta_B(0,0) \geq 1/\beta_p(B) . \tag{6.1}$$

Note that equality holds if $p = 2$ and B is a Hilbert space.

To prove this inequality, we shall use the following consequence of Remarks 5.2 and 5.3:

$$\zeta_B(0,0) = L_\delta(0,0) . \tag{6.2}$$

Let $Z \in \underset{\sim}{Z}_\delta(0,0)$. Then Z determines simple martingales f and g with jumps of size less than δ by

(6.3)
$$f_n = (X_n + Y_n)/2$$

and

(6.4)
$$g_n = (X_n - Y_n)/2 .$$

It is easy to check that g is the transform of f by a sequence ε of numbers in $\{-1,1\}$. Also, by (6.4) and (5.2),

(6.5)
$$|g_\infty| \geq 1 .$$

Let $\lambda = 1/\beta_p(B)(p-1)^{1/p}$,

$$F_n = \sum_{k=1}^{n} u_k d_k ,$$

$$G_n = \sum_{k=1}^{n} \varepsilon_k u_k d_k ,$$

where d is the difference sequence of f, as usual, and u_n is the indicator function of the set $\{\tau \geq n\}$ where

$$\tau = \inf\{n: |f_n| > \lambda\} .$$

Note that

$$\begin{aligned} 1 &= P(|g_\infty| \geq 1) \\ &\leq P(|g_\infty| \geq 1, \ \tau = \infty) + P(\tau < \infty) \\ &\leq P(|G_\infty| \geq 1) + P(f^* > \lambda) \end{aligned}$$

where $f^* = \sup_n |f_n|$, the maximal function of f. Since G is a $\pm$ 1-transform of the simple martingale F,

$$\begin{aligned} P(|G_\infty| \geq 1) &\leq \|G_\infty\|_p^p \\ &\leq \beta_p^p(B) \|F_\infty\|_p^p \\ &\leq \beta_p^p(B) \|F_\infty\|_\infty^{p-1} \|F_\infty\|_1 . \end{aligned}$$

Since $d^* \leq \delta$ and $F_\infty = f_\tau$, we have that $\|F_\infty\|_\infty \leq \lambda + \delta$. Applying Doob's optional stopping theorem to the nonnegative submartingale $|f_n|$, we obtain $\|F_\infty\|_1 \leq \|f_\infty\|_1$. Also, $P(f^* > \lambda) \leq \|f_\infty\|_1/\lambda$ where, by (6.3), $2\|f_\infty\|_1 = E|X_\infty + Y_\infty|$. Therefore,

$$2 \leq [\beta_p^p(B)(\lambda + \delta)^{p-1} + 1/\lambda]E|X_\infty + Y_\infty|$$

so that

$$2 \leq [\beta_p^p(B)(\lambda + \delta)^{p-1} + 1/\lambda]L_\delta(0,0) \ .$$

Replacing $L_\delta(0,0)$ by $\zeta_B(0,0)$ and letting $\delta \to 0$, we obtain

$$\begin{aligned} 2 &\leq [\beta_p^p(B)\lambda^{p-1} + 1/\lambda]\zeta_B(0,0) \\ &= q(p-1)^{1/p}\beta_p(B)\zeta_B(0,0) \ . \end{aligned}$$

Here we have used $q = p/(p-1)$ and our choice of λ. The expression $q(p-1)^{1/p}$ is maximized at $p = 2$ and (6.1) follows.

This completes the proof of the first half of Theorem 3.1. A completely different proof will be given in Remark 7.1.

7. A GENERAL BOUNDARY VALUE PROBLEM

In Section 5, we characterized the lower solution of the boundary value problem considered there in terms of zigzag martingales. In Section 6, we used this characterization to study B-valued martingales and their $\pm$ 1-transforms. We shall explore here, and in later sections, some of the further consequences of the underlying ideas.

Let S be a biconvex subset of $B \times B$ and S_∞ a nonempty subset of S:

$$S_\infty \subset S \subset B \times B \ .$$

Let $F: S_\infty \to \mathbb{R}$. We place no further conditions on F, not even the condition of measurability. The problem is: If there exists at least one biconcave function u on S such that $u \geq F$ on S_∞, find the least such function. There is of course the dual problem for biconvex functions.

A number of examples will given below. In Section 5, the biconvex set S is the whole space, $S_\infty = \{x \in S: |x - y| \geq 2\}$, and $F(x,y) = |x + y|$. But, as we have seen there, if $S_\infty = \{x \in S: |x| \vee |y| \geq 1\}$, then the problem has the same lower solution.

In this section let

$$(7.1) \qquad \underset{\sim}{Z}(x,y) = \underset{\sim}{Z}(x,y;\, S,S_\infty)$$

denote the set of all simple zigzag martingales Z, defined on the Lebesgue unit interval and with values in S, such that $Z_1 \equiv (x,y)$ and Z_∞ has all of its values in S_∞. We assume that $\underset{\sim}{Z}(x,y)$ is nonempty for all $(x,y) \in S$. This is usually easy to check, particularly in the examples of interest to us here. Let

$$(7.2) \qquad U_F(x,y) = \sup\{EF(Z_\infty) : Z \in \underset{\sim}{Z}(x,y)\} \,,$$

$$(7.3) \qquad L_F(x,y) = \inf\{EF(Z_\infty) : Z \in \underset{\sim}{Z}(x,y)\} \,.$$

THEOREM 7.1. <u>The function</u> U_F <u>is the least biconcave function</u> $u\colon S \to \mathbb{R}$ <u>such that</u> $u \geq F$ <u>on</u> S_∞ <u>provided at least one such function exists. The existence of such a function is assured if</u> $U_F(x,y) < \infty$ <u>for all</u> $(x,y) \in S$. <u>The function</u> L_F <u>is the greatest biconvex function</u> $u\colon S \to \mathbb{R}$ <u>such that</u> $u \leq F$ <u>on</u> S_∞ <u>provided at least one such</u> u <u>exists and this will be the case if</u> $L_F(x,y) > -\infty$ <u>for all</u> $(x,y) \in S$.

PROOF. The proof is similar to that of Lemma 5.1. For example, if $u\colon S \to \mathbb{R}$ is a biconvex function such that $u \leq F$ on S_∞ and $Z \in \underset{\sim}{Z}(x,y)$ where $(x,y) \in S$, then $u(Z_\infty) \leq F(Z_\infty)$ so that, by Lemma 4.1,

$$(7.4) \qquad u(x,y) = Eu(Z_1) \leq \cdots \leq Eu(Z_\infty) \leq EF(Z_\infty) \,.$$

This implies that $u(x,y) \leq L_F(x,y)$. If $(x,y) \in S_\infty$, the the constant martingale $Z_n \equiv (x,y)$ shows that $L_F(x,y) \leq F(x,y)$. By the splicing argument, L_F satisfies (5.4) whether or not L_F has its values in $\mathbb{R}$. But L_F is real-valued, hence is biconvex, if L_F does not assume $-\infty$ as one of its values or, by (7.4), if the lower class of functions u is nonempty.

The dyadic analogue of Theorem 7.1 is proved in exactly the same way. Replace $\underset{\sim}{Z}(x,y)$, as in (7.1), by

$$\underset{\sim}{Z}^0(x,y) = \{Z \in \underset{\sim}{Z}(x,y) : Z \text{ is dyadic}\} .$$

Let U_F^0 and L_F^0 be the resulting dyadic analogues of (7.2) and (7.3).

THEOREM 7.2. <u>Suppose that</u> $U_F^0(x,y) < \infty$ <u>for all</u> $(x,y) \in S$. <u>Then</u> U_F^0 <u>is the least midpoint biconcave function</u> $u: S \to \mathbb{R}$ <u>such that</u> $u \geq F$ <u>on</u> S_∞.

The dual statement for L_F^0 is of course also valid.

Here is one application of Theorem 7.2. Let $\beta_p^0(B)$ be the least $\beta \in [0,+\infty]$ such that if d is the difference sequence of a B-valued dyadic martingale, ϵ is a sequence in $\{-1,1\}$, and n is a positive integer, then

$$\|\sum_{k=1}^{n} \epsilon_k d_k\|_p \leq \beta \|\sum_{k=1}^{n} d_k\|_p .$$

Thus, $\beta_p^0(B)$ is the analogue for dyadic martingales of $\beta_p(B)$.

LEMMA 7.1. <u>If</u> B <u>is a Banach space and</u> $1 < p < \infty$, <u>then</u> $\beta_p(B) = \beta_p^0(B)$.

This can be seen in several different ways. One possible proof is suggested in [37]. Here we give an entirely different proof. Conceptually simple, it rests on the elementary fact that a midpoint concave function that is locally bounded from below is concave.

PROOF. It is clear that $\beta_p^0(B) \leq \beta_p(B)$. To show that

$$\beta_p(B) \leq \beta_p^0(B) , \tag{7.5}$$

we shall assume that $\beta_p^0(B)$ is finite and use Theorem 7.2 with $S_\infty = S = B \times B$ and

$$F(x,y) = \left|\frac{x+y}{2}\right|^p - \delta^p \left|\frac{x-y}{2}\right|^p$$

where $\delta = \beta_p^0(B)$, the dyadic constant. Then, for all $(x,y) \in S$,

$$F(x,y) \leq U_F^0(x,y) , \tag{7.6}$$

an inequality that follows from the definition of U_F^0, and

$$(7.7) \qquad U_F^0(x,y) < \infty .$$

To prove (7.7), we need to show only that $U_F^0(0,0)$ is finite since, by the dual of (5.4) for our problem here,

$$\begin{aligned} U_F^0(x,y) + U_F^0(x,-y) + U_F^0(-x,y) + U_F^0(-x,-y) \\ \leq 2[U_F^0(x,0) + U_F^0(-x,0)] \\ \leq 4\, U_F^0(0,0) . \end{aligned}$$

To show that $U_F^0(0,0)$ is finite, indeed, that $U_F^0(0,0) \leq 0$, we let $Z \in \underset{\sim}{Z}^0(0,0)$ and define the simple dyadic martingales f and g by $f_n = (X_n - Y_n)/2$ and $g_n = (X_n + Y_n)/2$. Then f and g satisfy

$$f_n = \sum_{k=1}^{n} d_k ,$$

$$g_n = \sum_{k=1}^{n} \epsilon_k d_k$$

for some sequence of signs $\epsilon_k \in \{-1,1\}$. By the definition of $\beta_p^0(B)$,

$$EF(Z_\infty) = \|g_\infty\|_p^p - \delta^p\|f_\infty\|_p^p \leq 0$$

so that $U_F^0(0,0) \leq 0$ and (7.7) is proved. Accordingly, by Theorem 7.2, U_F^0 is mid-point biconcave. Since F is locally bounded from below, so is U_F^0. Therefore, U_F^0 is biconcave and $U_F \leq U_F^0$. (The reverse inequality is clear.) In particular,

$$U_F(0,0) \leq 0 .$$

Now consider a simple (not necessarily dyadic) martingale f with $f_1 \equiv 0$ and a transform g of f by a sequence ϵ in $\{-1,1\}$. Let $Z_n = (X_n, Y_n)$ be the n-th term of the simple zigzag martingale $Z \in \underset{\sim}{Z}(0,0)$ defined by $X_n = g_n + f_n$ and $Y_n = g_n - f_n$. Then $f_n = (X_n - Y_n)/2$ and $g_n = (X_n + Y_n)/2$ so that

$$\|g_\infty\|_p^p - \delta^p\|f_\infty\|_p^p = EF(Z_\infty) \leq U_F(0,0) \leq 0 .$$

Therefore, at least for simple B-valued martingales started at the origin, (2.2) holds with $\beta = \delta$. But using a straightforward approximation argument and keeping in mind that the assumption that f starts at the origin is harmless (see [13] for both arguments), we can conclude that (2.2) holds in general with $\beta = \delta$. This completes the proof of (7.5) and the lemma.

REMARK 7.1. Suppose that $1 < p < \infty$ and $\beta_p(B) < \infty$. Then, by the proof of Lemma 7.1, the function F defined by

$$F(x,y) = \left|\frac{x+y}{2}\right|^p - \beta^p\left|\frac{x-y}{2}\right|^p$$

is majorized by the biconcave function U_F and $U_F(0,0) = 0$. Here $\beta = \beta_p(B)$. Let $\zeta\colon B \times B \to \mathbb{R}$ be defined by

$$\zeta(x,y) = \frac{1 - U_F(x,-y)}{p\beta^p} .$$

Then ζ is biconvex on $B \times B$ and

$$\zeta(0,0) = \frac{1}{p\beta^p} > 0 .$$

Furthermore, (3.1) is satisfied so B is ζ-convex. To check (3.1), let x and y satisfy $|x| = |y| = 1$. Set $t = |x+y|/2$ and note that

$$\left|\frac{x-y}{2}\right| \geq |x| - t = 1 - t .$$

Accordingly, since $F \leq U_F$,

$$\begin{aligned} \zeta(x,y) &\leq \frac{1 - F(x,-y)}{p\beta^p} \\ &\leq \frac{1 - (1-t)^p + \beta^p t^p}{p\beta^p} \\ &\leq 2t = |x+y| . \end{aligned}$$

Here we have used $0 \leq t \leq 1$ and $\beta \geq 1$.

So here we have another proof of the assertion that a UMD-space is ζ-convex but one that yields a smaller lower bound on $\zeta_B(0,0)$ than the proof in Section 6.

8. B IS ζ-CONVEX $\Rightarrow$ B $\in$ UMD

Suppose that $\zeta\colon B \times B \to \mathbb{R}$ is biconvex with $\zeta(0,0) > 0$ and

$$\zeta(x,y) \le |x + y| \quad \text{if} \quad |x| = |y| = 1 .$$

Then, as we shall show,

$$\beta_p(B) \le \frac{72}{\zeta(0,0)} \cdot \frac{(p+1)^2}{p-1}, \quad 1 < p < \infty . \tag{8.1}$$

Thus, a ζ-convex Banach space has the UMD property.

Even if $\zeta(0,0)$ is replaced by $\zeta_B(0,0)$, as it can be, the bound on the right is not sharp but it does have the best possible order of magnitude as p approaches either endpoint of the interval $(1,\infty)$. This follows from the inequality

$$\beta_p(B) \ge p^* - 1 \tag{8.2}$$

where $p^* = p \vee q$ and $q = p/(p-1)$. To obtain the lower bound $p^* - 1$, note that $\beta_p(B) \ge \beta_p(\mathbb{R})$ and use

$$\beta_p(\mathbb{R}) = p^* - 1 , \tag{8.3}$$

one of the results of [16].

We shall assume throughout this section that $u\colon B \times B \to \mathbb{R}$ is a biconvex function such that $u(0,0) > 0$,

$$u(x,y) \le |x + y| \quad \text{if} \quad |x| \vee |y| \ge 1 , \tag{8.4}$$

and, for all $x \in B$,

$$u(x,-x) \le u(0,0) . \tag{8.5}$$

At the end of this section, we shall construct such a function u from the function ζ described above.

If x and y are in B, then, by (8.5) and Lemma 3.2,

(8.6)
$$\begin{aligned} u(x,y) &= u(x,-x+x+y) \\ &\le u(x,-x) + |x+y| \\ &\le u(0,0) + |x+y| . \end{aligned}$$

Also, we can and do assume that

(8.7)
$$u(x,y) = u(-x,-y) ,$$

for if this does not already hold, we can replace u by the mapping $(x,y) \to u(x,y) \vee u(-x,-y)$. By (8.7) and the biconvexity of u,

(8.8)
$$u(0,0) \le [u(x,0) + u(-x,0)]/2 = u(x,0)$$

and, similarly, $u(0,0) \le u(0,y)$.

LEMMA 8.1. *Suppose that* Z *is a simple zigzag martingale with values in* $B \times B$ *such that* $X_1 = 0$ *or* $Y_1 = 0$. *Then*

(8.9)
$$u(0,0)P(|X_\infty| \vee |Y_\infty| \ge 1) \le \|X_\infty + Y_\infty\|_1 .$$

PROOF. By (8.4), the left-hand side of (8.9) is less than or equal to

$$u(0,0)P(|X_\infty + Y_\infty| - u(Z_\infty) + u(0,0) \ge u(0,0)) .$$

By (8.6), the function $|X_\infty + Y_\infty| - u(Z_\infty) + u(0,0)$ is nonnegative. Therefore, by Chebyshev's inequality, the left-hand side of (8.9) is less than or equal to

$$E(|X_\infty + Y_\infty| - u(Z_\infty) + u(0,0)) .$$

The desired inequality (8.9) now follows from $Eu(Z_\infty) \ge Eu(Z_1)$, which is a consequence of Lemma 4.1, and from $u(Z_1) \ge u(0,0)$, which is a consequence of (8.8) and the assumption that $Z_1 = (X_1,0)$ or $(0,Y_1)$.

LEMMA 8.2. *Let* f *be a simple* B-*valued martingale and* g *its transform by a sequence of numbers in* $\{-1,1\}$. *Then, for* $\lambda > 0$,

(8.10)
$$\lambda P(|g_\infty| \ge \lambda) \le \frac{2}{u(0,0)} \|f_\infty\|_1 .$$

PROOF. It is enough to prove this for $\lambda = 1$. We write $f_n = \sum_{k=1}^{n} d_k$ and $g_n = \sum_{k=1}^{n} \epsilon_k d_k$, as usual, and set

$$X_n = f_n + g_n = \sum_{k=1}^{n} (1 + \epsilon_k) d_k ,$$

$$Y_n = f_n - g_n = \sum_{k=1}^{n} (1 - \epsilon_k) d_k ,$$

and $Z_n = (X_n, Y_n)$. Then

$$f_n = (X_n + Y_n)/2 ,$$

$$g_n = (X_n - Y_n)/2 ,$$

$X_1 = 0$ or $Y_1 = 0$, and Z is a simple zigzag martingale. Therefore, by Lemma 8.1,

$$\begin{aligned} P(|g_\infty| \geq 1) &= P(|X_\infty - Y_\infty| \geq 2) \\ &\leq P(|X_\infty| \vee |Y_\infty| \geq 1) \\ &\leq \|X_\infty + Y_\infty\|_1 / u(0,0) \\ &= 2\|f_\infty\|_1 / u(0,0) \end{aligned}$$

and this completes the proof.

REMARK 8.1. Let f be any B-valued martingale and g its transform by a sequence of numbers in $\{-1,1\}$. Then the above lemma easily yields the inequality

$$\lambda P(g^* \geq \lambda) \leq \frac{2}{u(0,0)} \|f\|_1 , \tag{8.11}$$

where $g^* = \sup_n |g_n|$ and $\|f\|_1 = \sup_n \|f_n\|_1$. To deduce (8.11) from (8.10), use approximation and the natural stopping time argument; see [13] and [16].

LEMMA 8.3. <u>Let</u> f <u>be a dyadic</u> B-<u>valued martingale and</u> g <u>its transform by a sequence</u> ϵ <u>of numbers in</u> $\{-1,1\}$. <u>Then, for</u> $0 < p < \infty$,

$$\|g^*\|_p \leq \frac{38}{u(0,0)} \cdot \frac{(p+1)^2}{p} \|f^*\|_p . \tag{8.12}$$

PROOF. It is enough to prove (8.12) for $(f_1, \ldots, f_n, f_n, f_n, \ldots)$ so we can and do assume in the proof that f is also simple.

Let $\delta > 0$ and $\beta > 2\delta + 1$. The first step is to show that, for all $\lambda > 0$,

$$P(g^* > \beta\lambda,\ f^* \le \delta\lambda) \le \alpha P(g^* > \lambda) \tag{8.13}$$

where

$$\alpha = \frac{8\delta}{u(0,0)(\beta - 2\delta - 1)} . \tag{8.14}$$

If the constant function f_1 satisfies $|f_1| > \delta\lambda$, then the left-hand side of (8.13) vanishes and the inequality is true. From now on suppose that $|f_1| \le \delta\lambda$. Let

$$\mu = \inf\{n : |g_n| > \lambda\} ,$$
$$\nu = \inf\{n : |g_n| > \beta\lambda\} ,$$
$$\sigma = \inf\{n : |f_n| > \delta\lambda \ \text{ or } \ |d_{n+1}| > 2\delta\lambda\} ,$$

and u_n be the indicator function of the set $\{\mu < n \le \nu \wedge \sigma\}$. Note that, since f is dyadic, we have that F defined by

$$F_n = \sum_{k=1}^{n} u_k d_k$$

is a martingale. On the set $\{\sigma \le \mu\}$, all of the u_k vanish so $F_\infty = 0$. In particular, this is the case on $\{\mu = \infty\} = \{g^* \le \lambda\}$. On $\{\sigma > \mu\}$,

$$|F_\infty| = |f_{\nu \wedge \sigma} - f_\mu| \le 4\delta\lambda .$$

Therefore,

$$\|F_\infty\|_1 \le 4\delta\lambda\, P(g^* > \lambda) .$$

Now consider G defined by

$$G_n = \sum_{k=1}^{n} \epsilon_k u_k d_k .$$

By Lemma 8.2,

$$P(g^* > \beta\lambda,\ f^* \leq \delta\lambda) \leq P(\mu \leq \nu < \infty,\ \sigma = \infty)$$

$$\leq P(|G_\infty| > (\beta - 2\delta - 1)\lambda)$$

$$\leq \frac{2}{u(0,0)} \cdot \frac{\|F_\infty\|_1}{(\beta - 2\delta - 1)\lambda}$$

$$\leq \alpha P(g^* > \lambda) \ .$$

If $\alpha\beta^p < 1$, then

$$\|g^*\|_p^p \leq \frac{\beta^p}{\delta^p(1 - \alpha\beta^p)} \|f^*\|_p^p \tag{8.15}$$

as can be seen as follows:

$$\|g^*\|_p^p = \beta^p \|g^*/\beta\|_p^p$$

$$= \beta^p \int_0^\infty P(g^* > \beta\lambda)\, p\lambda^{p-1} d\lambda$$

$$\leq \alpha\beta^p \int_0^\infty P(g^* > \lambda)\, p\lambda^{p-1} d\lambda$$

$$+ \beta^p \int_0^\infty P(f^* > \delta\lambda)\, p\lambda^{p-1} d\lambda$$

$$= \alpha\beta^p \|g^*\|_p^p + \beta^p \delta^{-p} \|f^*\|_p^p \ .$$

The term $\alpha\beta^p\|g^*\|_p^p$ can now be subtracted from both sides since $\|g^*\|_p$ is finite by the simplicity of g.

Now substitute $\beta = 1 + 1/p$ and

$$\delta = \frac{u(0,0)\, p^p}{8(p + 1)^{p+1} + 2p^{p+1}}$$

in (8.14) and (8.15) to obtain (8.12). The number 38 can be replaced by 36 if $1 \leq p < \infty$.

LEMMA 8.4. *If* $1 < p < \infty$, *then*

$$\beta_p(B) \leq \frac{36}{u(0,0)} \cdot \frac{(p + 1)^2}{p - 1} \ . \tag{8.16}$$

PROOF. Since $\beta_p(B) = \beta_p^0(B)$, by Lemma 7.1, it will be enough to show that if f is a simple dyadic B-valued martingale and g is its transform by a sequence of numbers in $\{-1,1\}$, then

$$\|g_\infty\|_p \le \frac{36}{u(0,0)} \cdot \frac{(p+1)^2}{p-1} \|f_\infty\|_p . \tag{8.17}$$

Clearly, $\|g_\infty\|_p \le \|g^*\|_p$ and, by Doob's inequality, $\|f^*\|_p \le q\|f_\infty\|_p$ where $q = p/(p-1)$. Thus, (8.17) follows from (8.12) and the remark above (8.16).

LEMMA 8.5. *Suppose that* ζ *is a biconvex function on*

$$\{(x,y) \in B \times B: |x| \vee |y| \le 1\}$$

such that

$$\zeta(x,y) \le |x+y| \quad \text{if} \quad |x| = |y| = 1 .$$

Then there is a biconvex function $u: B \times B \to \mathbb{R}$ *such that* (8.4) *and* (8.5) *both hold as well as*

$$u(0,0) \ge \zeta(0,0)/(1+r) \tag{8.18}$$

where r *is any number in the interval* $[0,1]$ *satisfying*

$$\sup_{|x| \le 1} \zeta(x,-x) \le \sup_{|x| \le r} \zeta(x,-x) . \tag{8.19}$$

PROOF. We can and do assume that ζ is nonnegative. Then, by Lemma 3.1 and its proof, there is a nonnegative biconvex function $v: B \times B \to \mathbb{R}$ such that

$$v(x,y) \le |x+y| \quad \text{if} \quad |x| \vee |y| \ge 1$$

and $v(0,0) = \zeta(0,0)$, in fact,

$$v(x,-x) = \zeta(x,-x) \quad \text{if} \quad |x| \le 1 .$$

Now define u on $B \times B$ by

$$u(x,y) = \sup_{|b| \le r} \frac{v[(1+r)x + b,\ (1+r)y - b]}{1+r} .$$

Since v is locally bounded from above so is u. Also, u is biconvex and $u(0,0) \geq v(0,0)/(1+r) = \zeta(0,0)/(1+r)$, so (8.18) holds. Suppose that $|x| \vee |y| \geq 1$ and $|b| \leq r$; for example, suppose that $|x| \geq 1$. Then

$$|(1+r)x + b| \geq (1+r)|x| - |b| \geq (1+r) - r = 1$$

and, by the properties of v,

$$u(x,y) \leq \frac{|(1+r)x + b + (1+r)y - b|}{1+r} = |x+y| .$$

Furthermore,

$$\begin{aligned}(1+r)u(x,-x) &= \sup_{|b| \leq r} v[(1+r)x + b, -(1+r)x - b] \\ &\leq \sup_{b \in B} v(b,-b) = \sup_{|b| \leq 1} v(b,-b) \\ &= \sup_{|b| \leq 1} \zeta(b,-b) = \sup_{|b| \leq r} \zeta(b,-b) \\ &= \sup_{|b| \leq r} v(b,-b) = (1+r)u(0,0)\end{aligned}$$

so both (8.4) and (8.5) hold.

We can now complete the proof of (8.1). Let ζ be as in the above lemma with $\zeta(0,0) > 0$ and let u be as described. Then $u(0,0) \geq \zeta(0,0)/2$ and (8.1) follows from Lemma 8.4.

This completes the proof of Theorem 3.1. For a number of other martingale conditions equivalent to ζ-convexity (for example, the almost everywhere convergence of transforms of L^1_B-bounded martingales), see [13].

REMARK 8.2. If B satisfies

(8.20) $$\zeta_B(x,-x) \leq \zeta_B(0,0) , \quad x \in B ,$$

then we can let $u = \zeta_B$ and there is no need for Lemma 8.5. There are other desirable consequences; for example, the inequality (8.11) becomes sharp. Does (8.20) hold for every Banach space B? We can show that at least for many of the classical Banach spaces the inequality (8.20) does hold. For example, consider

$B = \ell^r$ with $1 < r < \infty$. Let n be a positive integer, T the linear isometry $x \to (0,\ldots,0,x_1,x_2,\ldots)$ in which the first n coordinates are 0, and M the range of T. Then $\zeta_M(Tx,Ty) \le \zeta_B(x,y)$ by the maximality of ζ_B. In particular, $\zeta_M(0,0) \le \zeta_B(0,0)$. Let $z \in B$ and $z^n = (z_1,\ldots,z_n,0,0\ldots)$. Define ζ on $M \times M$ by $\zeta(x,y) = \zeta_B(x + z^n, y - z^n)$. If $x,y \in M$ and $|x| \vee |y| \ge 1$, then $|x + z^n| \vee |y - z^n| \ge 1$ so that $\zeta(x,y) \le |x + y|$. Also, ζ is biconvex on $M \times M$. Therefore, by the maximality of ζ_M, we can conclude that $\zeta \le \zeta_M$ on $M \times M$. Letting $x = y = 0$, we obtain

$$\zeta_B(z^n,-z^n) \le \zeta_M(0,0) \le \zeta_B(0,0) .$$

The inequality $\zeta_B(z,-z) \le \zeta_B(0,0)$ now follows from Lemma 3.2 and the fact that $z^n \to z$ as $n \to \infty$. If B is ℓ^1 or ℓ^∞ or any other Banach space for which $\zeta_B(0,0) = 0$, then $\zeta_B(x,y) = |x + y|$ on $B \times B$ and (8.20) is obvious.

REMARK 8.3. Once an inequality of the form (8.13), with α suitably small, is proven for two nonnegative functions, then L^p-inequalities and more easily follow. See [20] and, in particular, Lemma 7.1 of [12]. For example, let Φ be any continuous nondecreasing function from $[0,\infty]$ into $[0,\infty]$ with $\Phi(0) = 0$ and

(8.21) $$\Phi(2\lambda) \le c\Phi(\lambda) , \quad \lambda > 0 .$$

Then, for f and g as in Lemma 8.3, we have that

(8.22) $$E\Phi(g^*) \le cE\Phi(f^*) .$$

The choice of $c = c(8.22)$ depends only on $c(8.21)$ and $\zeta_B(0,0)$.

9. $B \in UMD \Rightarrow B \in HT$

Itô integration with respect to Brownian motion can be used to prove this result, which is due to McConnell and the author, and that was the original approach. For a proof accessible to mathematicians not familiar with the Itô calculus, see [15] where the proof is essentially self-contained. It rests on (i) the square-function inequality for real martingales [11], an immediate consequence, via Khintchine's inequality, of $\mathbb{R} \in$ UMD [11], which is here an immediate consequence of the assumption that $B \in$ UMD, (ii) a decoupling inequality for Rademacher functions that also follows easily from the UMD condition, and (iii) Taylor's theorem with remainder. Both $\mathbb{R}$-valued and B-valued martingales appear in the proof, but the square-function inequality is used only in the real case. In fact, the square-function inequality does not hold for B-valued martingales, not even for the special case of Rademacher series with coefficients in B [35], unless B is isomorphic to a Hilbert space. If $B \in$ UMD, the decoupling inequality serves as an effective substitute.

Here we shall use a similar decoupling method for Brownian motion and thereby return to the underlying idea of the original approach.

Let $B \in$ UMD and $1 < p < \infty$. Suppose that N is a positive integer and $a_1,\dots,a_N$ and $b_1,\dots,b_N$ belong to B. Define B-valued functions u and v on $\mathbb{C}$ by

$$u(re^{i\theta}) = \sum_{k=1}^{N} (a_k \cos k\theta + b_k \sin k\theta) r^k ,$$

$$v(re^{i\theta}) = \sum_{k=1}^{N} (a_k \sin k\theta - b_k \cos k\theta) r^k .$$

To prove that $B \in$ HT, it is enough to show that

$$\int_0^{2\pi} |u(e^{i\theta})|^p d\theta \approx \int_0^{2\pi} |v(e^{i\theta})|^p d\theta . \tag{9.1}$$

Here the symbol "$\approx$" is to mean that the left-hand side is not greater than $c_p(B)$ times the right-hand side, and the right-hand side is not greater than $c_p(B)$ times the left-hand side, where the choice of the positive real number $c_p(B)$ depends only on B and p. The choice may vary from one use of the equivalence symbol to the next.

Let $Z = (Z_t)_{t \geq 0}$ be a complex Brownian motion and X and Y the real and imaginary parts of Z. We shall suppose that $Z_0 \equiv 0$ and that all of the paths of Z are continuous. If μ is a measurable function with values in $[0,\infty]$, we shall say that μ is a stopping time of Z if μ is a stopping time relative to the minimal right-continuous filtration with respect to which Z is adapted. We shall always let τ denote the stopping time of Z defined by

$$\tau = \inf\{t: |Z_t| = 1\} . \tag{9.2}$$

LEMMA 9.1. *Suppose that* μ *and* ν *are stopping times of* Z *such that* $\mu \leq \nu \leq \tau$. *Then*

$$\|u(Z_\nu) - u(Z_\mu)\|_p \approx \|v(Z_\nu) - v(Z_\mu)\|_p . \tag{9.3}$$

Note that (9.1) is an immediate application of this lemma. Let $\mu \equiv 0$ and $\nu = \tau$ to obtain

$$\|u(Z_\tau)\|_p \approx \|v(Z_\tau)\|_p . \tag{9.4}$$

Since Z_τ is uniformly distributed on the unit circle, (9.4) implies (9.1).

Another application of Lemma 9.1 is given in Section 11.

PROOF. Note that u and v and their partial derivatives are continuous on the closed unit disk. Thus, for example, $u(Z_\nu) - u(Z_\mu)$ is bounded. Since $u_{xx} + u_{yy} = 0$, Itô's formula gives

$$\|u(Z_\nu) - u(Z_\mu)\|_p = \| \int_\mu^\nu u_x(Z_t)dX_t + u_y(Z_t)dY_t\|_p . \tag{9.5}$$

Let Z' be another complex Brownian motion with the same distribution as Z. Moreover, suppose that Z and Z' are defined on the same probability space and that Z and Z' are independent. The next step is to show that

$$\| \int_\mu^\nu u_x(Z_t)dX_t + u_y(Z_t)dY_t\|_p \tag{9.6}$$

$$\approx \| \int_\mu^\nu u_x(Z_t)dX'_t + u_y(Z_t)dY'_t\|_p .$$

This is the analogue of the decoupling lemma in [15] and follows directly from the discussion of Garling in [32], but we shall give a different proof here. By approximation and scaling this can be reduced to showing that

$$(9.7)\qquad \Big\|\sum_{k=1}^{n} U_{2k}(X_{2k}-X_{2k-2})+V_{2k}(Y_{2k}-Y_{2k-2})\Big\|_p$$

$$\approx \Big\|\sum_{k=1}^{n} U_{2k}(X'_{2k}-X'_{2k-2})+V_{2k}(Y'_{2k}-Y'_{2k-2})\Big\|_p$$

where U_{2k} and V_{2k} are functions of $Z_0, Z_2, \ldots, Z_{2k-2}$. Let $d_k = X_k - X_{k-1}$ and $e_k = Y_k - Y_{k-1}$ so that

$$X_{2k} - X_{2k-2} = d_{2k} + d_{2k-1} ,$$

$$Y_{2k} - Y_{2k-2} = e_{2k} + e_{2k-1} ,$$

and U_{2k} and V_{2k} are functions of $d_2+d_1, \ldots, d_{2k-2}+d_{2k-3}, e_2+e_1, \ldots, e_{2k-2}+e_{2k-3}$ if $k \geq 2$, and are constant for $k = 1$. One can now replace $X'_{2k} - X'_{2k-2}$ by $d_{2k} - d_{2k-1}$ and $Y'_{2k} - Y'_{2k-2}$ by $e_{2k} - e_{2k-1}$ in the right-hand side of (9.7) and keep the same value for the norm: Note that

$$(1, d_2-d_1, e_2-e_1, d_2+d_1, e_2+e_1, \ldots)$$

is an orthogonal Gaussian sequence and is therefore an independent sequence with the same distribution as

$$(1, X'_2-X'_0, Y'_2-Y'_0, X_2-X_0, Y_2-Y_0, \ldots) .$$

With these replacements, the left-hand side of (9.7) becomes

$$(9.8)\qquad \Big\|\sum_{k=1}^{n} U_{2k}(d_{2k}+d_{2k-1})+V_{2k}(e_{2k}+e_{2k-1})\Big\|_p$$

and the right-hand side becomes

$$(9.9)\qquad \Big\|\sum_{k=1}^{n} U_{2k}(d_{2k}-d_{2k-1})+V_{2k}(e_{2k}-e_{2k-1})\Big\|_p .$$

But (9.8) can be written as $\|F_{4n}\|_p$ where $F = (F_1, F_2, \ldots)$ is the martingale with difference sequence

$$(U_2d_2, U_2d_1, V_2e_2, V_2e_1, \ldots)$$

and (9.9) can be written as $\|G_{4n}\|_p$ where G is a $\pm$ 1-transform of F. Therefore, by the UMD condition,

$$\|F_{4n}\|_p \approx \|G_{4n}\|_p$$

and (9.6) is proved.

Since $u_x = v_y$ and $u_y = -v_x$, the right-hand side of (9.6) equals

$$\| \int_\mu^\nu v_x(Z_t)d(-Y_t') + v_y(Z_t)dX_t'\|_p \ .$$

Because μ and ν are stopping times of Z and $(X,Y,-Y',X')$ has the same distribution as (X,Y,X',Y'), this is equal to

$$\| \int_\mu^\nu v_x(Z_t)dX_t' + v_y(Z_t)dY_t'\|_p \ .$$

By the result for u, this is equivalent to $\|v(Z_\nu) - v(Z_\mu)\|_p$ and completes the proof of Lemma 9.1.

10. $B \in HT \Rightarrow B \in UMD$

For the proof, see Bourgain [6].

11. EXTENSION OF THE M. RIESZ INEQUALITY

Much of what we have discussed up to this point can be summarized by

(11.1) $$B \text{ is } \zeta\text{-convex} \Leftrightarrow B \in UMD \Leftrightarrow B \in HT \ .$$

In the limited space that remains for us here, we shall look a little beyond (11.1).

Suppose that B is ζ-convex and $a_1,a_2,\ldots,b_1,b_2,\ldots$ belong to B. Let

(11.2) $$u(z) = \sum_{k=1}^{\infty} (a_k \cos k\theta + b_k \sin k\theta)r^k \ ,$$

(11.3) $$v(z) = \sum_{k=1}^{\infty} (a_k \sin k\theta - b_k \cos k\theta)r^k \ .$$

We assume that the two series converge at each point $z = re^{i\theta}$ in the open unit disk D of $\mathbb{C}$ or, equivalently, that

$$\limsup_{n\to\infty} |a_n|^{1/n} \le 1 \quad \text{and} \quad \limsup_{n\to\infty} |b_n|^{1/n} \le 1 \ .$$

We can take B to be, as usual, a Banach space over either the real or complex field.

Fix $0 < \alpha < 1$ and let $\Gamma_\alpha(\theta)$ be the interior of the smallest convex set containing the circle $|z| = \alpha$ and the point $e^{i\theta}$. The function $N_\alpha(u)$, defined on $[0,2\pi)$ by

$$N_\alpha(u)(\theta) = \sup_{z \in \Gamma_\alpha(\theta)} |u(z)| ,$$

is the nontangential maximal function of u.

THEOREM 11.1. *Let* $\Phi: [0,\infty] \to [0,\infty]$, *with* $\Phi(0) = 0$, *be a continuous and nondecreasing function satisfying the growth condition* (8.21). *Then*

$$\int_0^{2\pi} \Phi(N_\alpha(u))d\theta \approx \int_0^{2\pi} \Phi(N_\alpha(v))d\theta . \tag{11.4}$$

The constants in this double inequality may be chosen to depend only on α, *the growth constant* $c(8.21)$, *and the* ζ-*convexity constant* $\zeta_B(0,0)$.

This theorem extends to ζ-convex spaces one of the results for real-valued u and v of [21]. The above theorem easily implies (9.1) just as the corresponding result in the real case implies the classical M. Riesz inequality.

So the sizes of $N_\alpha(u)$ and $N_\alpha(v)$ are comparable in a dramatic way--but only if B is ζ-convex.

PROOF. Let $0 < \rho < 1$ and $U(z) = u(\rho z)$ for z in D. Since $N_\alpha(U) \uparrow N_\alpha(u)$ as $\rho \uparrow 1$, with a similar result for $N_\alpha(v)$, we can assume in the proof that u and v are defined by (11.2) and (11.3) on a disk with center at 0 and a radius greater than 1. In this case the two series converge absolutely on the closed unit disk, so in the proof we can and do assume that u and v are given by the finite sums of Section 9.

Let Z and τ be as in Section 9 and consider the Brownian maximal functions of u and v:

$$u^* = \sup_{0 \le t < \infty} |u(Z_{\tau \wedge t})| ,$$

$$v^* = \sup_{0 \le t < \infty} |v(Z_{\tau \wedge t})| .$$

Then, as we shall show,

(11.5) $$E\Phi(u^*) \approx E\Phi(v^*) .$$

For $\beta > 1$, $\delta > 0$, and $\lambda > 0$, consider the three stopping times of Z defined by

$$\mu = \inf\{t: |v(Z_{\tau \wedge t})| > \lambda\} ,$$

$$\nu = \inf\{t: |v(Z_{\tau \wedge t})| > \beta\lambda\} ,$$

$$\sigma = \inf\{t: |u(Z_{\tau \wedge t})| > \delta\lambda\} .$$

Then

$$\begin{aligned} P(v^* > \beta\lambda, \ u^* \leq \delta\lambda) &= P(\mu \leq \nu < \infty, \ \sigma = \infty) \\ &\leq P(|v(Z_{\tau \wedge \nu \wedge \sigma}) - v(Z_{\tau \wedge \mu \wedge \sigma})| \geq (\beta-1)\lambda) \\ &\leq \frac{\|v(Z_{\tau \wedge \nu \wedge \sigma}) - v(Z_{\tau \wedge \mu \wedge \sigma})\|_2}{(\beta-1)\lambda} . \end{aligned}$$

By Lemma 9.1 and its proof, the numerator is not greater than $\beta_2^2(B)$ times

$$\|u(Z_{\tau \wedge \nu \wedge \sigma}) - u(Z_{\tau \wedge \mu \wedge \sigma})\|_2 \leq 2\delta\lambda P(\mu < \infty) = 2\delta\lambda P(v^* > \lambda) .$$

By (8.1), $\beta_2(B) \leq 648/\zeta_B(0,0)$. Therefore

$$\begin{aligned} P(v^* > \beta\lambda, \ u^* \leq \delta\lambda) &\leq \frac{(648)^2}{\zeta_B^2(0,0)} \cdot \frac{2\delta}{\beta - 1} P(v^* > \lambda) \\ &= \varepsilon P(v^* > \lambda) . \end{aligned}$$

Since $\varepsilon \to 0$ as $\delta \to 0$, we can conclude from Lemma 7.1 of [12] that $E\Phi(v^*) \leq cE\Phi(u^*)$ where c depends only on $c(8.21)$ and on $\zeta_B(0,0)$. The double inequality (11.5) follows.

The next step is to observe that the Lebesgue measure of the set where $N_\alpha(u) > \lambda$ is comparable to $P(u^* > \lambda)$:

(11.6) $$m(N_\alpha(u) > \lambda) \approx P(u^* > \lambda) .$$

Here the constants depend only on the size of $\Gamma_\alpha(\theta)$, that is, only on α. The proof uses the fact that $N_\alpha(u) = N_\alpha(|u|)$ and that $|u|$ is subharmonic and continuous. The proof is then exactly the same as for real-valued u; see [21]. By (11.6),

$$\int_0^{2\pi} \Phi(N_\alpha(u))d\theta = \int_0^\infty m(N_\alpha(u) > \lambda)d\Phi(\lambda)$$
$$\approx \int_0^\infty P(u^* > \lambda)d\Phi(\lambda)$$
$$= E\Phi(u^*)$$

with a similar equivalence for v. The inequality (11.4) then follows from (11.5).

12. EXTENSION OF A THEOREM ON HARDY SPACES

Many theorems about Hardy spaces of scalar-valued functions carry over to ζ-convex spaces. Here is an example.

THEOREM 12.1. Let B be ζ-convex and u and v be given by (11.2) and (11.3). Then, for $0 < p < \infty$,

$$\|N_\alpha(u)\|_p^p \approx \sup_{0<r<1} \int_0^{2\pi} |u(re^{i\theta})|^p \vee |v(re^{i\theta})|^p d\theta . \tag{12.1}$$

For $B = \mathbb{R}$, the function $F = u + iv$ is analytic and $|u| \vee |v| \leq |F| \leq 2(|u| \vee |v|)$. Therefore, F belongs to the Hardy space H^p if and only if $N_\alpha(u)$ belongs to the Lebesgue space $L^p(0,2\pi)$. The "only if" part of this result is in [34], the "if" part in [21].

PROOF. With Φ as in Theorem 11.1, we have that

$$\Phi(|u(re^{i\theta})|) \vee \Phi(|v(re^{i\theta})|) \leq \Phi(N_\alpha(u)(\theta)) + \Phi(N_\alpha(v)(\theta)) .$$

Therefore, by Theorem 11.1,

$$\sup_{0<r<1} \int_0^{2\pi} \Phi(|u(re^{i\theta})|) \vee \Phi(|v(re^{i\theta})|)d\theta \leq c \int_0^{2\pi} \Phi(N_\alpha(u))d\theta . \tag{12.2}$$

This contains one side of the double inequality (12.1).

Even if $B = \mathbb{R}$, it is not true that the converse of (12.2) holds for all functions Φ as in Theorem 11.1; see the last paragraph of [21]. On the other hand, for all Banach spaces B, the converse does hold for powers $\Phi(\lambda) = \lambda^p$. To see this we can assume that u and v are given by finite sums as in Section 9. Let w be the bounded harmonic function on D with boundary function $|u(e^{i\theta})|^{p/2} \vee |v(e^{i\theta})|^{p/2}$. We shall show that

$$|u|^{p/2} \leq 2^p w \quad \text{on } D \tag{12.3}$$

so that $N_\alpha^p(u) = N_\alpha^2(|u|^{p/2}) \leq 2^{2p} N_\alpha^2(w)$. The desired result then follows from the classical fact [34] that

$$\int_0^{2\pi} N_\alpha^2(w)\,d\theta \leq c_\alpha \sup_{0<r<1} \int_0^{2\pi} w^2(re^{i\theta})\,d\theta = c_\alpha \int_0^{2\pi} w^2(e^{i\theta})\,d\theta \ .$$

To show (12.3), let $f(z) = \varphi(u(z)) + i\varphi(v(z))$ where $\varphi\colon B \to \mathbb{R}$ satisfies $\varphi(a+b) = \varphi(a) + \varphi(b)$, $\varphi(\beta b) = \beta\varphi(b)$, and $|\varphi(b)| \leq |b|$ for $a, b \in B$ and $\beta \in \mathbb{R}$. Then $f\colon \mathbb{C} \to \mathbb{C}$ is analytic. Consequently, $|f|^{p/2}$ is subharmonic on D with a boundary function majorized by the boundary function of $2^{p/2}w$. Therefore, for $z \in D$,

$$|\varphi(u(z))|^{p/2} \leq |f(z)|^{p/2} \leq 2^{p/2} w(z) \ .$$

This implies that $|u|^{p/2} \leq 2^{p/2} w$ if $\mathbb{R}$ is the scalar field of B. In the complex case, the real and imaginary parts of linear functionals $\varphi + i\psi$ can be handled separately to obtain (12.3).

For other results about Hardy spaces, Fourier multiplier transformations, and singular integrals of B-valued functions, see Berkson, Gillespie, and Muhly [3], Blasco [4], Bourgain [7], [9], Bourgain and Davis [10], McConnell [38], Rubio de Francia [44], Rubio de Francia, Ruiz, and Torrea [45], and Virot [49].

13. SHARP INEQUALITIES FOR STOCHASTIC INTEGRALS

The boundary methods that we have described here to study B-valued martingales and their transforms give, in principle, a way to calculate the best constants in a number of inequalities. In particular, if $B = \mathbb{R}$, it is possible to derive some of these constants explicitly; see [16], [17], and [18]. We shall describe here a few of these results in the context of stochastic integration. See [16], [18], and [19] for more information and the proofs.

Let $(\Omega, \mathfrak{F}_\infty, P)$ be a complete probability space and $\mathfrak{F} = (\mathfrak{F}_t)_{t \geq 0}$ a non-decreasing right-continuous family of sub-σ-fields of $\mathfrak{F}_\infty$ where $\mathfrak{F}_0$ contains all $A \in \mathfrak{F}_\infty$ with $P(A) = 0$. Suppose that $M = (M_t)_{t \geq 0}$ is a real martingale adapted to $\mathfrak{F}$ such that almost all of the paths of M are right-continuous on $[0,\infty)$ and have left-limits on $(0,\infty)$. Let $V = (V_t)_{t \geq 0}$ be a predictable process with values in $[-1,1]$ and denote by $N = V \cdot M$ the stochastic integral of V with respect to M: N is an adapted right-continuous process with left-limits such that

$$N_t = \int_{[0,t]} V_s \, dM_s \quad \text{a.s.}$$

For background, see [25].

If $1 \leq p \leq \infty$, let $\|M\|_p = \sup_t \|M_t\|_p$ and $N^*(\omega) = \sup_t |N_t(\omega)|$.

THEOREM 13.1. *Let* $1 < p < \infty$ *and* $p^* = p \vee q$ *where* $1/p + 1/q = 1$. *Then*

$$\|N\|_p \leq (p^* - 1)\|M\|_p \tag{13.1}$$

and the constant $p^* - 1$ *is best possible. Furthermore, strict inequality holds if* $p \neq 2$ *and* $0 < \|M\|_p < \infty$.

THEOREM 13.2. *If* $\|M\|_\infty \leq 1$ *and* $2 \leq p < \infty$, *then*

$$\|N\|_p^p \leq \frac{1}{2}\,\Gamma(p+1) . \tag{13.2}$$

If $1 \leq p \leq 2$, *then*

$$\sup_{\lambda > 0} \lambda^p P(N^* \geq \lambda) \leq 2\|M\|_p^p / \Gamma(p+1) . \tag{13.3}$$

The bounds on the right in (13.2) *and* (13.3) *are best possible.*

Note that if $P(N^* \geq 1) = 1$, then (13.3) implies that

$$\frac{1}{2}\,\Gamma(p+1) \leq \|M\|_p^p .$$

The following theorem extends (13.2). For the corresponding extension of (13.3), see Theorem 8.1 of [16].

THEOREM 13.3. *If* $\|M\|_\infty \leq 1$ *and* $\Phi\colon [0,\infty) \to [0,\infty)$ *is a convex function that has a convex first derivative on* $(0,\infty)$ *with* $\Phi'(0+) = \Phi(0) = 0$, *then*

$$\sup_t E\Phi(|N_t|) \leq \frac{1}{2}\int_0^\infty \Phi(t)e^{-t}dt$$

and the bound on the right is best possible.

If we let $\Phi(t) = e^{\alpha t} - 1 - \alpha t$, where $0 < \alpha < 1$, and use

$$\|N\|_1 \leq \|N\|_2 \leq \|M\|_2 \leq \|M\|_\infty \leq 1 ,$$

we obtain the following corollary, which is also an immediate consequence of Theorem 13.2.

COROLLARY 13.1. *If* $\|M\|_\infty \leq 1$ *and* $0 < \alpha < 1$, *then*

$$\sup_t Ee^{\alpha|N_t|} < \frac{2 - \alpha^2}{2(1 - \alpha)} .$$

Note that if $\|N\|_1 = 1$, then $|N_\infty| = 1$ a.s. and the left-hand side is e^α, which is strictly smaller than the right-hand side.

Up to now we have assumed that the predictable process V has its values in the interval $[-1,1]$. We now replace $[-1,1]$ by $[a,b]$ where $a \leq 0 \leq b$.

THEOREM 13.4. *Let* a *and* b *be real numbers satisfying* $a \leq 0 \leq b$ *and* V *a predictable process such that* $a \leq V_t(\omega) \leq b$. *Then the stochastic integral* $N = V\cdot M$ *satisfies*

$$\lambda P(N^* \geq \lambda) \leq (b - a)\|M\|_1 , \quad \lambda > 0 .$$

The constant $b - a$ *is best possible.*

See [18] for the proof.

Note that if $a = 0$ and $b = 1$, then the best constant is 1, the same as in Doob's classical inequality

$$\lambda P(M^* \geq \lambda) \leq \|M\|_1 .$$

14. MONOTONE BASES OF L^p

Recall that a sequence $e = (e_1, e_2, \ldots)$ in real $L^p(0,1)$ is a basis of $L^p(0,1)$ if, for every $f \in L^p(0,1)$, there is a unique sequence $a = (a_1, a_2, \ldots)$ in $\mathbb{R}$ such that $\|f - \sum_{k=1}^{n} a_k e_k\|_p \to 0$ as $n \to \infty$. Write

$$T_n f = \sum_{k=1}^{n} a_k e_k .$$

Then $T_1, T_2, \ldots$ is a nondecreasing sequence of contractive projections in $L^p(0,1)$: T_n is a linear transformation and $T_m T_n = T_n T_m = T_m$ if $1 \le m \le n$. A basis e is a monotone basis if the T_n are contractions:

$$\|T_n f\|_p \le \|f\|_p .$$

If e is a basis, let $K_p(e)$ be the least $K \in [1,\infty]$ such that

$$\|\sum_{k=1}^{n} \epsilon_k a_k e_k\|_p \le K \|\sum_{k=1}^{n} a_k e_k\|_p$$

for all choices of $\epsilon_k \in \{-1,1\}$, $a_k \in \mathbb{R}$, and $n \ge 1$. The basis e is an unconditional basis if the unconditional constant $K_p(e)$ is finite.

Pelczyński and Rosenthal [40] and Dor and Odell [28] discovered that, for $1 < p < \infty$, every monotone basis of $L^p(0,1)$ is unconditional. The following theorem from [16] makes this result more precise.

THEOREM 14.1. *Let $1 < p < \infty$ and $p^* = p \vee q$ where $1/p + 1/q = 1$. Then every monotone basis of $L^p(0,1)$ has the unconditional constant $p^* - 1$.*

Here is a related theorem.

THEOREM 14.2. *Suppose that $1 < p < \infty$ and $(\Omega, \mathcal{A}, \mu)$ is a positive measure space (not necessarily finite or σ-finite). Let $T_1, T_2, \ldots$ be any nondecreasing sequence of contractive projections in $L^p(\Omega, \mathcal{A}, \mu)$ and set $T_0 = 0$. If $\epsilon_k \in \{-1,1\}$, $k \ge 1$, and $f \in L^p(\Omega, \mathcal{A}, \mu)$, then*

$$\|\sum_{k=1}^{\infty} \epsilon_k (T_k - T_{k-1}) f\|_p \le (p^* - 1)\|f\|_p$$

where the series on the left is the limit in the strong operator topology of its partial sums. Furthermore, strict inequality holds if $\|f\|_p > 0$ *and* $p \neq 2$.

See [16] for further results and discussion.

REFERENCES

[1] D. J. Aldous, Unconditional bases and martingales in $L_p(F)$, Math. Proc. Cambridge Phil. Soc. 85 (1979), 117-123.

[2] A. Benedek, A. P. Calderón, and R. Panzone, Convolution operators on Banach space valued functions, Proc. Nat. Acad. Sci. 48 (1962), 356-365.

[3] E. Berkson, T. A. Gillespie, and P. S. Muhly, Théorie spectrale dans les espaces UMD, C. R. Acad. Sc. Paris (1986).

[4] O. Blasco, Espacios de Hardy de funciones con valores vectoriales, Publicaciones del Seminario Matematico, Universidad de Zaragoza, 1985.

[5] S. Bochner and A. E. Taylor, Linear functionals on certain spaces of abstractly-valued functions, Ann. Math. 39 (1938), 913-944.

[6] J. Bourgain, Some remarks on Banach spaces in which martingale difference sequences are unconditional, Ark. Mat. 21 (1983), 163-168.

[7] J. Bourgain, Extension of a result of Benedek, Calderón, and Panzone, Ark. Mat. 22 (1984), 91-95.

[8] J. Bourgain, On martingale transforms in finite dimensional lattices with an appendix on the K-convexity constant, Math. Nachr. 119 (1984), 41-53.

[9] J. Bourgain, Vector valued singular integrals and the H^1 - BMO duality, Probability Theory and Harmonic Analysis, J. A. Chao and W. A. Woyczynski, editors, Marcel Dekker, New York (1986), 1-19.

[10] J. Bourgain and W. J. Davis, Martingale transforms and complex uniform convexity, Trans. Amer. Math. Soc.

[11] D. L. Burkholder, Martingale transforms, Ann. Math. Statist. 37 (1966), 1494-1504.

[12] D. L. Burkholder, Distribution function inequalities for martingales, Ann. Probab. 1 (1973), 19-42.

[13] D. L. Burkholder, A geometrical characterization of Banach spaces in which martingale difference sequences are unconditional, Ann. Probab. 9 (1981), 997-1011.

[14] D. L. Burkholder, Martingale transforms and the geometry of Banach spaces, Proceedings of the Third International Conference on Probability in Banach Spaces, Tufts University, 1980, Lecture Notes in Mathematics, 860 (1981), 35-50.

[15] D. L. Burkholder, A geometric condition that implies the existence of certain singular integrals of Banach-space-valued functions, Conference on Harmonic Analysis in Honor of Antoni Zygmund, University of Chicago, 1981, Wadsworth International Group, Belmont, California, 1 (1983), 270-286.

[16] D. L. Burkholder, Boundary value problems and sharp inequalities for martingale transforms, Ann. Probab. 12 (1984), 647-702.

[17] D. L. Burkholder, An elementary proof of an inequality of R. E. A. C. Paley, Bull. London Math. Soc. 17 (1985), 474-478.

[18] D. L. Burkholder, An extension of a classical martingale inequality, Probability Theory and Harmonic Analysis, J. A. Chao and W. A. Woyczynski, editors, Marcel Dekker, New York (1986), 21-30.

[19] D. L. Burkholder, A sharp and strict L^p-inequality for stochastic integrals, Ann. Probab. 14 (1986).

[20] D. L. Burkholder and R. F. Gundy, Extrapolation and interpolation of quasi-linear operators on martingales, Acta Math. 124 (1970), 249-304.

[21] D. L. Burkholder, R. F. Gundy, and M. L. Silverstein, A maximal function characterization of the class H^p, Trans. Amer. Math. Soc. 157 (1971), 137-153.

[22] A. P. Calderón and A. Zygmund, On the existence of certain singular integrals, Acta Math. 88 (1952), 85-139.

[23] A. P. Calderón and A. Zygmund, On singular integrals, Amer. J. of Math. 78 (1956), 289-309.

[24] M. L. Cartwright, Manuscripts of Hardy, Littlewood, Marcel Riesz and Titchmarsh, Bull. London Math. Soc. 14 (1982), 472-532.

[25] C. Dellacherie and P.-A. Meyer, Probabilités et potentiel: Théorie des martingales, Hermann, Paris, 1980.

[26] J. Diestel and J. J. Uhl, Vector Measures, Math. Surveys 15, American Mathematical Society, Providence, Rhode Island, 1977.

[27] J. L. Doob, Stochastic Processes, Wiley, New York, 1953.

[28] L. E. Dor and E. Odell, Monotone bases in L_p, Pacific J. Math. 60 (1975), 51-61.

[29] P. Enflo, Banach spaces which can be given an equivalent uniformly convex norm, Israel J. Math. 13 (1972), 281-288.

[30] C. Fefferman and E. M. Stein, H^p spaces of several variables, Acta Math. 129 (1972), 137-193.

[31] L. Gårding, Marcel Riesz in memoriam, Acta Math. 127 (1970), i-xi.

[32] D. J. H. Garling, Brownian motion and UMD-spaces, Conference on Probability and Banach Spaces, Zaragoza, 1985, Lecture Notes in Mathematics.

[33] J. A. Gutiérrez, On the Boundedness of the Banach Space-Valued Hilbert Transform, Ph.D. Dissertation, University of Texas, Austin, 1982.

[34] G. H. Hardy and J. E. Littlewood, A maximal theorem with function-theoretic applications, Acta Math. 54 (1930), 81-116.

[35] S. Kwapień, Isomorphic characterizations of inner product spaces by orthogonal series with vector valued coefficients, Studia Math. 44 (1972), 583-595.

[36] J. Lindenstrauss and L. Tzafriri, Classical Banach Spaces I: Sequence Spaces, Springer, New York, 1977.

[37] B. Maurey, Système de Haar, Séminaire Maurey-Schwartz (1974-75), École Polytechnique, Paris, 1975.

[38] T. R. McConnell, On Fourier multiplier transformations of Banach-valued functions, Trans. Amer. Math. Soc. 285 (1984), 739-757.

[39] T. R. McConnell, A Skorohod-like representation in infinite dimensions, Probability in Banach Spaces V, Tufts University, 1984, Lecture Notes in Mathematics, 1153 (1985), 359-368.

[40] A. Pelczyński and H. P. Rosenthal, Localization techniques in L^p spaces, Studia Math. 52 (1975), 263-289.

[41] G. Pisier, Un exemple concernant la super-réflexivité, Séminaire Maurey-Schwartz (1974-75), École Polytechnique, Paris, 1975.

[42] M. Riesz, Les fonctions conjuguées et les séries de Fourier, C. R. Acad. Sci. Paris, 178 (1924), 1464-1467.

[43] M. Riesz, Sur les fonctions conjuguées, Math. Z. 27 (1927), 218-244.

[44] J. L. Rubio de Francia, Fourier series and Hilbert transforms with values in UMD Banach spaces, Studia Math. 81 (1985), 95-105.

[45] J. L. Rubio de Francia, F. J. Ruiz, and J. L. Torrea, Calderón-Zygmund theory for operator-valued kernels, Advances in Math.

[46] J. Schwartz, A remark on inequalities of Calderón-Zygmund type for vector-valued functions, Comm. Pure Appl. Math. 14 (1961), 785-799.

[47] E. M. Stein, Singular Integrals and Differentiability Properties of Functions, Princeton, New Jersey, Princeton University Press, 1970.

[48] S. Vági, A remark on Plancherel's theorem for Banach space valued functions, Ann. Scuola Norm. Sup. Pisa Cl. Sci. 23 (1969), 305-315.

[49] B. Virot, Quelques inégalitiés concernant les transformées de Hilbert des fonctions à valeurs vectorielles, C. R. Acad. Sc. Paris, 293 (1981), 459-462.

[50] A. Zygmund, Trigonometric Series I, II, New York, Cambridge University Press, 1959.

MARTINGALE THEORY : AN ANALYTICAL FORMULATION WITH SOME APPLICATIONS IN ANALYSIS.

S.D. Chatterji.

Département de Mathématiques

Ecole Polytechnique Fédérale

Lausanne.

Introduction.

The main theme of the lecture notes that follow is that martingale theory is a versatile tool in questions of classical analysis. This is, of course, nowadays, widely recognized. But, the recognition is based, mostly, on its applications to potential theory and partial differential equations which depend on fairly sophisticated techniques using Brownian motions and other more general Markov processes. My objective here is to show that even the most elementary facts of martingale theory, in its simplest formulation, used imaginatively, would yield a number of deep and difficult results of classical analysis and would lead to new questions of interest to probabilists as well as analysts.

The basic ingredients of the martingale theory needed are summarized in Chapter 1 where we have collected some notational and terminological conventions as well. A resumé of some necessary measure theoretic facts is also given there. It is assumed that the reader is more or less familiar with this sort of material. Some indications as to where one can find a more leisurely discussion of the topics is also given. The martingale theory needed is such that it appears, at least in the real-valued case, in almost any graduate level text. The vector-valued situation as well as some of the measure theory may seem a little esoteric to the beginners — although the general trend must be familiar to those who are acquainted with the standard facts. Chapters 2 and 3 deal with some generalizations of elementary martingale theory. The results of Chapter 2 are used in a very limited way later. Those of Chapter 3 are presented partly for their intrinsic interest but mostly for their applications in Chapter 4 where we take up some aspects of the Lebesgue differentiation theorem in $\mathbb{R}^n$. Chapter 5 proves a fixed point theorem due to Ryll-Nardzweski by using martingale theory. The inclusion of this chapter was mostly intended to show how martingale theory may be useful in a problem which has apparently nothing to do with probability or measure theory. Chapter 6 deals with the determination of the absolute continuity or singularity of two probability measures, a topic where martingale theory plays a known and natural role. We have treated there the case of product measures and Gaussian measures because of their general importance. Finally, in Chapter 7 we discuss a new, relatively recent, development in probability theory which started out by using martingale methods.

The chapters are written in a such a way that they can be read almost independantly of each other. No attempt has been made to cover any topic in an encyclopaedic manner. However, each chapter ends with a section which gives some historical background and a choice of further works to be consulted.

Some years ago (1973), I published, in a Lecture Notes volume, a text (cf. references) whose (French) title would seem to indicate considerable overlap with the present notes. In fact, although the topics treated here form almost a subset of those treated in the previous volume, much of the presentation in the present one is quite different and some of the theorems and proofs involved were found only after the publication of the previous volume. The list of references at the end has been restricted to books and monographs only; in general, these have substantial bibliographies. References to articles are briefly given in the Notes and Remarks of different chapters; indications are to be found there as to how further relevant literature can be looked up.

Chapter 1.

Notation, terminology, preliminaries.

§1. General notation.

In general, we follow the set-theoretical and functional notations of Bourbaki except for occasional deviations to facilitate typing. Thus $\mathbb{N} = \{0,1,2,\ldots\}$, $\mathbb{Z} = \{0,\pm 1,\pm 2,\ldots\}$, $\mathbb{R}$ = the set of real numbers, $\mathbb{C}$ = the set of complex numbers. If A,B are two sets, we write $A \setminus B = A \cap B^c$ where B^c = the complement of B and $A \setminus B$ is the set of points in A but not in B. Also $A \Delta B = (A \setminus B) \cup (B \setminus A)$. If $f : X \to Y$ and $A \subset Y$, we write $\{f \in A\}$ for $f^{-1}(A) = \{x \in X \mid f(x) \in A\}$ — as is customary in probabilistic literature. If $f : X \to Y$, $g : Y \to Z$ then we often denote by $g(f)$ the map $g \circ f$: $X \to Z$ if that is convenient for typographical reasons and if it does not cause any confusion.

The phrase "if and only if" will be abbreviated to "iff".

§2. Measure-theoretical notation, terminology and preliminaries.

Let Ω be a set. An algebra of subsets of Ω is a family A of subsets of Ω such that (i) $\Omega \in A$ and (iii) finite unions and complements of sets in A are again in A. A sigma-algebra (of subsets of Ω) A is an algebra which contains countable unions of its sets. A measurable space is a set Ω endowed with a sigma-algebra A — denoted generally by the ordered pair (Ω,A). A measure space is a triple (Ω,A,μ) where (Ω,A) is a measurable space and $\mu : A \to \overline{\mathbb{R}}_+ = [0,\infty]$ is a countably additive, non-negative set function (often simply called a measure). If $\mu(\Omega) < \infty$, we talk of a finite measure space; if $\mu(\Omega) = 1$, we say that (Ω,A,μ) is a probability space.

If F is any family of sets in Ω, by $\sigma(F)$ we denote the sigma-algebra generated be F. Further, if $\{f_t\}_{t\in T}$ is any collection of measurable functions from Ω to some other measurable spaces then $\sigma\{f_t, t \in T\}$ will denote the smallest sigma-algebra in Ω with respect to which all the f_t's are measurable; this sigma-algebra is also termed as one generated by f_t, $t \in T$. The expressions "measurable function" and "random variable" are used interchangeably.

Let A be an algebra of subsets of Ω and let E be a normed space (over $\mathbb{R}$ or $\mathbb{C}$); the norm in E will be noted by $|\cdot|$. Let $\varphi : A \to E$ by a finitely additive set function. The total variation $|\varphi| : A \to [0,\infty]$ of φ is defined by the formula

$$|\varphi|(A) = \sup \sum_{j=1} |\varphi(A_j)|$$

where the sup is taken over all finite A-partitions $\{A_j\}$ of A i.e. $A_j \in A$ for $1 \leq j \leq n$, A_j's are disjoint, $A = \bigcup_{j=1}^{n} A_j$. Then $|\varphi|$ is a finitely additive set function also. If $|\varphi|(\Omega) < \infty$ then φ is called a finitely additive (E-valued, if we need to emphasize that) set function of bounded variation on (Ω, A). The class of all such will be denoted by $M(A,E)$; endowed with the map $\varphi \mapsto |\varphi|$, this latter is a Banach space. (cf. Dunford and Schwartz or Diestel and Uhl for this and much else in these notes). If E is $\mathbb{R}$ or $\mathbb{C}$ we abbreviate $M(A,E)$ to $M(A)$; the non-negative elements of the latter are indicated by $M_+(A)$. Thus, if $\varphi \in M(A,E)$ then $|\varphi| \in M_+(A)$.

If $\varphi \in M(A,E)$ and $\lambda \in M_+(A)$ we say that φ is λ-absolutely continuous or absolutely continuous with respect to λ (in symbols : $\varphi << \lambda$) if, for any $\varepsilon > 0$, there is a $\delta > 0$ such that $|\varphi|(A) < \varepsilon$ whenever $\lambda(A) < \delta$. We say that φ is λ-singular or φ is singular (or orthogonal) to λ (in symbols : $\varphi \perp \lambda$) if, for any $\varepsilon > 0$, there is a set $N \in A$ such that

$$|\varphi|(N) + \lambda(N^c) < \varepsilon.$$

If (Ω, A) is a measurable space (i.e. A is a sigma-algebra) and P,Q are two probability measures there (or, more generally, finite measures) then it is known that $P << Q$ iff $Q(A) = 0$ implies $P(A) = 0$. If $P << Q$ and $Q << P$, we say that P and Q are equivalent (in symbols : $P \sim Q$). We shall also use the fact that, with the above hypotheses on P,Q, $P \perp Q$ iff there is $N \in A$ such that $P(N) = 0$ and $Q(N^c) = 0$.

Let Ω,E be two sets, A a family of subsets of Ω and $f : \Omega \to E$. We say that f is a simple function if $f(\Omega)$ is a finite subset $\{x_1, \ldots, x_n\}$ of E; f is called A-simple if further the sets $\{f = x_j\}$, $1 \leq j \leq n$, are in A.

Let (Ω, A, μ) be a finite measure space, E a Banach space and $f : \Omega \to E$. We say that f is strongly measurable (also called measurable in the Bochner sense) if f is the limit μ a.e. of a sequence of A-simple functions. If E is $\mathbb{R}^n$, then this corresponds to the usual measurability of f. In these notes, we shall only consider strong measurability; hence, we shall usually say that $f : \Omega \to E$ is measurable (or A-measurable, if we wish to emphasize A) and omit the adjective "strong".

Let Ω be a set equipped with an algebra of subsets A and let E be a Banach space. If $f : \Omega \to E$ and $\mu \in M(A)$ then $\int f \, d\mu$ is taken in the sense of Dunford and Schwartz. If (Ω, A, μ) happens to be a finite measure space (as would be generally the case in the following) then $\int f \, d\mu$ exists iff (i) f is strongly measurable and (ii) $\int |f| \, d\mu < \infty$. By $L_E^p(\Omega, A, \mu)$ or more simply $L_E^p(\mu)$, $1 \leq p \leq \infty$, we denote the usual Banach spaces of μ-equivalence classes of E-valued functions associated with (Ω, A, μ). However, we shall have few occasions to distinguish a function f from its μ-equivalence class. Further, when no confusion is possible we shall say that f is in L^1 or that f_n's converge in L^p to f instead of writing out $\int |f|\, d\mu < \infty$ or $\lim_n \int |f_n - f|^p\, d\mu = 0$. Also, $L^p_E(\mu)$ will be written as L^p if there is only one μ in question and only one E in view or that E is $\mathbb{R}$ or $\mathbb{C}$; further $\|f\|_p$ will stand for $(\int |f|^p\, d\mu)^{1/p}$, $1 \leq p \leq \infty$ in all cases and $\|f\|_\infty$ the usual L^∞- norm of f.

In chapters 3 and 4, we need the space of functions f which are such that $\int |f|(\overset{+}{\ln}|f|)^\varepsilon\, d\mu < \infty$ for some $\varepsilon \geq 1$. This class will be variously denoted by $L^1_E(\overset{+}{\ln} L^1_E)^\varepsilon$ or simply $L^1(\overset{+}{\ln} L^1)^\varepsilon$ and the integral above will be abbreviated to $\|f\|_{L^1(\overset{+}{\ln} L^1)^\varepsilon}$. Here $\overset{+}{\ln}$ is the function $\overset{+}{\ln} : [0,\infty[\to [0,\infty[$ such that $\overset{+}{\ln}\, x = 0$ if $0 \leq x \leq 1$ and $\overset{+}{\ln}\, x = \ln x$ if $x > 1$. The conventions regarding L^p will be used for $L^1(\overset{+}{\ln} L^1)^\varepsilon$ also.

The integral $\int f\, d\mu$ will be sometimes written as $\mathbb{E}_\mu(f)$ or even $\mathbb{E}(f)$ — particularly in probabilistic contexts. Various other symbols for $\int f\, d\mu$ will be used according to convenience and as customary in analysis e.g. $\int f(x)\, \mu(dx)$ etc.

On several occasions, we shall write $\int f\, d\mu$ in a situation where μ is a countably additive non-negative set function on an algebra of subsets A of a set Ω with $\mu(\Omega) < \infty$ but $f : \Omega \to E$, is an E-valued, $\sigma(A)$-measurable function (E a Banach space). The understanding is that the integral is taken with respect to the unique countably additive extension of μ to $\sigma(A)$.

If (Ω, A) is as in the preceding paragraph, $\varphi \in M(A,E)$, E some Banach space, it may be the case that $\varphi(A) = \int_A f\, d\mu$, $A \in A$, for some $\mu \in M(A)$. We shall write this relation as $\varphi = f \cdot \mu$. It is known that $|\varphi| = |f| \cdot |\mu|$.

§3. R N P

A Banach space E is said to have the Radon-Nikodym Property (RNP) if given any finite measure space (Ω, A, μ) and any $\varphi \in M(A,E)$ which is μ-absolutely continuous (so that φ is also countably additive, μ being so), φ is of the form $f \cdot \mu$. For E to have RNP, it is enough to verify the above for $\Omega = [0,1]$ and μ = Lebesgue measure on the Borel subsets (or $\mu = |\varphi|$ on the Borel subsets). The function $f \in L^1_E(\mu)$ is called the Radon-Nikodym derivative of φ with respect to μ and noted as $d\varphi | d\mu$ or $D_\mu \varphi$. It is known that any reflexive space (e.g. Hilbert space) and any separable dual space (e.g ℓ^1 but not L^1 over [0,1] with Lebesgue measure) has RNP.

If E has RNP and (Ω, A) is a space Ω equipped with an algebra of subsets A then it can be shown that for any $\varphi \in M(A,E)$ and any $\mu \in M_+(A)$ which is countably additive

$$\varphi = f \cdot \mu + \Theta.$$

where $\Theta \in M(A,E)$, $\Theta \perp \mu$ and $f \in L^1_E(\mu)$, f being in general only $\sigma(A)$-measurable. Further, f and Θ are uniquely determined by φ and μ; we write also $f = D_\mu \varphi$ and call it the generalized derivative of φ with respect to μ. Actually, f is the Radon-Nikodym derivative of the μ-absolutely continuous part of φ and Θ is the μ-singular part of φ. (cf. Diestel and Uhl, p. 30-31). Further,

$$|\varphi| = |f| \cdot \mu + \Theta$$

holds. Although not quite accurate historically, we shall call the above decomposition of φ, its Lebesgue decomposition with respect to μ.

§4. Martingales.

Let (Ω, Σ, P) be a probability space and E a Banach space. If $f \in L^1_E(P)$ and Σ' is any sigma-sub-algebra of Σ, it can be shown that there is a unique (upto P-equivalence) $g \in L^1_E(P)$ which is Σ'-measurable and which satisfies

$$\int_A f \, dP = \int_A g \, dP, \qquad A \in \Sigma'$$

(cf. Neveu). We call g the (P-)conditional expectation of f given the sigma-algebra Σ' and write $g = \mathbb{E}_P\{f|\Sigma'\}$ or simply $\mathbb{E}\{f|\Sigma'\}$ when P needs no emphasizing. If Σ' is of the form $\sigma\{\xi_t, t \in T\}$ then we write $\mathbb{E}\{f|\xi_t, t \in T\}$ (if necessary along with a subscript indicating P). We shall not recall the properties of conditional expectations (for which see Neveu, Doob) except to underline the following :

$$|\mathbb{E}\{f|\Sigma'\}| \leq \mathbb{E}\{|f| \, |\Sigma'\}$$

and (Jensen's inequality):

$$\varphi\{\mathbb{E}(f|\Sigma')\} \leq \mathbb{E}\{\varphi(f)|\Sigma'\}$$

for $f \geq 0$ and φ convex on $\mathbb{R}_+$ to $\mathbb{R}_+$.

Let $\{\Sigma_n\}_{n\in\mathbb{N}}$ be an increasing sequence of sigma-algebras in a probability space (Ω,Σ,P) with $\Sigma_n \subset \Sigma$, $n \in \mathbb{N}$. Let $f_n : \Omega \to E$, E a Banach space, be Σ_n-measurable. The sequence $\{f_n, \Sigma_n\}_{n\in\mathbb{N}}$ is called a martingale if $f_m = \mathbb{E}\{f_n|\Sigma_m\}$ for $m \leq n$. In other words $\{f_n, \Sigma_n\}_{n\in\mathbb{N}}$ is a martingale if

$$\int_A f_m \, dP = \int_A f_n \, dP, \quad A \in \Sigma_m$$

for any $m \leq n$. We write this out to indicate that the notion of martingales can be defined without defining conditional expectations — although, of course, the latter are indispensable in most serious work in probability theory.

An important example of martingales is given by $f_n = \mathbb{E}\{f|\Sigma_n\}$ where f is some function in $L^1_E(P)$. For ease of reference, we give a list of properties of martingales which we shall need in the sequel in the form of some theorems.

Theorem 1.

Let (Ω,Σ, P) be a probability space, E an arbitrary Banach space and $\{f_n,\Sigma_n\}_{n\in\mathbb{N}}$ an E-valued martingale.

(i) If $f_n = \mathbb{E}\{f|\Sigma_n\}$ then $f_n \to f_\infty = \mathbb{E}\{f|\Sigma_\infty\}$ a.e. and in L^1 where $\Sigma_\infty = \sigma(\bigcup_n \Sigma_n)$; if $f \in L^p_E$, $1 < p < \infty$, then the convergence holds in L^p as well.

(ii) If E has RNP and $\sup_n \|f_n\|_1 < \infty$ (we say that $\{f_n,\Sigma_n\}_{n\in\mathbb{N}}$ is a L^1-bounded martingale) then $\lim_n f_n$ exists a.e.

(iii) The sequence $\{g_n = |f_n|, \Sigma_n\}_{n\in\mathbb{N}}$ forms a non-negative submartingale i.e.

$$\mathbb{E}\{g_n|\Sigma_m\} \geq g_m \qquad \text{a.e.}$$

for all $n \geq m$.

Corollary :

If $\{f_n,\Sigma_n\}_{n\in\mathbb{N}}$ is a real-valued martingale sequence such that the associated difference sequence $d_n = f_n - f_{n-1}$, $n \geq 1$, $d_0 = f_0$ is L^2-bounded (i.e. $\sup_n \|d_n\|_2 < \infty$) then for any sequence $\{c_n\}_{n\in\mathbb{N}}$ of real numbers in ℓ^2 (i.e. $\sum_n |c_n|^2 < \infty$), the series $\sum_n c_n d_n$ converges a.e.

Theorem 2.

Let (Ω,Σ,P) be a probability space and $\{g_n,\Sigma_n\}_{n\in\mathbb{N}}$ a non-negative submartingale.

(i) If $\varphi : \mathbb{R}_+ \to \mathbb{R}_+$ is non-decreasing and convex then $\varphi(g_n) = h_n$ is such that $\{h_n,\Sigma_n\}_{n\in\mathbb{N}}$ is also a non-negative submartingale; this applies, in particular, to $\varphi(x) = x^p$ or $(\overset{+}{\ln}\, x)^p$, $1 \leq p < \infty$.

(ii) (Maximal inequalities). Let $g_n^* = \max_{0 \leq k \leq n} g_k$.

(a) For any $t > 0$,

$$P\left\{g_n^* \geq t\right\} \leq \frac{1}{t} \int_{\{g_n^* \geq t\}} g_n \, dP \leq \frac{1}{t} \mathbb{E}\{g_n\}$$

(b)
$$\| g_n^* \|_p \leq \left(\frac{p}{p-1}\right) \|g_n\|_p \quad , \quad 1 < p < \infty$$

(c)
$$\| g_n^* \|_1 \leq \frac{e}{e-1} \left[1 + \mathbb{E}\left\{g_n \overset{+}{\ln}\, g_n\right\}\right] .$$

The notion of martingales generalizes immediately to the case of families $\{\Sigma_\alpha\}_{\alpha\in I}$ where I is a general, partially ordered, directed index set. Unfortunately, the a.e. convergence theory of martingales $\{f_\alpha,\Sigma_\alpha\}_{\alpha\in I}$ is not simple — even if I is countable. Also the maximal inequalities for the associated non-negative submartingales $\{|f_\alpha|, \Sigma_\alpha\}_{\alpha\in I}$ are, in general, invalid. This point is discussed a little more in §5 where some relevant literature is cited. Here we state a simple, general theorem which will be of later use to us. To avoid some unnecessary measure-theoretical technicalicaties, we consider the case of countable I only.

Theorem 3.

Let (Ω,Σ,P) be a probability space, E a Banach, I a denumerable, directed set, $\{\Sigma_\alpha\}_{\alpha\in I}$ an increasing family of sigma-sub-algebras of Σ and $\{f_\alpha,\Sigma_\alpha\}_{\alpha\in I}$ and E-valued martingale (i.e. fo all $\alpha\leqslant\beta$, $f_\alpha = \mathbb{E}\{f_\beta|\Sigma_\alpha\}$). Let $\Sigma_\infty = \sigma\{\Sigma_\alpha, \alpha\in I\}$.

(a) If $f_\alpha = \mathbb{E}\{f|\Sigma_\alpha\}$ for some $f\in L^p_E$, $1\leqslant p<\infty$, then $f_\alpha \to f_\infty = \mathbb{E}\{f|\Sigma_\infty\}$ in L^p.

(b) If E has RNP then $f_\alpha = \mathbb{E}\{f|\Sigma_\alpha\}$ for some $f\in L^p_E$, $1<p<\infty$ iff $\sup_\alpha\|f_\alpha\|_p<\infty$.

(c) If E has RNP then $f_\alpha = \mathbb{E}\{f|\Sigma_\alpha\}$ for some $f\in L^1_E$ iff $\{|f_\alpha|\}_{\alpha\in I}$ is uniformly integrable, a necessary and sufficient condition for which is the existence of some measurable $\varphi : \mathbb{R}_+ \to \mathbb{R}_+$ with $\lim_{t\to\infty}\varphi(t)/t = \infty$ such that $\sup_\alpha \mathbb{E}\{\varphi(|f_\alpha|)\} < \infty$. ($\varphi$ can then be chosen to be also convex and increasing with $\varphi(0) = 0$).

We conclude this section with a few remarks concerning the proofs of theorem 1-3. Further discussion and bibliographical indications are given in §5. A simple proof of part (i) of theorem 1 is outlined in the proof of theorem 2 of Chapter 2; part (iii) is even more elementary. Part (ii) of theorem 1 is deep; it will be used only when $E = \mathbb{R}$ via the corollary which follows theorem 1. The proof of the corollary is as follows. Let $g_n = \sum_{j=0}^n c_j d_j$; then $\{g_n,\Sigma_n\}_{n\in\mathbb{N}}$ is a martingale. Further

$$\mathbb{E}\left\{|g_n|^2\right\} = \sum_{j=0}^n |c_j|^2\ \mathbb{E}\left\{|d_j|^2\right\} \quad ;$$

this proves that $\{g_n\}$ is L^2-bounded — hence, a fortiori, L^1-bounded. We now conclude that $\lim_n g_n$ exists a.e. which completes the proof of the corollary. Theorem 2 is quite elementary — although as pointed out in §5, the maximal inequalities have far-reaching consequences. Part (a) of theorem 3 is an immediate consequence of the L^p-convergence part of theorem 1(i) (which in itself is quite elementary) since convergence of a directed family in a metric space is equivalent to the convergence of subsequences. For (b) and (c), we need the most elementary properties of uniform integrability of $|f_\alpha|$, $\alpha \in I$, to conclude that the E-valued set function $\varphi(A) = \lim_\alpha \int_A f_\alpha \, dP$, $A \in \mathcal{A} = \bigcup_\alpha \Sigma_\alpha$, defined on the algebra $\mathcal{A}$ (the existence of φ is a trivial reinterpretation of the martingale property) is countably additive, of bounded variation and P-absolutely continuous. The RNP of E then gives that $\varphi = f \cdot P$ and this f will do for parts (b) and (c). Finally, we remark that theorem 1 of Chapter 2 gives a more general form of the basic martingale convergence theorem as contained in part (ii) of theorem 1 above.

§5. Notes and Remarks.

Finitely additive set functions are often neglected in many expositions of measure theory — although that is certainly not the case with the two references given in §2 viz. Diestel and Uhl, Dunford and Schwartz. Besides their intrinsic interest and usefulness in many questions of functional analysis, they arise naturally in martingale theory. A martingale $\{f_n, \Sigma_n\}_{n \in \mathbb{N}}$ is indeed equivalent to the presence of a finitely additive set function φ defined on the algebra $\mathcal{A} = \bigcup_n \Sigma_n$ ($\varphi(A) = \lim_n \int_A f_n dP$) whose restriction to Σ_n possesses P-Radon-Nikodym derivative f_n; the martingale is L^1-bounded iff φ is of bounded variation. Many properties of $\{f_n\}$ can be viewed through φ and indeed, this led us to the generalization of martingales studied in Chapter 2. As long as no comprehensive exposition of all aspects of measure theory is written, the reader will have to pick up his facts about the finitely additive theory from different sources; those given here should suffice for the present notes. What we have called the Lebesgue decomposition above is due to the combined efforts of many of whom at least Hewitt and Yoshida as well as Rickart must be mentioned. In our References we donot give the exact location of these since they can be found in both of the references already cited.

This is not the place to discuss the history of RNP; this has been done competently by others and the matter can be looked up in Diestel and Uhl which abounds in RNP's, their multifarious avatars and teeming descendants. The terminology (some contend that it is highly unsuitable) was introduced, rather unsuspectingly, in Chatterji (Math. Scand. 22 (1968), p. 21-41), to simplify the statement of a vector-valued martingale convergence theorem (about which see below). However, it was pointed out there that an equivalent property (called (D) — for derivation, I imagine) had been introduced many years ago by Bochner and Taylor (Ann. Math. (2) 39 (1938), p. 913-944); to my knowledge, that is its first explicit occurrence, although the matter must have been in the minds of Clarkson (Trans. A.M.S. 40 (1936), p. 396-414), Dunford, Pettis and others around the 30's.(For further references on RNP and related vector-valued integration cf. Diestel and Uhl, Dunford and Schwartz). It should be emphasized that I donot claim any part in the success story of RNP which is due to many first rate workers whose contributions can be gauged from the monograph of Diestel and Uhl.

Martingale theory which goes back to the work of Lévy and Ville (in the late 30's) is, in its modern form, the uncontested creation of Doob. Although much has been written on it since, Doob [1953] still remains, in my opinion, the best introduction to the subject. It contains a full treatment of theorems 1 and 2 (in the real-valued case) — and naturally, much else. We cite Doob [1984] also in the References — one reason for that being the desire to indicate a rich source where martingale theory of a more elaborate nature is applied to classical potential theory and many other related matters. Other excellent references are Dellacherie and Meyer, Neveu; the latter is the most appropriate monographic reference to the three theorems of this chapter in the vector-valued case. Two other references are Chatterji and Hunt.

Theorem 1 (ii) in its present form appears in my Math. Scand. (1968) paper referred to above. However, an earlier publication of A. and C. Ionescu Tulcea (Trans. A.M.S. 107 (1963), p. 313-337) contains the result also — even though it is not formulated explicitly in the theorem concerning it. There, the theorem is stated for separable, dual spaces and the authors rightly point out, in various remarks, that their proof yields a more general result which is our theorem 1 (ii). This fact has been overlooked by almost every one — including the present author. Hence, although this part of the theorem has often been cited (by others) as "Chatterji's theorem" since the appearance of my 1968 paper, it is clear that this is not justified chronologically. It should be pointed out that the proof used by the Tulceas is different. Before closing this historical discussion, it should be added that the vector-valued martingale theory started with a paper of Scalora(Pacific J.Math.11(1961),p.347-374) and one of mine (Bull. A.M.S. 66 (1960), p. 395-398); our works were independent and actually Scalora's work predated mine by a few months. Again, our methods were different. The main difference between our results was that mine had a few simple but useful results (like theorem 1 (i)) valid for general Banach spaces. There is now a big bibliography touching on various aspects of the vector-valued martingale theory. The references to these (as well as those mentioned above) can be looked up in Neveu (cf. also Egghe).

We now pass to a discussion of martingales indexed by general directed sets I other than $\mathbb{N}$. If I is uncountable some elementary measure theoretical circumspection is needed for the validity of even such a simple theorem as theorem 3 (a). The latter would be false, as stated, if we maintain our present definitions, for the same reason that the Lebesgue dominated convergence theorem does not go through for general index sets I. In fact a standard (almost trivial) counter-example for the last can be used as a counter-example here also. Amongst those who have not been careful on this point, I shall cite only myself (Chatterji). One easy way out is to assume that all the Σ_α's contain all the null sets of Σ but a more general discussion is contained in the works of Chow and Krickeberg (for references to which cf. Neveu). That a.e. convergence in theorem 3 (a) does not hold appears first in a counter-example of Dieudonné in 1950 (cf. Neveu). A much simpler counter-example due to Chow can be found in Chatterji p. 48-49. These counter-examples would also imply that the maximal inequalities of theorem 2 cannot hold. Indeed, it is easy to see that if the maximal inequalities were valid for some indexed (say denumerable) family $\{\Sigma_\alpha\}_{\alpha \in I}$ then theorems like theorems 1 and 3 would follow immediately. In fact, this is the modern approach to the a.e. convergence questions : one proves the a.e. convergence theorem for a dense class of objects and then passes to the general case via the maximal inequalities. Thus, these latter occupy the centre of the stage. In Chatterji p. 27 - 28 it is shown that the maximal inequalities applied even to entirely trivial martingales give rise to substantial classical analytical inequalities due to Hardy. In my paper in Lecture Notes in Maths. No. 794, p. 361-364, Springer-Verlag (1980), it is further pointed out how the failure of the maximal inequalities for a general index set can be seen extremely simply. Much work has been done in recent years in order to understand what structure the Σ_α's must have (for $\alpha \in I$, a general index set) in order for various martingale and submartingale (as well as amarts, introduced in Chapter 2) convergence theorems to hold in an a.e. sence. They often involve what are called "Vitali type" conditions, initiated in this area by Krickeberg. The paper by A. Millet and L. Sucheston in the Lecture Notes volume 794 (referred to above) contains an useful discussion of these conditions and references to some recent developments (cf. also the monograph of Hayes and Panc).

Finally, it should be pointed out that most of martingale theory can be (and has been) generalized easily to the case of sigma-finite measures P. This is formally useful, in certain situations (cf. Chapter 4, for exemple). Cf. also Chatterji and Hunt for this.

Chapter 2.

A basic convergence theorem.

§1. Statement of theorem.

Let Ω be an abstract set, E a Banach space and A_n, $n \in \mathbb{N}$ an increasing sequence of algebras of subsets of Ω; we write $A_\infty = \bigcup_n A_n$. Clearly, A_∞ also is an algebra of subsets. If μ is any set function defined on a class of sets containing A_n, we shall write $\pi_n \mu$ for the restriction of μ to A_n. These notations will be held fast throughout this chapter.

If $\varphi_n \in M(A_n, E)$, $n \in \mathbb{N}$, we say that $\{\varphi_n\}$ converges (to φ) if for each $A \in A_\infty$

$$\varphi(A) = \lim_n \varphi_n(A)$$

exists (briefly : $\varphi_n \to \varphi$); note that each A in A_∞ belongs to all A_n for n sufficienttly large so that it makes sense to talk about $\lim_n \varphi_n(A)$. Note further that $\varphi : A_\infty \to E$ is additive although not necessarily of bounded variation. We say that $\{\varphi_n\}$ converges regularly to φ (briefly : $\varphi_n \to \varphi$ regularly) if there is a sequence $\nu_n \in M_+(A_n)$ which decreases to 0 (i.e. $\pi_n \nu_{n+1} \leqslant \nu_n$ and $\nu_n(\Omega) \to 0$) and $|\pi_n \varphi - \varphi_n| \leqslant \nu_n$. The sequence $\{\nu_n\}$ will be sometimes called an associated sequence for the regularly converging sequence $\{\varphi_n\}$; clearly an associated sequence $\nu_n \to 0$ regularly also. Notice that if $\varphi_n \to \varphi$ regularly then $\pi_n \varphi \in M(A_n, E)$ even though φ need not be of bounded variation.

If $\lambda \in M_+(A_\infty)$ is countably additive and E has RNP then we have seen that each $\varphi_n \in M(A_n, E)$ can be written as

$$\varphi_n = f_n \cdot \lambda + \Theta_n$$

where $f_n = D_\lambda \varphi_n$ and $\Theta_n \perp \lambda$. A natural question to ask is the following : under what conditions does $\lim_n f_n$ exist a.e (λ) ? The basic convergence theorem of this chapter states that the limit exists a.e. (λ) if $\{\varphi_n\}$ converges regularly to φ and $\varphi \in M(A_\infty, E)$ (i.e. that φ is of bounded variation.)

Theorem 1.

Let A_n, A_∞ be as above and E be a Banach space with RNP. If $\varphi_n \in M(A_n, E)$, $n \in \mathbb{N}$, converges regularly to $\varphi \in M(A_\infty, E)$ then for any countably additive $\lambda \in M_+(A_\infty)$ we have

$$\lim_n D_\lambda \varphi_n(\omega) = D_\lambda \varphi(\omega) \qquad \omega \text{ a.e. } (\lambda)$$

Before passing to the proof of the theorem in the next section let us make a few remarks. First, note that the existence a.e. (λ) of $\lim_n f_n$, $f_n = D_\lambda \varphi_n$, is neither necessary nor sufficient for $\{\varphi_n\}$ to converge. Choosing, for instance, $\varphi_n = f_n \cdot \lambda$ (with $\Omega = [0,1]$, λ = Lebesgue measure, $A_n = A$ = Borel sets of Ω, $f_n : \Omega \to \mathbb{R}$) we may easily convince ourselves of this. If $f_n \to 0$ in $L^1(\lambda)$ but $f_n \not\to 0$ a.e. (λ) then we have a sequence $\{\varphi_n\}$ which converges to 0 but $f_n = D_\lambda \varphi_n$ does not converge a.e. (λ). If, on the other hand, $f_n \to 0$ a.e. (λ) but $\int_0^1 f_n \, d\lambda$ has no limit then we obtain a sequence $\{\varphi_n\}$ not converging at all although $f_n = D_\lambda \varphi_n \to 0$ a.e. (λ). However, it is interesting to observe that if $\varphi_n \to \varphi$ regularly (a condition independent of any measure λ) and if φ is of bounded variation then, according to our theorem, $D_\lambda \varphi_n \to D_\lambda \varphi$ a.e. (λ) for <u>any</u> countably additive $\lambda \in M_+(A_\infty)$. Secondly, it is very easy to give important examples of regularly converging sequences $\{\varphi_n\}$. The following are perhaps the most useful ones :

(1) Let φ be any additive E-valued set function defined on A_∞ such that $\varphi_n = \pi_n \varphi \in M(A_n, E)$. Clearly $\varphi_n \to \varphi$ regularly; here an associated sequence ν_n is given by $\nu_n = 0$. We shall call such sequences $\{\varphi_n\}$, $\varphi_n = \pi_n \varphi$, <u>martingale type sequences</u>.

(2) Let $\varphi_n \in M(A_n, E)$ be such that

$$\sum_n \|\pi_n \varphi_{n+1} - \varphi_n\| < \infty.$$

It is easy to see that $\{\varphi_n\}$ is a regularly converging sequence. Indeed, for $A \in A_n$, define

$$\nu_n(A) = \sum_{k \geq n} |\pi_k \varphi_{k+1} - \varphi_k|(A);$$

clearly, $\nu_n \in M_+(A_n)$ and ν_n decreases to 0 in the sense described above. Further, if $A \in A_n$ and $m > n$,

$$\begin{aligned} |\varphi_m(A) - \varphi_n(A)| &\leq \sum_{k=n}^{m-1} |\varphi_{k+1}(A) - \varphi_k(A)| \\ &\leq \sum_{k=n}^{m-1} |\pi_k \varphi_{k+1} - \varphi_k|(A) \\ &\leq \nu_n(A) \leq \nu_n(\Omega) \quad ; \end{aligned}$$

this proves that $\lim_n \varphi_n(A) = \varphi(A)$ exists for all $A \in A_\infty$ and $\varphi_n \to \varphi$ regularly. We shall call such sequences $\{\varphi_n\}$ <u>semi-martingale type</u> sequences. Obviously, these are more general than the martingale type sequences.

(3) If E is a Banach lattice and $\varphi_n \in M(A_n,E)$ is such that $\pi_n\varphi_{n+1} \geq \varphi_n$ for all n, we call $\{\varphi_n\}$ an E-valued submartingale type sequence. Similarly $\{\varphi_n\}$ is called an E-valued supermartingale type sequence if $\pi_n\varphi_{n+1} \leq \varphi_n$ for all n. If $E = \mathbb{R}$ and we suppose further that $\lim_n \varphi_n(A) = \varphi(A)$ exists for all $A \in A_\infty$ and $\pi_n\varphi \in M(A_n,E)$ then we can see that $\varphi_n \to \varphi$ regularly. To see this in the submartingale case we take $\nu_n = \pi_n\varphi - \varphi_n$ and verify that $\{\nu_n\}$ is a suitable associated sequence; in the supermartingale case we take $\nu_n = \varphi_n - \pi_n\varphi$. For a general Banach lattice E, further conditions are needed on a sub- orsuper-martingale type sequence to converge regularly.

(4) A type of sequence $\{\varphi_n\}$ which converges comes up naturally in the recent development of amart theory. To define this type of sequence we need to introduce bounded stopping times τ relative to $\{A_n\}$ i.e. $\tau : \Omega \to \mathbb{N}$ such that $\sup_\omega \tau(\omega) < \infty$ and $\{\tau = j\} \in A_j$, $j \in \mathbb{N}$. Let T be the class of all bounded stopping times relative to $\{A_n\}$. We define A_τ, $\tau \in T$, to be the class all sets $A \in A_\infty$ such that $\{\tau=j\} \cap A \in A_j$, $j \in \mathbb{N}$. It is easy to verify that A_τ is an algebra of sets and that $\sigma \leq \tau$ in T implies that $A_\sigma \subset A_\tau$. Given $\varphi_n \in M(A_n,E)$, $n \in \mathbb{N}$ and $\tau \in T$ we define $\varphi_\tau \in M(A_\tau,E)$ as follows :

$$\varphi_\tau(A) = \sum_{j=1}^{N} \varphi_j(A \cap \{\tau=j\})$$

for $A \in A_\tau$, $N = \sup_\omega \tau(\omega)$. It is easy to verify that φ_τ is indeed additive and of bounded variation. Let us now say that $\{\varphi_n\}$ is a sequence of amart type if $\lim_\tau \varphi_\tau(\Omega)$ exists where τ runs through the directed set T. It turns out then (see bibliographical remarks in §3 below) that $\{\varphi_n\}$ converges to some E-valued additive φ defined on A_∞. Further, if $E = \mathbb{R}$ or $\mathbb{C}$, $\varphi_n \to \varphi$ regularly; also, in this case, if $\varphi_n \to \varphi$ regularly, it can be shown that $\{\varphi_n\}$ is an amart type sequence. Thus, if $E = \mathbb{R}$ or $\mathbb{C}$, there is complete identity between regularly converging sequences and amart type sequences. For general spaces E, it is fairly easy to show that regularly converging sequences are of amart type but the converse need not hold. (cf. §3 for bibliographical remarks).

Of course, the motivation for proving our theorem 1 and the justification for introducing the terminology above (in (1) - (4)) come from the theory of classical stochastic processes where the φ_n are of the form $f_n \cdot \lambda$, λ being a probability measure on the sigma-algebra $\sigma(A_\infty)$ and the A_n's form an increasing sequence of sigma-algebras. In this situation, the sequence $\{\varphi_n\}$is of martingale type (respectively semi-, sub-, super-martingale or amart type) if and only if the E-valued stochastic process $\{f_n\}$, f_n being A_n-measurable and λ-integrable, is a martingale (respectively semi-, sub-, super-martingale or amart). Thus our theorem gives a convenient common generalization for the convergence a.e. of L^1-bounded (i.e. $\sup_n \int |f_n| d\lambda < \infty$) martingales (semi-, sub-, super-martingales, amarts) since the condition that the limiting φ is of bounded variation is exactly equivalent to the L^1-boundedness of $\{f_n\}$ (if $\varphi_n = f_n \cdot \lambda$). However, all our applications will be based on classical martingales $\{f_n\}$ and therefore we refrain from any further development of the theory of regularly converging sequences $\{\varphi_n\}$ here.

§2. Proof of theorem 1.

The proof will be given in two steps. First, we shall consider the case $\varphi = f \cdot \lambda$ and $\varphi_n = \pi_n \varphi$. In this case, $D_\lambda \varphi_n = \mathbb{E}\{f|\sigma(A_n)\}$ and theorem 1 reduces to the usual martingale convergence theorem (for the Banach-valued case) of an elementary type as given in theorem 1 (i), Chapter 1. We reformulate it here as theorem 2. Second, we treat the case of $\varphi_n \to \varphi$ regularly where φ is λ-singular and we show that here $D_\lambda \varphi_n \to 0$ a.e. (λ) (cf. theorem 3). The proof of theorem 1 then follows easily from a combination of theorems 2 and 3.

Theorem 2.

Let A_n, A_∞ be as above and let E be an arbitrary Banach space. Let $\lambda \in M_+(A_\infty)$ be countably additive and f an E-valued, $\sigma(A_\infty)$-measurable, λ-integrable function. Then

$$\lim_n \mathbb{E}_\lambda\{f|\sigma(A_n)\} = f \quad \text{a.e. } (\lambda).$$

The convergence takes place in L^p as well if $f \in L^p_E(\lambda)$ $1 \leq p < \infty$.

Proof of theorem 2 :

Since the proof is well-known, we shall be very brief. The convergence statements are obvious if f is a simple function measurable with respect to some $\sigma(A_n)$. Such functions are dense in $L^p_E(\lambda)$ whence the result follows by a standard reasoning.

Theorem 3.

Let $\varphi_n \in M(A_n, E)$, $n \in \mathbb{N}$, $\varphi \in M(A_\infty, E)$, E an arbitrary Banach space, $\lambda \in M_+(A_\infty)$ countably additive. Suppose that $\varphi_n = f_n \cdot \lambda + \Theta_n$, $\Theta_n \perp \lambda$, f_n being $\sigma(A_n)$-measurable and λ-integrable. If $\varphi_n \to \varphi$ regularly and $\varphi \perp \lambda$ then $f_n \to 0$ a.e. (λ).

Proof of theorem 3 :

Choose $g_n : \Omega \to E$ such that g_n is A_n-simple and $\sum_n \int |f_n - g_n| d\lambda < \infty$. It is enough to show that $g_n \to 0$ a.e. (λ).

Suppose $\{\nu_n\}$ is an associated sequence for $\{\varphi_n\}$.

Since $\varphi \perp \lambda$, for all positive ε_1, ε_2, we can find $A \in A_\infty$ such that

$$|\varphi|(A) < \varepsilon_1, \quad \lambda(A^c) < \varepsilon_2.$$

Given $\varepsilon_3 > 0$ we can find N such that $A \in A_N$ and

$$\sum_{n \geq N} \int |f_n - g_n| d\lambda < \varepsilon_3, \quad \nu_N(\Omega) < \varepsilon_3.$$

Now, for any $\varepsilon > 0$,

$$\lambda\left\{A; \max_{N \leq n \leq N'} |g_n| > \varepsilon\right\}$$

$$= \sum_{n=N}^{N'} \lambda(B_n), \quad \begin{cases} B_n = \{A; |g_n| > \varepsilon, |g_j| \leq \varepsilon, N \leq j < n\} \in A_n & \text{if } n > N \text{ and} \\ B_N = \{A; |g_N| > \varepsilon\} \in A_N \end{cases}$$

$$\leq \frac{1}{\varepsilon} \sum_{n=N}^{N'} \int_{B_n} |g_n| d\lambda$$

$$\leq \frac{1}{\varepsilon} \left\{ \sum_{n=N}^{N'} \int |f_n - g_n| d\lambda + \sum_{n=N}^{N'} \int_{B_n} |f_n| d\lambda \right\}$$

$$\leq \frac{1}{\varepsilon} \left\{ \varepsilon_3 + \sum_{n=N}^{N'} |\varphi_n|(B_n) \right\}$$

(since $\varphi_n = f_n \cdot \lambda + \Theta_n$, $\Theta_n \perp \lambda$ implies that for any $B \in A_n$, $|\varphi_n|(B) = |f_n \cdot \lambda|(B) + |\Theta_n|(B)$

$$= \int_B |f_n| d\lambda + |\Theta_n|(B) \quad)$$

$$\leq \frac{1}{\varepsilon} \left\{ \varepsilon_3 + \sum_{n=N}^{N'} \left[|\pi_n \varphi|(B_n) + \nu_n(B_n) \right] \right\}$$

(since $\varphi_n = \varphi_n - \pi_n\varphi + \pi_n\varphi$ gives $|\varphi_n| \leq |\varphi_n - \pi_n\varphi| + |\pi_n\varphi| \leq \nu_n + |\pi_n\varphi|$,

ν_n being an associated sequence for $\{\varphi_n\}$)

$$\leq \frac{1}{\varepsilon} \left\{ \varepsilon_3 + \sum_{n=N}^{N'} |\pi_{N'}\varphi|(B_n) + \sum_{n=N}^{N'} \nu_n(B_n) \right\}$$

$$\leq \frac{1}{\varepsilon} \left\{ \varepsilon_3 + |\pi_{N'}\varphi|(\tilde{A}) + \sum_{n=N}^{N'} \nu_n(B_n) \right\}$$

($\tilde{A} = \bigcup_{n=N}^{N'} B_n$; note that the B_n's are disjoint

and $|\pi_n\varphi|(B_n) \leq |\pi_{N'}\varphi|(B_n)$ if $n \leq N'$)

$$\leq \frac{1}{\varepsilon} \left\{ \varepsilon_3 + |\varphi|(A) + \sum_{n=N}^{N'} \nu_n(B_n) \right\}$$

$$\leq \frac{1}{\varepsilon}\left\{\varepsilon_3 + \varepsilon_1 + \nu_N(\Omega)\right\}.$$

To obtain the last inequality we have to show that

$$\sum_{n=N}^{N'} \nu_n(B_n) \leq \nu_N(\Omega).$$

Now

$$\nu_N(\Omega) = \nu_N(B_N) + \nu_N(B_N^c)$$

$$\geq \nu_N(B_N) + \nu_{N+1}(B_N^c)$$

$$= \nu_N(B_N) + \nu_{N+1}(B_{N+1}) + \nu_{N+1}(B_N^c \setminus B_{N+1})$$

$$\geq \nu_N(B_N) + \nu_{N+1}(B_{N+1}) + \nu_{N+2}(B_N^c \setminus B_{N+1})$$

... (by induction)

$$\geq \sum_{n=N}^{N'} \nu_n(B_n) + \nu_{N'+1}\Big(\bigcap_{n=N}^{N'} B_N^c\Big)$$

$$\geq \sum_{n=N}^{N'} \nu_n(B_n)$$

which proves the needed inequality; we have used the positivity of the associated sequence ν_n and its decreasing nature.

Hence,

$$\lambda\left\{\max_{N\leq n\leq N'} |g_n| > \varepsilon\right\}$$

$$\leq \lambda(A^c) + \frac{1}{\varepsilon}\left\{\varepsilon_3 + \varepsilon_1 + \varepsilon_3\right\}$$

$$\leq \varepsilon_2 + \frac{1}{\varepsilon}\left\{2\varepsilon_3 + \varepsilon_1\right\}.$$

Thus, given $\varepsilon > 0$, $\delta > 0$, we can first choose ε_1, ε_2, ε_3 such that

$$0 < \varepsilon_2 < \delta/4, \; 0 < \varepsilon_3 < \frac{\varepsilon\delta}{4}, \; 0 < \varepsilon_1 < \frac{\varepsilon\delta}{4}.$$

If N is chosen as above, we shall then have

$$\lambda\left\{\max_{N\leq n\leq N'} |g_n| > \varepsilon\right\} < \frac{\delta}{4} + \left\{\frac{2\delta}{4} + \frac{\delta}{4}\right\} = \delta$$

for any $N' \geq N$. This means that $g_n \to 0$ a.e. (λ).

Proof of theorem 1 :

First, consider the case $\varphi_n = \pi_n \varphi$, $\varphi \in M(A_\infty, E)$. Write $\varphi = f \cdot \lambda + \Theta$, $\Theta \in M(A_\infty, E)$, $\Theta \perp \lambda$, f being $\sigma(A_\infty)$-measurable and λ-integrable. then

$$\varphi_n = \pi_n \varphi = \pi_n(f \cdot \lambda) + \pi_n \Theta$$

and

$$D_\lambda \varphi_n = D_\lambda\{\pi_n(f \cdot \lambda)\} + D_\lambda \pi_n \Theta$$

$$= \mathbb{E}_\lambda\{f | \sigma(A_n)\} + D_\lambda \pi_n \Theta$$

$$\to f = D_\lambda \varphi \quad \text{a.e. } (\lambda)$$

(by theorems 2 and 3; theorem 3 gives that $D_\lambda \pi_n \Theta \to 0$ a.e. (λ)).

In the general case, write $\psi_n = (\pi_n \varphi - \varphi_n)$; then $\psi_n \to 0$ regularly and $0 \perp \lambda$. By theorem 3, $D_\lambda \psi_n \to 0$ a.e. (λ). Thus

$$D_\lambda \varphi_n = D_\lambda(\pi_n \varphi) - D_\lambda \psi_n$$

$$\to D_\lambda \varphi \qquad \text{a.e. } (\lambda).$$

This proves theorem 1.

Remark : It is easy to see that in theorems 1 - 3 we can replace the countably additive finite measure λ by one which is sigma-finite on the algebra A_0 i.e. $\Omega = \bigcup_n A_n$, $A_n \in A_0$, $\lambda(A_n) < \infty$ and A_n's disjoint.

§3. Notes and Remarks.

As pointed out in the Notes and Remarks of Chapter 1, the origin of theorem 1 of this chapter lies in a method of proof for the vector-valued martingale convergence theorems. The relevant paper was published by me in Manuscripta Math. 4 (1971), p. 213-224 where the real-valued case is given; the vector-valued case was dealt with briefly in my article in Lectures Notes in Maths. vol. 541, p. 173-179, Springer-Verlag (1976) and a missing detail is supplied in my paper in J. Mult. Analysis 15 (1984), p. 410-413 . The relationship with amarts is discussed in my article in Lecture Notes in Maths. vol. 1089, p. 272-287, Springer-Verlag (1984) where the works of Austin, Bellow, Edgar, Lamb, Sucheston and others are referred to. I take this occasion to rectify a mistake in the last-mentioned article; lemma 1 (p. 284) there is not exact in that the set function β there is not additive but only super-additive on A. However, it is additive on A_1 and that is all that is needed in the argument given in the article.

There is a vast literature on amarts of quite recent vintage for which cf. Egghe, Gut and Schmidt. Our own interest in the matter is via theorem 1 of this chapter and is only methodological. It seems that this theorem (along with Lamb's further generalization of it for which see my article in Lecture Notes in Math., vol. 1089) brings to a natural conclusion a line of thought explicitly appearing in two well-known papers of Andersen and Jessen published in 1946 and 1948 (cf. Doob [1953]). Some works of S. Johansen and Karush, referred to in my 1971 Manuscripta Math. paper push the Andersen and Jessen thinking further; in fact, Lamb's proof is entirely in the Andersen-Jessen spirit. It seems to me however, that the method adopted here suits better the situation of a general Banach space where order relations are absent. It is possible to push the theory further by considering regularly converging nets of vector-valued set-functions and even taking the underlying probability measure a finitely additive sigma-finite set function. When the indexing set is $\mathbb{R}_+$, we may also obtain sample path regularity theorems for the stochastic process formed by the Radon-Nikodym derivatives. The latter is known from the works of Edgar and Sucheston in the amarts context (cf. Egghe for references). Indeed, it seems that the study of these regularly converging families donot lead to any really novel phenomena or new applications; they constitute a methodological stream-lining. As pointed out before, all our applications in the sequel are based on the most orthodox type of martingales (as listed in Chapter 1) supplemented by the considerations of Chapter 3 which are needed only in some parts of Chapter 4.

Chapter 3.

Convergence of multiparameter martingales.

§1. Statement of a theorem.

As explained in Chapter 1, L^1-bounded martingales $\{f_\alpha, A_\alpha\}_{\alpha\in I}$, indexed by a general directed set I, do not necessarily converge a.e. Here we state a theorem giving the convergence a.e. of f_α with $\alpha \in \mathbb{N}^d$ provided that the A_α's and f_α's are of a certain type.

If $\alpha = (\alpha_i)$, $\beta = (\beta_i)$ are in $\mathbb{N}^d$, we write $\alpha \leq \beta$ if $\alpha_i \leq \beta_i$ for all i. If $d = d' + d''$ and $\alpha' \in \mathbb{N}^{d'}$, $\alpha'' \in \mathbb{N}^{d''}$ then $\alpha = (\alpha', \alpha'') \in \mathbb{N}^d$ is defined in the obvious way.

Theorem 1.

Let (Ω, A, P) be a probability space and E a Banach space. Let $\{f_\alpha, A_\alpha\}_{\alpha\in\mathbb{N}^d}$ be an E-valued martingale such that

$$\mathbb{E}\left\{f_{(\beta,\gamma)} \mid A_{(\delta,\infty)} = f_{(\delta,\gamma)}\right\} \qquad \ldots \quad (1)$$

for all β,δ in $\mathbb{N}^{d'}$, $\delta \leq \beta$, $\gamma \in \mathbb{N}^{d''}$, $d' + d'' = d$, $d', d'' \geq 1$ and where $A_{(\delta,\infty)}$ is the sigma-algebra generated by $\{A_{(\delta,t)}, \ t \in \mathbb{N}^{d''}\}$.

If $f_\alpha = \mathbb{E}\{f|A_\alpha\}$ with $f \in L^1(\overset{+}{\ell n} L^1)^{d-1}$

then $\lim_\alpha f_\alpha = \mathbb{E}\{f|A_\infty\}$, $A_\infty = \sigma(\bigcup_\alpha A_\alpha)$

exists a.e. (and in L^1, if $d \geq 2$).

Remarks :

1. Note that in case E has RNP then any E-valued martingale $\{f_\alpha, A_\alpha\}_{\alpha\in I}$, for an arbitrary set I, is of the form $f_\alpha = \mathbb{E}\{f|A_\alpha\}$ if the following condition holds :

$$\sup_\alpha \mathbb{E}\left\{|f_\alpha|(\overset{+}{\ell n}|f_\alpha|)^\varepsilon\right\} < \infty \qquad \ldots \quad (2)$$

for some $\varepsilon \geq 1$. This follows from theorem 3 of Chapter 1 and the fact that $\Phi_\varepsilon(x) = x(\overset{+}{\ell n} x)^\varepsilon$, $\varepsilon \geq 1$ is a monotone increasing, convex function of $x \geq 0$ with $\lim_{x\to\infty} \Phi_\varepsilon(x)/x = \infty$.

Note also that if $f_\alpha = \mathbb{E}\{f|A_\alpha\}$, $\alpha \in I$ and f is in $L^1(\overset{+}{\ell n} L^1)^\varepsilon$, $\varepsilon \geq 1$, then $\{f_\alpha\}$ satisfies (2). This follows from Jensen's inequality.

2. Condition (1) is verified for any martingale $\{f_\alpha, A_\alpha\}_{\alpha\in\mathbb{N}^d}$ provided that Ω, P, A_α have the following product structure :

$$\Omega = \Omega_1 \times \ldots \times \Omega_d$$

$$A_\alpha = A^{(1)}_{\alpha_1} \otimes \ldots \otimes A^{(d)}_{\alpha_d} \quad , \quad \alpha = (\alpha_i) \in \mathbb{N}^d$$

$$P = P_1 \otimes \ldots \otimes P_d$$

where $(\Omega_i, A^{(i)}, P_i)$, $1 \leq i \leq d$, are probability spaces and, for each i, $\left\{A^{(i)}_n\right\}_{n\in\mathbb{N}}$ is an increasing sequence of sigma-algebras in $A^{(i)}$. To see this, it is enough to verify that

$$\int_{\Omega''} \varphi''(\omega'')\left\{\int_{\Omega'} \varphi'(\omega') f_{(\beta,\gamma)}(\omega',\omega'') P'(d\omega')\right\} P''(d\omega'')$$

$$= \int_{\Omega''} \varphi''(\omega'')\left\{\int_{\Omega'} \varphi'(\omega') f_{(\delta,\gamma)}(\omega',\omega'') P'(d\omega')\right\} P''(d\omega'') \quad \ldots \quad (2)$$

for all bounded (real-valued) A'_δ-measurable φ' and all bounded (real-valued) A''_t-measurable φ'', for all $t \geq \gamma$, t, γ in $\mathbb{N}^{d''}$ where $A'_\delta = A^{(1)}_{\delta_1} \otimes \ldots \otimes A^{(d')}_{\delta_{d'}}$, $A''_t = A^{(d'+1)}_{t_1} \otimes \ldots \otimes A^{(d'')}_{t_{d''}}$, $P' = P_1 \otimes \ldots \otimes P_{d'}$, $P'' = P_{d'+1} \otimes \ldots \otimes P_{d''}$, $\omega = (\omega',\omega'') \in \Omega' \times \Omega'' = \Omega$, $\Omega' = \Omega_1 \times \ldots \times \Omega_{d'}$, $\Omega'' = \Omega_{d'+1} \times \ldots \times \Omega_{d''}$. But the martingale property of $\{f_\alpha, A_\alpha\}_{\alpha\in\mathbb{N}^d}$ gives the validity of (2) for $t = \gamma$ whence we obtain the equality P'' - a.e. of the two ω''-functions appearing inside $\{\ldots\}$ in (2). From this follows the validity of (2) for all $t \geq \gamma$ and condition (1) of theorem 1 is established in case of product structures.

§2. Proof of Theorem 1.

If d = 1, the theorem is contained in theorem 1 of Chapter 1. Let d > 1 and assume the validity of the theorem for (d - 1). Let c(E) be the Banach space of all convergent sequences $x = (x_n) \in E \times E \times \ldots$ endowed with the norm $|x|_c = \sup_n |x_n|$. Define

$$A_{(\infty,n)} = \sigma\left\{A_{(\beta,n)} \quad , \quad \beta \in \mathbb{N}^{d-1}\right\}$$

$$f_{(\infty,n)} = \mathbb{E}\left\{f \middle| A_{(\infty,n)}\right\} \quad , \quad n \in \mathbb{N}$$

$$g = (f_{(\infty,n)})_{n\in\mathbb{N}}$$

$$h_n = \mathbb{E}\left\{|f| \,\middle|\, A_{(\infty,n)}\right\} .$$

According to theorem 1 of Chapter 1, g is a c(E)-valued random variable; also g is integrable since

$$\mathbb{E}\left\{|g|_c\right\} = \mathbb{E}\left\{\sup_n |f_{(\infty,n)}|\right\}$$

$$\leqslant \mathbb{E}\left\{\sup_n h_n\right\}$$

$$\leqslant a + a\,\mathbb{E}\left\{|f|\,\ell\overset{+}{n}\,|f|\right\} < \infty$$

(where $a = e/(e-1)$; cf. theorem 2, Chapter 1)

since $f \in L^1_E\,(\ell\overset{+}{n}\,L^1_E)^{d-1}$ and $d \geqslant 2$.

Let us now verify that

$$\mathbb{E}\left\{g|A_{(\beta,\infty)}\right\} = g_\beta = (f_{(\beta,n)})_{n\in\mathbb{N}} \quad , \quad \beta \in \mathbb{N}^{d-1}$$

i.e.

$$\mathbb{E}\left\{f_{(\infty,n)}|A_{(\beta,\infty)}\right\} = f_{(\beta,n)} \quad , \quad n \in \mathbb{N} .$$

This last follows from the fact that

$$f_{(\infty,n)} = \lim_{\beta'\to\infty} \mathbb{E}\left\{f|A_{(\beta',n)}\right\} \qquad \text{(limit in } L^1\text{)}$$

whence

$$\mathbb{E}\left\{f_{(\infty,n)}|A_{(\beta,\infty)}\right\}$$

$$= \lim_{\beta'\to\infty} \mathbb{E}\left\{\mathbb{E}\left\{f|A_{(\beta',n)}\right\} A_{(\beta,\infty)}\right\}$$

$$= \lim_{\beta'\to\infty} \mathbb{E}\left\{f_{(\beta',n)}|A_{(\beta,\infty)}\right\}$$

$$= f_{(\beta,n)} \qquad \text{(according to condition (1) of theorem 1).}$$

We now verify that g is not only (a c(E)-valued) integrable function but also that it is in the class $L^1(\ell\overset{+}{n}L^1)^{d-2}$.

This is obvious if $d = 2$ for then we simply have to have g in L^1. If $d \geqslant 3$ and $\Phi_\varepsilon(x) = x(\ell\overset{+}{n}x)^\varepsilon$, $x \geqslant 0$, we have

$$\Phi_{(d-2)}\,(\sup_n h_n) = \sup_n \Phi_{(d-2)}\,(h_n)$$

$$\leqslant \sup_n \mathbb{E}\left\{\Phi_{(d-2)}(|f|)\mid A_{(\infty,n)}\right\}$$

(by Jensen's inequality for conditional expectations)

whence

$$\mathbb{E}\left\{\Phi_{(d-2)}\left(\sup_n h_n\right)\right\}$$

$$\leqslant \mathbb{E}\left[\sup_n \mathbb{E}\left\{\Phi_{(d-2)}(|f|) \mid A_{(\infty,n)}\right\}\right]$$

$$\leqslant a + a\,\mathbb{E}\left\{\Phi_{(d-2)}(|f|)\ \ell n^{+}\ \Phi_{(d-2)}(|f|)\right\}$$

($a = e/(e-1)$; cf. theorem 2, Chapter 1)

$$\leqslant a + a\cdot(d-1)\ \mathbb{E}\left\{\Phi_{(d-1)}(|f|)\right\} < \infty$$

where we have used the elementary inequality

$$\Phi_{\varepsilon}(x)\ \ell n^{+}\ \Phi_{\varepsilon}(x) \leqslant (\varepsilon+1)\cdot\Phi_{(\varepsilon+1)}(x)\ ,\quad x \geqslant 0,\quad \varepsilon \geqslant 1$$

which can be verified easily by considering the three intervals $0 \leqslant x \leqslant 1$, $1 < x < e$, $x \geqslant e$ separately. Finally,

$$\mathbb{E}\left\{\Phi_{(d-2)}(|g|_c)\right\}$$

$$\leqslant \mathbb{E}\left\{\Phi_{(d-2)}\left(\sup_n h_n\right)\right\} \qquad (\text{since } |g|_c \leqslant \sup_n h_n)$$

$$\leqslant a + a\cdot(d-1)\ \mathbb{E}\left\{\Phi_{(d-1)}(|f|)\right\} < \infty \qquad \ldots \quad (1)$$

Now we use our induction hypothesis on the c(E)-valued martingale $\{g_\beta, A_{(\beta,\infty)}\}_{\beta\in\mathbb{N}^{d-1}}$ with $g_\beta = \mathbb{E}\{g|A_{(\beta,\infty)}\}$. Note that we have g in $L^1(\ell n^{+} L^1)^{d-2}$ and that $\{A_{(\beta,\infty)}\}_{\beta\in\mathbb{N}^{d-1}}$ satisfies condition (1) of Theorem 1, as can be verified easily. Thus $\lim_\beta g_\beta$ exists in c(E) a.e. Since $g_\beta = (f_{(\beta,n)})_{n\in\mathbb{N}}$ we have that $\lim_\beta f_{(\beta,n)}$ exists (in E) uniformly in n a.s. This proves the a.s. existence of $\lim_\alpha f_\alpha$, $\alpha \in \mathbb{N}^d$. The identification of the limit (and its existence in L^1 if $d \geqslant 2$) follows from the general fact given in theorem 1 (i) of Chapter 1.

§3. Maximal inequalities for multiparameter martingales.

Theorem 2.

With the notation and hypotheses of theorem 1, we have :

(a) $$P\left\{\sup_\alpha |f_\alpha| > t\right\} \leqslant \frac{1}{t}\left[b(d) + c(d)\,\mathbb{E}\left\{|f|\,(\ell n^{+}|f|)^{d-1}\right\}\right]$$

for all $t > 0$ where b(d), c(d) are positive constants independent of everything except d.

(b) $$\mathbb{E}\left\{(\sup_{\alpha}|f_\alpha|)^p\right\} \leqslant (\frac{p}{p-1})^d \cdot \mathbb{E}\left\{|f|^p\right\}$$

(c) $$\mathbb{E}\left\{\sup_{\alpha}|f_\alpha|\right\} \leqslant A(d) + B(d)\ \mathbb{E}\left\{|f|\,(\ell n^+ f)^d\right\}$$

where A(d), B(d) are positive constant depending only on d.

Remark : The constants A(d), B(d), b(d), c(d) above can be calculated explicitly from the recurrence relations obtained below.

Proof : (Notation as in the proof of theorem 1 in §2).

The proof is based again on induction on d. If d = 1, theorem 2 is contained in the maximal inequalities given in theorem 2 of Chapter 1 (with b(1) = 0, c(1) = 1 , A(1) = B(1) = e/(e-1) = a). Suppose that the theorem has been proved for (d-1). Then (for $\alpha \in \mathbb{N}^d$)

$$P\left\{\sup_{\alpha}|f_\alpha| > t\right\} \leqslant P\left\{\sup_{\beta}\ \mathbb{E}(|g|_c|A_{(\beta,\infty)}) > t\right\}$$

$$\leqslant \frac{1}{t}\left[b(d-1) + c(d-1)\ \mathbb{E}\left\{\Phi_{d-2}(|g|_c)\right\}\right]$$

(since here $\beta \in \mathbb{N}^{d-2}$, the induction hypothesis applies)

$$\leqslant \frac{1}{t}\left[b(d-1) + c(d-1)\left\{a + (d-1)\ a\ \mathbb{E}\left[\Phi_{d-1}(|f|)\right]\right\}\right]$$

(cf. (1) of §2)

$$= \frac{1}{t}\left[b(d) + c(d)\ \mathbb{E}\left\{\Phi_{d-1}(|f|)\right\}\right]$$

with b(d) = b(d-1) + a c(d-1) and
c(d) = (d-1) a c(d-1), b(1) = 0, c(1) = 1.

This proves (a). For (b) and (c) we again proceed by induction on d and the fact (already used) that

$$\sup_{\alpha}|f_\alpha| \leqslant \sup_{\beta}\ \mathbb{E}(|g|_c|A_{(\beta,\infty)})$$

where $\alpha \in \mathbb{N}^d$, $\beta \in \mathbb{N}^{d-1}$. Thus

$$\mathbb{E}\left\{(\sup_{\alpha}|f_\alpha|)^p\right\} \leqslant \mathbb{E}\left[\left[\sup_{\beta}\ \mathbb{E}(|g|_c|A_{(\beta,\infty)})\right\}^p\right]$$

$$\leqslant (\frac{p}{p-1})^{d-1}\ \mathbb{E}\left\{|g_c|^p\right\}$$

(by the induction hypothesis for (d-1) applied to $\sup_{\beta}\ \mathbb{E}\left\{|g|_c|A_{(\beta,\infty)}\right\}$, $\beta \in \mathbb{N}^{d-1}$)

$$\leqslant (\frac{p}{p-1})^{d-1}\ \mathbb{E}\left\{(\sup_{n} h_n)^p\right\}$$

(since $|g|_c \leqslant \sup_n h_n$, $h_n = \mathbb{E}\{|f| \,|A_{(\infty,n)}\})$

$$\leqslant (\frac{p}{p-1})^d \ \mathbb{E}\left\{|f|^p\right\}$$

according to the 1-parameter result. This proves (b).

Similarly, to prove (c), we have

$$\mathbb{E}\left\{\sup_\alpha |f_\alpha|\right\} \leqslant \mathbb{E}\left\{\sup_\beta \ \mathbb{E}\,(|g|_c | A_{(\beta,\infty)})\right\}$$

$$\leqslant A(d-1) + B(d-1)\ \mathbb{E}\left\{\Phi_{d-1}(|g|_c\right\}$$

(by induction hypothesis for (d-1) as used above)

$$\leqslant A(d-1) + B(d-1)\left[a+d\cdot a\ \mathbb{E}\left\{\Phi_d(|f|)\right\}\right]$$

(using (1) of §2)

$$\leqslant A(d) + B(d)\ \mathbb{E}\left\{\Phi_d(|f|)\right\}$$

with $A(d) = A(d-1) + a\cdot B(d-1)$ and
$B(d) = d\cdot a\cdot B(d-1)$, $A(1) = B(1) = a$

This proves (c) and completes the proof of Theorem 2.

§4. Notes and Remarks.

Theorems 1 and 2 in the real-valued case are due to Cairoli's work (cf. references in Neveu, Dellacherie and Meyer) published in 1970; Cairoli envisaged the case of a product structure there. In his ensuing studies (some with Walsh, cf. reference in Dellacherie and Meyer, he (and Walsh) use a condition often referred to as F4 which is what we have as condition (1) of theorem 1 in the case of d-dimensional parameters. Of course, the work of Cairoli and Walsh is essentially concerned with $\mathbb{R}_+^2$ i.e. with the continuous parameter case and theorems 1 and 2 are mere stepping stones to them. Theorem 1, as stated here, was given by I. Fazekas ("Convergence of vector-valued martingales with multidimensional indices" Publ. Math. Debrecen 30 (1983) p. 157-164); its proof, based on the reduction of the theorem to one concerning an one-parameter martingale with values in c or c(E), follows a method given in my article in Lecture Notes in Maths. vol. 526, p. 33-51, Springer-Verlag (1976).

Chapter 4.

Lebesgue's differentiation theorem.

§1. Statement of the theorem.

In this chapter, λ will be the Lebesgue measure on the Borel subset Σ of $\mathbb{R}^d$ and E a fixed Banach space. An open cube C in $\mathbb{R}^d$ with centre $a \in \mathbb{R}^d$ and side length $2\varepsilon > 0$ is the set of $x \in \mathbb{R}^d$ such that $\max_i |x_i - a_i| < \varepsilon$; here $a = (a_1, \dots, a\ )$ etc. A closed cube is the closure of an open cube. A sequence of Borel sets $A_n \in \Sigma$ is said to converge to $a \in \mathbb{R}^d$ if $a \in A_n$ for all n and diam $(A_n) \to 0$, diam(S) being the Euclidean diameter of the set S. The sequence $\{A_n\}$ converges regularly to $A \in \mathbb{R}^d$ if A_n converges to a, $0 < \lambda(A_n) < \infty$ and there exists a sequence of cubes C_n (open or closed) such that $A_n \subset C_n$ and $\sup_n \frac{\lambda(C_n)}{\lambda(A_n)} < \infty$; clearly, then C_n converges to a as well.

Theorem 1.

Let $\varphi \in M(\Sigma, E)$ i.e. φ is an E-valued additive function of bounded variation on the Borel sets Σ of $\mathbb{R}^d$, E being an arbitrary Banach space.

(a) If $\varphi \perp \lambda$ (Lebesgue measure in $\mathbb{R}^d$) then there exists a null set $N \in \Sigma$ (i.e. $\lambda(N) = 0$) such that for any $x \notin N$ and any sequence of Borel sets A_n converging regularly to x, we have
$$\lim_n \frac{\varphi(A_n)}{\lambda(A_n)} = 0.$$

(b) If $\varphi = f \cdot \lambda$, $f \in L^1_E(\lambda)$, then there exists a null set $N \in \Sigma$ such that for any $x \notin N$ and any sequence of Borel sets A_n converging regularly to x, we have $\lim_n \frac{\varphi(A_n)}{\lambda(A_n)} = f(x)$.

Corollary : If the Banach space E has RNP and $\varphi \in M(\Sigma, E)$ then there exists a null set $N \in \Sigma$ such that for any $x \notin N$ and any sequence of Borel sets A_n converging regularly to x,
$$\lim_n \frac{\varphi(A_n)}{\lambda(A_n)} = f(x)$$
exists and defines a version of $D_\lambda \varphi$.

§2. Proof of the theorem.

Let P^0_n be the partition of $\mathbb{R}^1$ given by the intervals $[m/3^n, (m+1)/3^n[$, $m \in \mathbb{Z}$ and P^1_n be that given by $[(m+\frac{1}{2})/3^n, (m+\frac{3}{2})/3^n[$, $m \in \mathbb{Z}$; we take $n \in \mathbb{N}$. The partition points of P^1_n are the mid-points of the contiguous partition points of P^0_n. Let $\varepsilon = (\varepsilon_i)$ with $\varepsilon_i = 0$ or 1, $i = 1, 2, \dots, d$ and let $\delta^\varepsilon_n = P^{\varepsilon_1}_n \times \dots \times P^{\varepsilon_d}_n$ be the product partition of

$\mathbb{R}^d$ defined in the obvious way.

Let Σ_n^ε be the sigma-algebra of subsets of $\mathbb{R}^d$ generated by the sets of the partition δ_n^ε. Clearly, for each ε, $\{\Sigma_n^\varepsilon\}_{n\in\mathbb{N}}$ is an increasing sequence of sigma-algebras; it is known and easy to prove that the algebra $\Sigma_\infty^\varepsilon = \bigcup_n \Sigma_n^\varepsilon$ generates Σ.

We start with the proof of (a). We shall apply, for each fixed $\varepsilon = (\varepsilon_i)$, theorem 3 of Chapter 2 (cf. also its obvious extension to sigma-finite λ as given in the remark at the end of Chapter 2) to $\varphi_n \in M_+(\Sigma_n^\varepsilon)$ and λ where φ_n is the restriction of $|\varphi|$ to Σ_n^ε. Clearly $\varphi_n = f_n \cdot \lambda$ where $f_n(x) = \frac{|\varphi|(B)}{\lambda(B)}$, B being the unique set of δ_n^ε containing x. Since $|\varphi| \perp \lambda$ (as $\varphi \perp \lambda$ on Σ and hence on $\Sigma_\infty^\varepsilon$) we have that $f_n(x) \to 0$ for $x \notin N$, $\lambda(N_\varepsilon) = 0$. Let $N = \bigcup_\varepsilon N_\varepsilon$. Then N is such that $\lambda(N) = 0$ and for $x \notin N$,

$$\lim_x \frac{|\varphi|(B_n)}{\lambda(B_n)} = 0$$

for any sequence of sets $B_n \in \bigcup_{k,\varepsilon} \delta_k^\varepsilon$ such that B_n converges to x. We shall show that this set N will have the property indicated in (a). For this we need a simple geometrical lemma (in $\mathbb{R}^1$).

Lemma.

Let I be the family of intervals of the type $]m/3^n, (m+1)/3^n[$, $m \in \mathbb{Z} \cup \frac{1}{2}\cdot\mathbb{Z}$, $n \in \mathbb{N}$. If $0 < (b-a) < \frac{1}{2}$ then $\exists\, I \in I$ such that $[a,b] \subset I$ and $(b-a) < \text{length}(I) \leq 6(b-a)$.

Proof of lemma :

Note that if $h > 0$ and $0 < (b-a) < h/2$ then $[a,b] \subset\,]mh, (m+1)h[$ for some m in $\mathbb{Z}$ or $\frac{1}{2}\cdot\mathbb{Z}$. Indeed, if $nh \in [a,b]$ for some n in $\mathbb{Z}$, then

$$[a,b] \subset\,](n-\tfrac{1}{2})h,\ (n+\tfrac{1}{2})h[$$

as is easily verified. If $nh \notin [a,b]$ for all $n \in \mathbb{Z}$ then we can find $m \in \mathbb{Z}$ such that $mh < a < (m+1)h$; in this case, $(m+1)h > b$ (since $(m+1)h \notin [a,b]$) so that $[a,b] \subset\,]mh, (m+1)h[$, $m \in \mathbb{Z}$.

Now take $h = 1/3^n$ where $n \in \mathbb{N}$ is such that $1/3^{n+1} \leq 2(b-a) < 1/3^n$. Then, according to the foregoing, there is an $m \in \mathbb{Z} \cup \frac{1}{2}\cdot\mathbb{Z}$ such that

$$[a,b] \subset\,]mh,(m+1)h[\ = I \in I$$

and

$$(b-a) \geq 1/(2\cdot 3^{n+1}) = (1/6)\ \text{length}\ (I).$$

This proves the lemma.

Going back to the proof of (a) of theorem 1, if C_n is any sequence of cubes converging to $x \notin N$ then, according to the lemma, we can find, for each n, a set $B_n \in \cup_{k,\varepsilon} \delta_k^\varepsilon$ such that $C_n \subset B_n$ and $\lambda(B_n) \leq 6^d \lambda(C_n)$. Clearly, B_n converges to x and from the choice of N and B_n we conclude that

$$\frac{|\varphi|(C_n)}{\lambda(C_n)} \leq \frac{|\varphi|(B_n)}{\lambda(C_n)} \leq 6^d \cdot \frac{|\varphi|(B_n)}{\lambda(B_n)} \to 0$$

Since $|\varphi(C_n)| \leq |\varphi|(C_n)$ we deduce that

$$\frac{\varphi(C_n)}{\lambda(C_n)} \to 0$$

for any sequence of cubes C_n converging to $x \notin N$. The same reasoning gives then that

$$\frac{\varphi(A_n)}{\lambda(A_n)} \to 0$$

for any sequence of Borel sets A_n converging regularly to $x \notin N$. This proves (a) of theorem 1.

To prove (b) we note first that here we may (and shall) suppose that E is separable; indeed, f being strongly measurable, we may suppose, without loss of generality that all the values of $f(\cdot)$ lie in a separable subspace so that $\varphi = f \cdot \lambda$ will imply that φ takes its values in some separable subspace too. Let $D \subset E$ be a denumberable dense subset. Applying theorem 3 of Chapter 2 to $\varphi_n \in M_+(\Sigma_n^\varepsilon)$ where φ_n is the restriction of $|\varphi - t \cdot \lambda|$ to Σ_n^ε, $t \in D$, $\varepsilon \in \{0,1\}^d$ we obtain, as before, the existence of a λ-null set $N \in \Sigma$ such that

$$\lim_n \frac{|\varphi - t\cdot\lambda|(B_n)}{\lambda(B_n)} = |f(x) - t| \qquad \dots (1)$$

holds for any sequence $B_n \in \cup_{k,\varepsilon} \delta_k^\varepsilon$ converging to $x \notin N$ and any $t \in D$. Here we have used the fact that $\varphi = f \cdot \lambda$ implies that $|\varphi - t\cdot\lambda| = |f - t| \cdot \lambda$.

We now show that (1) holds equally for all $t \in E$. This uses the density of D in E and a classical reasoning. Indeed, if we call $R_n(t)$ the ratio on the left hand side of (1), then we see that

$$|R_n(t) - R_n(s)| \leq \Big| |\varphi - t\cdot\lambda|(B_n) - |\varphi - s\cdot\lambda|(B_n) \Big| \,/\, \lambda(B_n)$$

$$\leq \Big| |\varphi - t\cdot\lambda| - |\varphi - s\cdot\lambda| \Big| (B_n) \,/\, \lambda(B_n)$$

$$\leq |(\varphi - t\cdot\lambda) - (\varphi - s\cdot\lambda)|(B_n) \;/\; \lambda(B_n)$$

$$= |t-s|$$

which shows that (1) holds for all $t \in E$ if it holds for t in some dense set D.

Substituting $t = f(x)$, $x \notin N$, in (1), we obtain

$$\lim_n \frac{|\varphi - f(x)\cdot\lambda|(B_n)}{\lambda(B_n)} = 0 \qquad \ldots \quad (2)$$

for any sequence $B_n \in \bigcup_{k,\varepsilon} \delta_k^\varepsilon$ converging to x. By the same reasoning as in the proof of (a), we now see that (2) holds also if we replace B_n there by A_n where A_n is any sequence in Σ converging regularly to $x \notin N$. Since

$$\left| \frac{\varphi(A_n)}{\lambda(A_n)} - f(x) \right| = |\varphi(A_n) - f(x)\cdot\lambda(A_n)| \;/\; \lambda(A_n)$$

$$\leq |\varphi - f(x)\cdot\lambda|(A_n) \;/\; \lambda(A_n)$$

we have completed the proof of (b) and of theorem 1.

The proof of the corollary follows immediately from the fact that if E has RNP then any $\varphi \in M(\Sigma, E)$ can be written as $f\cdot\lambda + \Theta$, $\Theta \perp \lambda$ so that theorem 1 applies separately to $f\cdot\lambda$ and Θ.

§3. Strong derivation.

In connexion with Theorem 1, it is natural to inquire whether $\lim_n \varphi(A_n)/\lambda(A_n)$ exists a.e. without the assumption that the sets A_n converge regularly. A detailed and somewhat subtle theory has been developed around the question; some references are given in §5. Here, let us note simply that the limit in question may not exist anywhere if we donot impose the condition of regular convergence on A_n, even if the set function φ is an indefinite integral (i.e. $\varphi = f\cdot\lambda$). In this context, one defines the notion of strong derivation. We say that $\varphi \in M(\Sigma, E)$ is strongly differentiable or derivable at $x \in \mathbb{R}^d$ if $\lim_n \varphi(A_n)/\lambda(A_n)$ exists for any sequence of bounded intervals A_n (i.e. sets of the form $I_1 \times \ldots \times I_d$, each I_j being a bounded non-degenerate interval of $\mathbb{R}^1$) converging to x. We shall limit ourselves to the case of φ of the form $f\cdot\lambda$.

Theorem 2.

Let $\varphi = f\cdot\lambda$ where f is E-valued (E an arbitrary Banach space) and

$$\int_{\mathbb{R}^d} |f| (\ell n^+ f)^{d-1} \, d\lambda < \infty \qquad \ldots \quad (1)$$

Then φ is strongly differentiable a.e. (with strong derivative equal to f).

Proof :

It is enough to establish the theorem for f which vanish outside some fixed cube of side length 1. In other words, we may (and do) consider φ, f, λ on some unit cube Q of $\mathbb{R}^d$. We do so in order to be able to apply Theorem 1 of Chapter 3 in the probabilistic context in which it was stated and proven.

To fix ideas, take $Q = [0,1]^d$. Let $\{P_n^0\}$, $\{P_n^1\}$ be any two sequences of finite partitions of [0,1] into intervals such that each interval of P_n^0 (or of P_n^1) is a union of those of P_{n+1}^0 (respectively, of P_{n+1}^1) and the maximum lengths of the intervals in P_n^0 or P_n^1 tend to 0 as $n \to \infty$. Let $\varepsilon = (\varepsilon_1, \ldots, \varepsilon_d)$ be some fixed choice of 0's and 1's (i.e. $\varepsilon_i = 0$ or 1, $1 \leq i \leq d$); for any such ε, let $\delta_\alpha^\varepsilon$, $\alpha = (\alpha_i) \in \mathbb{N}^d$, be the (product) partition of Q obtained from $P_{\alpha_i}^{\varepsilon_i}$, $1 \leq i \leq d$. In other words, a set in $\delta_\alpha^\varepsilon$ is an interval in Q of the form $I_1 \times \ldots \times I_d$ with $I_i \in P_{\alpha_i}^{\varepsilon_i}$, $1 \leq i \leq d$. Further, let A_α^ε be the sigma-algebra generated by the sets in $\delta_\alpha^\varepsilon$.

We proceed now as in the proof of Theorem 1 (b). We suppose, as we may, that the Banach space E is separable; let D be a denumerable dense set in E. For each ε and $t \in D$, we define the non-negative real-valued martingale $\{f_\alpha^{\varepsilon,t}, A_\alpha^\varepsilon\}_{\alpha \in \mathbb{N}^d}$ by the formula :

$$f_\alpha^{\varepsilon,t} = \mathbb{E}\left\{ |f-t| \,\middle|\, A_\alpha^\varepsilon \right\} .$$

Clearly,

$$f_\alpha^{\varepsilon,t}(x) = |\varphi - t\cdot\lambda|(A) \,/\, \lambda(A) = \frac{1}{\lambda(A)} \int_A |f-t| \, d\lambda$$

where A is the unique interval of $\delta_\alpha^\varepsilon$ which contains x.

Note that $Q, \lambda, A_\alpha^\varepsilon$ have a product structure and that f is in the integrability class needed for the application of theorem 1, Chapter 3. Thus $\lim_\alpha f_\alpha^{\varepsilon,t}(x)$ exists for $x \notin N$, $\lambda(N) = 0$ for any choice of ε (2^d possibilities) and any choice of $t \in D$. This means that

$$\lim_{n\to\infty} \frac{1}{\lambda(A_n)} \int_{A_n} |f(y)-t| \, dy = |f(x)-t| \qquad \ldots \quad (2)$$

for any sequence A_n converging to $x \notin N$ provided that the A_n's are in $\bigcup_{\alpha,\varepsilon} \delta_\alpha^\varepsilon$ and $t \in D$. Using the same argument as in the proof of theorem 1, we may conclude that (1) remains true for all $t \in E$.

We now specialize P_n^0, P_n^1 to be the special partitions used in the proof of theorem 1, restricted to Q. If B_n is any sequence of intervals in Q converging to x, we can, by using the lemma of §2, find a sequence of intervals $A_n \supset B_n$ such that A_n's are in our partitions $\bigcup_{\alpha,\varepsilon} \delta_\alpha^\varepsilon$ and A_n's converge to x in such a way that

$$\lambda(A_n) < 6^d \lambda(B_n).$$

The argument of §2 applies as before giving us that

$$\lim_{n\to\infty} \frac{1}{\lambda(B_n)} \int_{B_n} |f(y) - t| \, dy = |f(x)-t|$$

for any sequence of intervals B_n in Q converging to $x \notin N$ and any t in E. In particular, taking $t = f(x)$ we conclude as before that $\varphi = f \cdot \lambda$ posseses the strong derivative f(x) for $x \notin N$. This completes the proof of our theorem 2.

Remark :

Note that the same type of proof gives other results. Thus if f is in the class $L^1_E(\ell n^+ L^1_E)^{d-k-1}$ for some k in $\{1,2,\dots,(d-2)\}$ then $\int_{A_n} f \, d\lambda / \lambda(A_n) \to f(x)$ for any sequence of bounded intervals converging to $x \notin N$, $\lambda(N) = 0$, provided that (k+1) side lengths of the A_n are equal. The case k=0 corresponds to Theorem 2 and k = (d-1) corresponds to Theorem 1.

§4. Maximal inequalities.

Theorem 3.

Let $Q = [0,1]^d$, E a Banach space, $f : Q \to E$. For f λ-integrable, define Mf(x) (respectively $M_s f(x)$) by

$$\sup \frac{1}{\lambda(A)} \int_A |f(y)| dy$$

where the sup is taken over all cubes (respectively, over all intervals) in Q containing x.

(a) For any $t > 0$,

$$\lambda\{x \mid Mf(x) > t\} \leq \frac{\text{const.}}{t} \|f\|_1$$

$$\lambda\{x \mid M_s f(x) > t\} \leq \frac{1}{t} \left\{\text{const.} + \text{const.} \|f\|_{L^1(\ell n^+ L^1)^{d-1}}\right\}$$

(b) For any $p \in]1,\infty[$,

$$\|Mf\|_p \leq \text{const.} \|f\|_p$$

$$\|M_s f\|_p \leq \text{const. } \|f\|_p$$

(c)

$$\|Mf\|_1 \leq \text{const.} + \text{const. } \|f\|_{L^1(\ell n^+ L^1)}$$

$$\|M_s f\|_1 \leq \text{const.} + \text{const. } \|f\|_{L^1(\ell n^+ L^1)^d}$$

where const. stands for a constant (possibly different from one occurrence to another) independent of f and E (but depending on p in (b) and d).

Proof :

The proof depends on the corresponding maximal inequalities for martingales (Chapter 1 and Chapter 3). Those for Mf follow from the inequality

$$f^*(x) \leq Mf(x) \leq 6^d f^*(x)$$

where $f^*(x) = \max_\varepsilon \cdot M^\varepsilon|f|(x)$, $\varepsilon \in \{0,1\}^d$, $M^\varepsilon|f|$ being the maximal function of the positive martingale $\mathbb{E}\{|f|\,|\Sigma_n^\varepsilon\}$, $n \in \mathbb{N}$ where Σ_n^ε is the trace in Q of the corresponding sigma-algebra defined in §2. Indeed, the first part of the inequality is obvious since $f^*(x)$ is sup $\frac{1}{\lambda(A)} \int_A |f(y)|\, dy$ where sup is taken over all cubes A in Q containing x of a "special" type. The second part follows from the fact, explained in §2, that given any cube A containing x, there is a "special" cube $B \supset A$ with $\lambda(B) < 6^d \lambda(A)$; this gives

$$\frac{1}{\lambda(A)} \int_A |f|\, d\lambda \leq 6^d \cdot \frac{1}{\lambda(B)} \int_B |f|\, d\lambda$$

and leads to $Mf(x) \leq 6^d f^*(x)$. Now, from the maximal inequalities in Chapter 1, we see that

$$\lambda\left\{x \mid Mf(x) > t\right\} \leq \lambda\left\{x \mid f^*(x) > t/6^d\right\}$$

$$\leq \sum_\varepsilon \lambda\left\{x \mid M^\varepsilon|f|(x) > t/6^d\right\}$$

$$\leq \sum_\varepsilon \frac{6^d}{t} \cdot \|f\|_1$$

$$\leq \frac{\text{const.}}{t} \cdot \|f\|_1 \qquad (\text{const.} = (12)^d)$$

This proves (a) for Mf; (b) and (c) for Mf follow similarly from those concerning $M^\varepsilon|f|$.

The inequalities for $M_s f$ are deduced from the inequality (obtained exactly as above)

$$f_s^*(x) \leq M_s f(x) \leq 6^d f_s^*(x)$$

where $f_s^*(x) = \max_{\varepsilon} M_s|f|(x)$, $\varepsilon \in \{0,1\}^d$, $M_s^{\varepsilon}|f|$ being defined by

$$M_s^{\varepsilon}|f| = \sup_{\alpha} \mathbb{E}\left\{|f| \;\middle|\; A_{\alpha}^{\varepsilon}\right\}, \quad \alpha \in \mathbb{N}^d .$$

where A_{α}^{ε} are as defined in §3. We now use the preceding argument supplemented by the maximal inequalities for multiparameter martingales as in §3 of Chapter 3.

Remark : Explicit expressions for the constants can be easily written down from those occurring in the martingale maximal inequalities. However, it should be remembered that the constants in Theorem 3 will change if we replace $Q = [0,1]^d$ by some other bounded interval. Parts (a) and (b) remain valid if we replace Q by $\mathbb{R}^d$; a proof of the same type as above can be given once the martingale results have been extended to sigma-finite measure spaces. This is possible to do without much difficulty.

5. Notes and Remarks.

Theorem 1 in the real-valued case is essentially what is known as Lebesgue's differentiation theorem; its usual proof is via an important covering theorem called Vitali's covering theorem and can be found in many good texts (cf. Dunford and Schwartz). Some texts include also a statement like part (b) of theorem 1 for vector-valued functions (e.g. Dunford and Schwartz). Our presentation using martingales appears in my paper in Manuscripta Math. 4 (1971), p. 213-224. The approach however is not new; it goes back to de la Vallée Poussin to whom is due the elegant technique of "complementary nets" i.e. the use of the intervals P_n^0, P_n^1 in combination. The key to its success is the simple but luminous geometrical fact contained in the lemma of §2. This is the crucial step which permits the passage from the martingale result (limited to very special intervals) to the result using any cubes (Lebesgue's theorem). It permits the translation of most of general martingale theory (whose sets involved are somewhat arbitrary but once decided upon are rigidly fixed) to the classical case of $\mathbb{R}^d$ where we have to deal with general cubes and intervals (with no restrictions on end points); this is illustrated by the results of §4. Some authors seem to ignore the fact that an important passage is necessary to go from the general martingale results to those concerning $\mathbb{R}^d$. An accessible account of de la Vallée Poussin's method is given, in the classical framework, in the well-known text-book of Riesz and Sz.-Nagy (p. 84-89) where a reference to the original work of de la Vallée Poussin, may be found.

Theorem 2 on strong derivation is due to Jessen, Marcinkiewicz and Zygmund; it appeared first in a paper of theirs in 1935. Much work had been done on the theory of strong derivation in the 30's. The questions treated have gained new importance and depth due to the need for such theorems in the theory of singular integral operators and some questions of harmonic analysis. The two volumes of Guzman ([1975], [1981]) contain a readable exposition as well as references to relevant literature (including the paper of Jessen, Marcinkiewicz and Zygmund). In some ways, this last — mentioned paper has played a role in the theory of multiparameter martingales as well; this is clear from a reading of Cairoli's original article on the subject referred to in Chapter 3. Theorem 3 is also contained in the Jessen, Marcinkiewicz and Zygmund paper and is exposed from a different point of view in Guzman's books mentioned above. The martingale approach given here for theorems 2 and 3 seems new; it must be considered to be the most direct method once one has at ones disposal the martingale results of Chapters 1 and 3 and the geometrical lemma of §2.

I have tried in vain for several years to prove another differentiation theorem (due to Besicovitch, 1945-46) by methods analogous to those given here. In Besicovitch's theorem, the denominator measure λ of our theorem 1 is any arbitrary non-negative measure (and not just Lebesgue measure) and the converging sets are restricted to centred cubes. With these provisos, theorem 1 remains valid. I find this theorem truly remarkable. Regrettably, most standard measure theory text-books omit this theorem. Its proof is generally based on another elegant covering theorem (due to Besicovitch). An account can be found in Guzman [1975] mentioned above. Cf. also Hayes and Pauc and references therein for other material concerning this chapter.

Finally, more general strong derivation theorems, known as Ward's theorem (for which see Saks' book), which in the case of $\mathbb{R}^1$ go back to Denjoy and Young, can also be treated by martingale methods using stopping times. For this cf. Chow's work in Illinois J. Math. 9 (1965), p. 569-576.

Chapter 5.

Ryll-Nardzweski's fixed point theorem.

§1. Statement of a theorem.

Let E be a vector space and $\mathcal{T}$ some Hausdorff topology on E which makes E a topological vector space. If $K \subset E$ and $\mathcal{I}$ is a family of maps $\Theta : K \to K$ we say that $\mathcal{I}$ is a $\mathcal{T}$-distal (or, simply, distal) family if for every pair x,y in K, $x \neq y$, the set

$$S(x,y) = \{\Theta(x) - \Theta(y) \mid \Theta \in \mathcal{I}\}$$

is such that 0 is not in the $\mathcal{T}$-closure of $S(x,y)$.

Let $\Omega = [0,1]$ with Lebesgue measure λ on its Borel subsets. If $K \subset E$, we say that $\{f_n\}$ is a simple, K-valued, martingale sequence (defined on Ω) if

(i) $f_n : \Omega \to K$ is simple and measurable for all $n \in \mathbb{N}$ and

(ii) (martingale property)

$$\int_A f_{n+1} d\lambda = \int_A f_n d\lambda$$

for all $A \in \sigma\{f_1,\ldots,f_n\}$, $n \in \mathbb{N}$.

Note that f_n's being simple, the integrals are merely finite sums.

A subset K of E is said to have $\mathcal{T}$-MCP (or merely MCP - martingale convergence property) if any simple, K-valued, martingale sequence $\{f_n\}$ (defined on Ω) quasi-converges a.e. in the sense that for any $\mathcal{T}$-neighbourhood V of $0 \in E$ there is a λ-null set N (possibly depending on V) such that if $\omega \notin N$, $\{f_m(\omega) - f_n(\omega)\} \in V$ for m,n exceeding some $n_0(\omega)$.

Theorem 1.

Let E be a vector space and $\mathcal{T}_1, \mathcal{T}_2$ two Hausdorff topologies on E such that $(E,\mathcal{T}_i)$, $i = 1,2$, are topological vector spaces. Let K be a non-empty convex subset of E which is $\mathcal{T}_1$-compact and $\mathcal{T}_2$-MCP. Let $\mathcal{I}$ be a semi-group (under composition) of $\mathcal{T}_1$-continuous affine maps of K into K which is $\mathcal{T}_2$-distal. Then there exists a point $p \in K$ such that $\Theta(p) = p$ for all $\Theta \in \mathcal{I}$.

2. Proof of Theorem 1.

If $\Theta \in \mathcal{I}$, $\{x \in K \mid \Theta(x) = x\}$ is a $\mathcal{T}_1$-closed set contained in the $\mathcal{T}_1$-compact set K. Thus, to prove that

$$\bigcap_{\Theta\in I}\left\{x\in K \mid \Theta(x) = x\right\} \neq \emptyset$$

we verify that the family $\{x\in K|\Theta(x) = x\}$, $\Theta\in I$, has the finite intersection property.

Let $\Theta_0,\ldots,\Theta_{k-1}$ be any k distinct elements of I. We need to show that for some $p\in K$, $\Theta_j(p) = p$ for $0\leq j\leq (k-1)$.

If $\bar{\Theta} = (\Theta_0+\ldots+\Theta_{k-1})/k$ then clearly $\bar{\Theta} : K\to K$ is affine and T_1-continuous. We now use the following version of the Markov-Kakutani theorem : if K is a compact, convex subset of a Hausdorff topological vector space and $\bar{\Theta} : K\to K$ is a continuous, affine map then there is a point $p\in K$ such that $\bar{\Theta}(p) = p$. The following lemma now asserts that $\Theta_j(p) = p$ also for all j.

Lemma 1 :

Let E be a Hausdorff topological vector space, $K\subset E$ a convex set with MCP, G a semi-group of (not necessarily continuous) affine maps of K into K generated by $\Theta_0,\Theta_1,\ldots,\Theta_{k-1}$. We suppose that G is a distal family. If $\bar{\Theta}(p) = p$ for some $p\in K$ where $\bar{\Theta} = (\Theta_0+\ldots+\Theta_{k-1})/k$, then $\Theta(p) = p$ for all $\Theta\in G$.

Proof of lemma 1 :

Let $\Omega = [0,1]$ be endowed with Lebesgue measure λ and let $\omega = \sum_{n\geq 1}\alpha_n(\omega)k^{-n}$ be a k-adic expansion of ω. Thus $\alpha_n : \Omega\to\{0,1,\ldots,k-1\}$ are independent, identically, distributed random variables with $\lambda\{\alpha_n = j\} = 1/k$ for all n,j. Define

$$f_n(\omega) = \Theta_{\alpha_1(\omega)}\cdots\Theta_{\alpha_n(\omega)}(p).$$

We claim that $\{f_n\}$ is a simple, K-valued, martingale sequence. Indeed, if $A = \{\alpha_1 = j_1, \ldots, \alpha_n = j_n\}$, then

$$\int_A f_{n+1}d = \sum_{j=0}^{k-1}\int_{A\cap\{\alpha_{n+1}=j\}} f_{n+1}\, d\lambda$$

$$= \sum_{j=0}^{k-1}\lambda(A)\cdot\frac{1}{k}\cdot\Theta_{j_1}\Theta_{j_2}\cdots\Theta_{j_n}\Theta_j(p)$$

$$= \lambda(A)\,\Theta_{j_1}\cdots\Theta_{j_n}(\bar{\Theta}(p))$$

$$= \lambda(A)\,\Theta_{j_1}\cdots\Theta_{j_n}(p)$$

$$= \int_A f_n \, d\lambda$$

which proves the martingale property of $\{f_n\}$. Since K has MCP, $\{f_n\}$ quasi-converges a.e. In particular, for any neighbourhood V of $0 \in E$, there is a null set N such that if $\omega \notin N$, $f_{n+1}(\omega)-f_n(\omega) \in V$ for all large n. Note that

$$f_{n+1}(\omega)-f_n(\omega) = \Theta_{\alpha_1(\omega)} \cdots \Theta_{\alpha_n(\omega)} \Theta_{\alpha_{n+1}(\omega)}(p) - \Theta_{\alpha_1(\omega)} \cdots \Theta_{\alpha_n(\omega)}(p) .$$

We now show that $\Theta_j(p) = p$ for all j. Indeed, if for some j, $\Theta_j(p) \neq p$ then there is a neighbourhood V of 0 such that

$$V \cap \{\Theta(\Theta_j(p)) - \Theta(p) \mid \Theta \in G\} = \emptyset \qquad \dots \quad (1)$$

(since G is a distal family). Now, for this V, choose a null set $N \subset [0,1]$ such that if $\omega \notin N, f_{n+1}(\omega)-f_n(\omega) \in V$ for all large n. We can now take an $\omega \in [0,1] \setminus N$ such that $\{\alpha_n(\omega)\}$ contains j infinitely often; if $f_{n+1}(\omega)-f_n(\omega) \in V$ for $n \geq n(\omega)$, we may choose an $m \geq n(\omega)$ such that $\alpha_m(\omega) = j$ which will then have the property that $\Theta(\Theta_j(p))-\Theta(p) \in V$ with $\Theta = \Theta_{\alpha_1(\omega)} \cdots \Theta_{\alpha_{m-1}(\omega)}$. This contradicts (1) and proves that $\Theta_j(p) = p$ for all j. The proof of lemma 1 is complete.

An application of the lemma to E endowed with the topology T_2 completes the proof of theorem 1.

§ 3. On MCP (see §1 for definition)

Proposition 1.

Let E be any locally convex topological vector space. Then any weakly compact subset K of E has MCP.

Proof :

(a) We consider first the case of a normed space E. We have to prove that if $f_n : \Omega \to K$ ($\Omega = [0,1]$ with Lebesgue measure λ) is a simple, martingale sequence then $\{f_n\}$ quasi-converges a.e. We shall actually prove in this case that $\lim_n f_n(\omega)$ exists a.e.

Let $F_n(\omega)$ be the weak closure of $\{f_k(\omega) \mid k \geq n\}$; since $F_n(\omega)$ is a decreasing sequence of non-empty weakly compact subsets of K, $\bigcap_n F_n(\omega) \neq \emptyset$; choose $f(\omega) \in \bigcap_n F_n(\omega)$. For each $x' \in E'$, $\{\langle f_n, x' \rangle\}_{n \in \mathbb{N}}$ is a real-valued martingale sequence which is bounded (since $\sup_{x \in K} |\langle x, x' \rangle| < \infty$, K being weakly compact). Hence $\lim_n \langle f_n, x' \rangle$ exists a.e. We now verify that for all $x' \in E'$ $\lim_n \langle f_n, x' \rangle = \langle f, x' \rangle$ a.e. Indeed, for any ω, $x' \in E'$ and any $\varepsilon > 0$, the inequality

$$|\langle f_n(\omega) - f(\omega), x' \rangle| < \varepsilon$$

holds for infinitely many n — because of the choice of $f(\omega)$. So if $\lim_n \langle f_n(\omega), x' \rangle$ exists for some ω then the latter limit equals $\langle f(\omega), x' \rangle$. This shows that $\langle f_n(\omega), x' \rangle \to \langle f(\omega), x' \rangle$ for ω not in some λ-null set $N_{x'}$. Incidentally, this proves also that f is scalarly measurable with respect to the Lebesgue measurable sets and that f is separably valued. So f is strongly λ-integrable (λ being taken now on the Lebesgue measurable sets of $\Omega = [0,1]$). We now verify that if $\Sigma_n = \sigma\{f_1, \ldots, f_n\}$ then

$$f_n = \mathbb{E}\{f | \Sigma_n\}, \quad n \in \mathbb{N}.$$

Indeed, if $A \in \Sigma_n$ and $x' \in E'$,

$$\int_A \langle f_n, x' \rangle d\lambda = \int_A \langle f_{n+k}, x' \rangle d\lambda, \qquad k \geq 0$$

$$\to \quad (k \to \infty) \int_A \langle f, x' \rangle d\lambda$$

so that $\int_A f_n d\lambda = \int_A f \, d\lambda$ which proves our assertion above. Now, we can conclude from the vector-valued martingale convergence theorem (theorem 1 of Chapter 1) that $\lim_n f_n$ exists a.e. in the norm topology of E. This proves that K has MCP in this case.

(b) We consider now the case of a general locally convex space E. Let $f_n : \Omega \to K$ be a simple, martingale sequence, as before. It is enough to show that if V is any balanced, convex, open neighbourhood of $0 \in E$ then $f_m(\omega) - f_n(\omega) \in V$ for m,n sufficiently large ($\geq n_0(\omega)$) a.e. (ω).

Let p be the continous seminorm associated with V (i.e. $V = \{x \in E \mid p(x) < 1\}$) and let F be the (closed) subspace of E where p vanishes. Then if $\pi : E \to E/F$ is the quotient map, we know that π is linear and continuous [E having its initial topology E/F its quotient topology]. Hence π is also weakly continuous [E having topology $\sigma(E,E')$ and E/F having the topology $\sigma(E/F, (E/F)')$]. Thus if $K \subset E$ is weakly compact so is $\pi(K) \subset (E/F)$. Let us now consider E/F under the norm topology induced by p : $|x+F| = p(x)$. We assert that $\pi(K)$ is compact also in the weak topology of $(E/F, |\cdot|)$. Using obvious notation,

$$(E/F, |\cdot|)' \subset (E/F)'$$

so that $(E/F,|\cdot|)'$ - topology on E/F is smaller than the $(E/F)'$ - topology on E/F. Since $\pi(K)$ is compact in $(E/F)'$ - topology (as argued above) it is compact in $(E/F,|\cdot|)'$ - topology. Thus $g_n = \pi \circ f_n : \Omega \to \pi(K)$ is a simple, martingale sequence where $\pi(K)$ is a weakly compact subset of the normed space $(E/F,|\cdot|)$. By case (a), $\lim_n g_n$ exists a.e. Thus, there is a λ-null set N such that if $\omega \notin N$,

$$|g_n(\omega) - g_m(\omega)| = p\{f_n(\omega) - f_m(\omega)\} \to 0$$

as $m,n \to \infty$; this means $\{f_n(\omega) - f_m(\omega)\} \in V$ for $m,n \geq$ some $n_0(\omega)$ (since $p(x) < 1$ implies $x \in V$). We have thus established that K has MCP and the proof of Proposition 1 is complete.

§4. Ryll-Nardzweski's theorem.

Theorem 2.

Let E be a locally convex Hausdorff topological vector space and K a non-empty weakly compact convex subset of E. If G is a semigroup of continuous linear operators from E to E which maps K into K and which is distal on K then G has a fixed point in K.

Proof : Take T_1 to be the weak topology of E and T_2 the initial topology of E. Recall that any continous linear map from E to E is also continuous if E has weak topology. We can now use theorem 1 along with proposition 1 to deduce theorem 2.

Many other versions are possible and can be proven by combining theorem 1 with a suitable variation of proposition 1.

§5. Notes and Remarks.

Ryll-Nardzweski's theorem as in §4 appears in his article in the Fifth Berkley Symposium, 1967, vol. II, Part I, p. 55-61. His proof uses a "Monte-Carlo" method based on a vector-valued differentiation theorem. This was adapted to martingale arguments in Chatterji. The present exposition elaborates on the martingale argument by introducing MCP — a concept much used now in the context of Banach spaces. In the latter context, MCP of a set is equivalent to its having RNP — a notion easy to define but quite useful as a substitute for compactness as evidenced by Edgar's theorem (cf. Diestel and Uhl p. 145 for one version) which gives a non-compact version of Choquet's theorem for such sets. A scrutiny of our proofs here shows that we could have weakened our MCP to one where we demand only that $f_{n+1}(\omega)-f_n(\omega) \in V$ for $n \geq n_0(\omega)$, $\omega \notin N(V)$, $N(V)$ being a null set. This generalization is illusory in the case of Banach spaces (cf. Diestel and Uhl p. 216 : "trees in Banach spaces") but may be useful in the context of general topological vector spaces treated here.

Other expositions of the theorem are to be found in Bourbaki (p. IV 41-44), Dugundji and Granas (p. 100) where the proofs given are non-probabilistic. Many applications of the theorem are also indicated there. Dunford and Schwartz (p. 456) contains the version of the Markov-Kakutani theorem needed in this chapter as well as all the necessary information concerning topological vector spaces (ibid, Chapter V).

One novelty of Ryll-Nardzweski's theorem lies in its adroit mixture of weak and strong topologics. In our version (theorem 1), this is reflected in the choice of the topologies T_1 and T_2. The weaker the topology T_1, the easier it is for K to be T_1-compact. The stronger the topology T_2, the easier it is for the semi-group to be T_2-distal but more difficult for it to have T_2-MCP. Useful applications of it must strike a balance between T_1 and T_2. No doubt, the original statement of Ryll-Nardzweski is the simplest such balance which is at the same time very useful.

Chapter 6.

Absolute continuity and singularity.

§1. Preliminary.

In this chapter, we shall be concerned mostly with probability measures. If P,Q are two such measures on a measurable space (Ω,Σ), by D_PQ we shall denote, as before, the Radon-Nikodym derivative of the absolutely continuous part of Q with respect to P.

Theorem 1.

Let P,Q be two probability measures on the measurable space (Ω,Σ) and let $f = D_PQ$. Then :

(a) (i) $P \perp Q$ iff $P\{f = 0\} = 1$

(ii) $P \ll Q$ iff $P\{f > 0\} = 1$.

(b) Let $\{\Sigma_n\}_{n\in\mathbb{N}}$ be an increasing sequence of sigma-algebras such that $\sigma(\bigcup_n \Sigma_n) = \Sigma$; let P_n, Q_n be the restrictions of P,Q to Σ_n and let $f_n = D_{P_n} Q_n$. Then

(i) $\{f_n, \Sigma_n\}_{n\in\mathbb{N}}$ is a non-negative supermartingale with respect to P and $f_n \to f$ a.e. (P)

(ii) $P \perp Q$ iff

$$\lim_n \int f_n \, dP = \inf_n \int f_n^\beta \, dP = 0$$

for some $\beta \in]0,1[$ (e.g. $\beta = 1/2$)

(iii) $P \ll Q$ iff $\forall \varepsilon > 0, \exists \beta(\varepsilon) \in]0,1[$ such that

$$\int f_n^\beta \, dP > (1-\varepsilon)$$

for all $n \in \mathbb{N}$ and $0 < \beta < \beta(\varepsilon)$.

Proof :

(a) Write $Q(A) = \int_A f \, dP + Q(A \cap N)$

where $P(N) = 0$. Then (i) follows immediately from the uniqueness of the Lebesgue decomposition. For (ii), note that $f > 0$ a.e. (P) gives that $Q(A) = 0$ implies $P(A) = 0$ i.e. $P \ll Q$. Conversely, if $P \ll Q$ and $M = \{f = 0\}$,

$$\begin{aligned} P(M) &= P(M \cap N) + P(M \setminus N) \\ &\leq P(N) + P(M \setminus N) = 0 \end{aligned}$$

since $Q(M) = Q(M\cap N)$ implies that $Q(M\setminus N) = 0$ so that $P(M\setminus N) = 0$. This proves (a).

(b) Let $m \leq n$; for all $A \in \Sigma_n$,

$$Q(A) = \int_A f_n \, dP + Q'(A)$$

where Q' is a non-negative measure on Σ_n such that $Q' \perp P_n$. If now $A \in \Sigma_m$ then

$$Q(A) = \int_A \mathbb{E}_P \{f_n|\Sigma_m\} dP + \int_A g \, dP + Q''(A)$$

where $g \cdot P_m + Q''$ is the Lebesgue decomposition of $Q'|\Sigma_m$ with respect to P_m. By the uniqueness of the Lebesgue decomposition, we have

$$f_m = \mathbb{E}_P \{f_n|\Sigma_m\} + g \geq \mathbb{E}_P \{f_n|\Sigma_m\}$$

Thus $\{f_n, \Sigma_n\}_{n\in\mathbb{N}}$ is a non-negative P-supermartingale. Its convergence a.e. and the identification of the limit follow from the results in Chapters 1 and 2. This proves (i).

For each $\beta \in]0,1[$, $\{f_n\}_{n\in\mathbb{N}}$ is a uniformly P-integrable sequence since $(1/\beta) > 1$ and

$$\int (f_n^\beta)^{1/\beta} dP = \int f_n dP \leq 1$$

Hence, for $0 < \beta < 1$,

$$\lim_n \int f_n^\beta \, dP = \int f^\beta \, dP$$

Since, for $0 < \beta < 1$, $x \mapsto x^\beta$, $x \geq 0$ is a concave function, $\{f_n^\beta, \Sigma_n\}_{n\in\mathbb{N}}$ is also a supermartingale if $0 < \beta < 1$. This gives that $\int f_n^\beta \, dP$ is decreasing in n so that

$$\lim_n \int f_n^\beta \, dP = \inf_n \int f_n^\beta \, dP.$$

The proof of b (ii) is immediate using (a) (i).

From (a) (ii) we know that $P \ll Q$ iff $f > 0$ a.e. (P). Now

$$\lim_{\beta\to 0+} f^\beta(\omega) = \begin{cases} 1 & \text{if} \quad f(\omega) > 0 \\ 0 & \text{if} \quad f(\omega) = 0 \end{cases}$$

and

$$f^\beta(\omega) \leq \max \{f(\omega), 1\}, \quad 0 < \beta < 1$$

so that

$$\lim_{\beta\to 0+} \int f^\beta \, dP = \int \lim_{\beta\to 0+} f^\beta \, dP$$
$$= P\{f > 0\}$$

Hence, by what precedes, $P \ll Q$ iff

$$\lim_{\beta \to 0+} \inf_n \int f_n^\beta \, dP = 1$$

Since

$$\int f_n^\beta \, dP \leq \left(\int f_n dP\right)^\beta \leq 1, \qquad 0 < \beta < 1$$

statement (b) (iii) is only a reformulation of the above limit relation. This completes the proof of theorem 1.

§2. Kakutani's theorem on product measures.

Theorem 2.

Let p_j, q_j be two probability measures on (Ω_j, σ_j), $j \in \mathbb{N}$ and let $P = \otimes_j p_j$, $Q = \otimes_j q_j$ be the corresponding product measures on $(\Omega = \Pi_j \Omega_j, \ \Sigma = \otimes_j \sigma_j)$. Suppose $p_j \ll q_j$ for all j and let $h_j = dp_j/dq_j$. Then either $P \ll Q$ or $P \perp Q$. Also if $c = \prod_j \int h_j^{\frac{1}{2}} \, dq_j$ then $0 \leq c \leq 1$; $P \perp Q$ iff $c = 0$ and $P \ll Q$ iff $c > 0$.

Proof :

Let us start with the following simple remark : if p,q are two probability measures on some space and $p \ll q$, $h = dp/dq = D_q p$ then $g = 1/h = D_p q$ a.e.(p). Indeed,

$$p\{h = 0\} = \int_{\{h=0\}} h \, dq = 0$$

implies that g is well-defined and >0 a.e. (p). If

$$q(A) = \int_A \varphi \, dp + q(A \cap N), \quad p(N) = 0$$

is the Lebesgue decomposition of q with respect to p then

$$\begin{aligned} p(A) &= p(A \setminus N) \\ &= \int_{A \setminus N} h \, d \, q \\ &= \int_{A \setminus N} h \, \varphi \, dp = \int_A h \, \varphi \, dp \end{aligned}$$

so that $h \varphi = 1$ a.e. (p) whence $\varphi = 1/h = g$ a.e. (p). Note also that

$$\int g^{\frac{1}{2}} \, dp = \int g^{\frac{1}{2}} \cdot h \, dq = \int h^{\frac{1}{2}} \, dq \quad .$$

Now let Σ_n be the sigma-algebra in Ω determined by sets of the type $\prod_j A_j$ with

$A_j = \Omega$, $j \geq n+1$. Then Σ_n is an increasing sequence of sigma-algebras whose union generates Σ. Let $P_n = P|\Sigma_n$, $Q_n = Q|\Sigma_n$ and apply Theorem 1. To do so, introduce $f_n = D_{P_n}Q_n$, $g_n = D_{p_n}q_n$. As remarked above, $h_n = dp_n/dq_n = 1/g_n$ a.e. (p_n). We now verify that

$$f_n(\omega) = \prod_{j=1}^{n} g_j(\omega_j) \quad , \quad \omega = (\omega_j)_{j\in\mathbb{N}}$$

a.e. (P_n). This is easily done by induction on n. For n = 1, there is nothing to prove. Suppose the formula for f_n valid for some n. Then, for $A \in \Sigma_n$,

$$Q_n(A) = \int_A f_n \, dP_n + Q_n(A \cap N)$$

for some $N \in \Sigma_n$, $P(N) = 0$; let, for $B \in \sigma_{n+1}$

$$q_{n+1}(B) = \int_B g_{n+1} \, dp_{n+1} + q_{n+1}(B \cap M)$$

be the Lebesgue decomposition of q_{n+1} with respect to p_{n+1} (so that $M \in \sigma_{n+1}, p_{n+1}(M)=0$).

We see immediately that

$$\begin{aligned} Q_{n+1}(A \times B) &= Q_n(A) \times q_{n+1}(B) \\ &= \int_{A\times B} f_n \cdot g_{n+1} \, dP_{n+1} + Q'(A\times B) \end{aligned}$$

where

$$Q'(A\times B) = \left(\int_A f_n dP_n\right) q_{n+1}(B \cap M) + Q_n(A \cap N) \int_B g_{n+1} \, dp_{n+1} + Q_n(A \cap N) \, q_{n+1}(B \cap M)$$

i.e. Q' is a non-negative measure on Σ_{n+1} concentrated in the P_{n+1}-null set

$$\left\{\Omega_1 \times \ldots \times \Omega_n \times M \times \Omega_{n+2} \times \ldots\right\} \cup \left\{N \times \Omega_{n+1} \times \ldots\right\}$$

This proves that $f_{n+1}(\omega) = f_n(\omega) \cdot g_{n+1}(\omega_{n+1})$ and the induction is complete.

Now, according to Theorem 1 ((b) (ii)), $P \perp Q$ iff $\int f_n^{\frac{1}{2}} \, dP \to 0$. But

$$\int f_n^{\frac{1}{2}} \, dP = \prod_{j=1}^{n} \int g_j^{\frac{1}{2}} \, dp_j = \prod_{j=1}^{n} \int h_j^{\frac{1}{2}} \, dP$$

i.e. $P \perp Q$ iff $c = 0$.

Further, $$\int h_j^{\frac{1}{2}} \, dP \leq \int h_j \, dP = 1$$

so that $c = \lim_n \prod_{j=1}^{n} \int h_j^{\frac{1}{2}}\, dP$ exists and $0 \leq c \leq 1$. Thus if $c > 0$, P,Q are not mutually singular and the set $A = \{\omega | f(\omega) > 0\}$ is such that $P(A) > 0$. However since each $g_j > 0$ a.e. (P), the set A a.s. "does not depend on $\omega_1 \ldots \omega_n$" for any n so that by the 0-1 law of probability theory, $P(A) = 0$ or 1. Hence if $c > 0$, $P(A) = 1$ and $P \ll Q$. This completes the proof of theorem 2.

§3. Gaussian processes.

Theorem 3.

Let P,Q be two probability measures on the measurable space (Ω,Σ) and let $\{\xi_t\}_{t\in T}$ be a real-valued stochastic process (i.e. $\xi_t : \Omega \to \mathbb{R}$ is measurable for each $t \in T$, T being an arbitrary index set) such that (i) $\Sigma = \sigma\{\xi_t, t \in T\}$ and (ii) $\{\xi_t\}_{t\in T}$ is Gaussian for both P and Q (i.e. the P-law or the Q-law of any finite linear combination of the ξ_t's is a Gaussian measure in $\mathbb{R}^1$ — proper or degenerate). Then either $P \perp Q$ or $P \sim Q$.

The proof of the theorem is preceded by a number of elementary technical lemmas.

If Σ_1, Σ_2 are two sigma-algebras and μ is some non-negative measure on some sigma-algebra containing Σ_1 and Σ_2, we write $\Sigma_1 \overset{\mu}{\subset} \Sigma_2$ if for any $A \in \Sigma_1$ there is a $B \in \Sigma_2$ such that $\mu(A \Delta B) = 0$. We write $\Sigma_1 \overset{\mu}{=} \Sigma_2$ if $\Sigma_1 \overset{\mu}{\subset} \Sigma_2$ and $\Sigma_2 \overset{\mu}{\subset} \Sigma_1$.

Lemma 1 :

Let (Ω,Σ,μ) be any measure space, $\{A_t\}_{t\in T}$ a family of sigma-algebras of subsets of Ω such that $\Sigma = \sigma\{A_t, t \in T\}$. For any subset S of T let Σ_S denote the sigma-algebra $\sigma\{A_t, t \in S\}$. If for all $t \in T$, $A_t \overset{\mu}{\subset} \Sigma_S$ then $\Sigma_S \overset{\mu}{=} \Sigma$.

Proof :

Let Σ' be the family of $A \in \Sigma$ which is such that there is a $B \in \Sigma_S$ with $\mu(A \Delta B)=0$. We verify that Σ' is a sigma-algebra. Clearly, $\Omega \in \Sigma'$ and $A \Delta B = A^c \Delta B^c$ gives immediately that Σ' is stable for complementation. If $A_n \in \Sigma'$ and $B_n \in \Sigma_S$, $n \in \mathbb{N}$, such that $\mu(A_n \Delta B_n) = 0$ then $\mu(A \Delta B) = 0$ where $A = \cup_n A_n$, $B = \cup_n B_n$ since

$$A_m \cap B^c \subset A_m \cap B_m^c \quad , \quad B_m \cap A^c \subset B_m \cap A_m^c$$

imply that $\mu(A \cap B^c) = \mu(B \cap A^c) = 0$.
Now, by hypothesis, each A_t is contained in Σ'; hence, $\Sigma = \sigma\{A_t, t \in T\} \subset \Sigma'$ i.e. $\Sigma = \Sigma'$. This is equivalent to the affirmation of the lemma.

Lemma 2 :

Let the notation and hypotheses be as in Theorem 3 and suppose further that P and Q are not mutually singular.

(i) If for some $t, t_1, \dots, t_n$ in T and constants $a, c_1, \dots, c_n$ in $\mathbb{R}$

$$\xi_t = a + \sum_{i=1}^{n} c_i \, \xi_{t_i}$$

holds a.e. (P) then the same holds a.e. (Q).

(ii) Let $S \subset T$ be such that for any $t \in T \setminus S$, ξ_t is a.e. (P) (and hence a.e. (Q)) a finite linear combination of 1 and $\{\xi_t\}_{t \in S}$. Then

$$\Sigma \overset{P+Q}{=} \sigma\{\xi_s, \, s \in S\} \quad (=\Sigma_0)$$

This last relation implies that :

$$(P|\Sigma_0) \perp (Q|\Sigma_0) \quad \text{iff} \quad P \perp Q$$

and

$$(P|\Sigma_0) \sim (Q|\Sigma_0) \quad \text{iff} \quad P \sim Q.$$

Proof :

(i) Let $A = \{\xi_t = a + \sum_i c_i \xi_{t_i}\}$. Since both the P-law and the Q-law of $\xi_t - \sum_i c_i \xi_{t_i}$ is Gaussian, P(A) and Q(A) are either 0 or 1. If $P(A) = 1$ and $Q(A) < 1$ then $Q(A) = 0$ and $P \perp Q$, contrary to hypothesis. Hence $P(A) = 1$ implies $Q(A) = 1$.

(ii) Fix $t \in T$; let $\xi_t = a + \sum_{i=1}^{n} c_i \xi_{t_i} = \eta$ a.e. (P) (and hence a.e. (Q)) for some choice of $t_1 \dots t_n$ in S and numbers $a, c_1, \dots, c_n$ in $\mathbb{R}$. Then, for any Borel set $A \subset \mathbb{R}$,

$$\left\{\xi_t^{-1}(A) \, \Delta \, \eta^{-1}(A)\right\} \subset \left\{\xi_t \neq \eta\right\}.$$

This proves that if $A_t = \sigma\{\xi_t\}$ then $A_t \subset^{(P+Q)} \Sigma_0$ where $\Sigma_0 = \sigma\{\xi_s, s \in S\} = \sigma\{A_s, \, s \in s\}$. Lemma 1 now proves that $\Sigma_0 \overset{P+Q}{=} \Sigma$.

Since $\Sigma_0 \subset \Sigma$, it is trivial that $(P|\Sigma_0) \perp (Q|\Sigma_0)$ implies $P \perp Q$. To prove the converse, let $N \in \Sigma$ such that $P(N)=0$, $Q(N)=1$. Take $M \in \Sigma_0$ such that $(P+Q)(M \Delta N) = 0$. Then, $P(M) = P(M \cap N) + P(M \cap N^c) = 0$

and

$$\begin{aligned} Q(M) &= Q(M \cap N) + Q(M \cap N^c) \\ &= Q(M \cap N) \\ &= Q(N) - Q(N \cap M^c) = Q(N) = 1 \end{aligned}$$

i.e. $(P|\Sigma_0) \perp (Q|\Sigma_0)$ if $P \perp Q$.

Let us now prove that $(P|\Sigma_0) \ll (Q|\Sigma_0)$ implies $P \ll Q$, the converse being immediate since $\Sigma_0 \subset \Sigma$. If $A \in \Sigma$ and $Q(A) = 0$, find $B \in \Sigma_0$ such that $(P+Q)\ (A \Delta B) = 0$. Now, $Q\ (B) = Q(B \cap A) + Q(B \cap A^c) = 0$ so that $P(B) = 0$ if $(P|\Sigma_0) \ll (Q|\Sigma_0)$ thus $P(A) = P(A \cap B + P(A \cap B^c) = 0$. So we have proven that $P \ll Q$ if $(P|\Sigma_0) \ll (Q|\Sigma_0)$. This is sufficient to complete the proof of lemma 2.

Lemma 3 : (Notation and hypotheses same as in theorem 3).

If theorem 3 is valid for any denumerable set T then it is valid for an arbitrary T.

Proof :

For any subset S of T let Σ_S denote the sigma-algebra generated by ξ_t, $t \in S$. Then it is known (and easy to prove) that each set A in Σ is in some Σ_S, $S \subset T$, S denumerable. Write $P_S = P|\Sigma_S$. If $P_S \perp Q_S$ for some denumerable S then it is trivial that $P \perp Q$. If $P_S \sim Q_S$ for any denumerable $S \subset T$, we have to show that $P \sim Q$. Indeed, if $P(A) = 0$ for some $A \in \Sigma$ then $A \in \Sigma_S$ for some denumerable sub-set S of T and so $Q(A) = 0$. From this follows the proof of lemma 3.

Lemma 4.

Theorem 3 is valid for T finite.

Proof :

Suppose P and Q are not singular. Then, according to lemma 2, there is a finite subset S of T such that $\{1;\xi_t, t \in S\}$ is a linearly independent set in $L^2(P)$ as well as in $L^2(Q)$ and that each ξ_t is a linear combination of 1 and $\xi_{t'}$, $t' \in S$ a.e. (P) and a.e. (Q). Also, according to lemma 2, $P \sim Q$ iff $(P|\Sigma_S) \sim (Q|\Sigma_S)$ where $\Sigma_S = \sigma\{\xi_t, t \in S\}$. Now, if $S = \{t_1,\dots,t_n\}$ and $\eta = (\xi_{t_1},\dots,\xi_{t_n})$ then $\eta : \Omega \to \mathbb{R}^n$ is not only measurable but also $\eta^{-1}(\mathcal{B}) = \Sigma_S$ where $\mathcal{B}$ is the class of Borel sets in $\mathbb{R}^n$. Thus $(P|\Sigma_S) \sim (Q|\Sigma_S)$ iff $P' \sim Q'$ where $P' = P\eta^{-1}$, $Q' = Q\eta^{-1}$ are the laws of η in $\mathbb{R}^n$ under P and Q respectively. Now, P', Q' are Gaussian measures in $\mathbb{R}^n$ such that every proper affine subspace of $\mathbb{R}^n$ has P',Q' measures equal to zero. Then, it is known (and elementary) that each of P',Q' posseses a continuous density with respect to Lebesgue measure in $\mathbb{R}^n$. In particular, $P' \sim Q'$ whence, according to the preceding argument, $P \sim Q$. This proves lemma 4.

Lemma 5.

Let P,Q be two Gaussian measures in $\mathbb{R}^n$ which are absolutely continuous with respect to Lebesgue measure in $\mathbb{R}^n$. Suppose that the mean vectors of P,Q are p,q and their covariance matrices are C,D respectively. Then, for $0 < \beta < 1$, we have the formula ·

$$\int \left(\frac{dQ}{dP}\right)^{\beta} dP = \prod_{j=1}^{n} \left\{\frac{\lambda_j^{\beta}}{\beta\lambda_j+(1-\beta)}\right\}^{\frac{1}{2}} \cdot \exp\left[\frac{-\beta(1-\beta)}{2} \sum_{j=1}^{n} \frac{\lambda_j m_j^2}{\beta\lambda_j+(1-\beta)}\right]$$

where $\lambda_1,\ldots,\lambda_n$ are the eigen-values of the positive definite matrix $C^{\frac{1}{2}} D^{-1} C^{\frac{1}{2}}$ and $m = (m_j)$ is such that $m = V^{-1} C^{-\frac{1}{2}}(q-p)$ for a certain orthogonal matrix V.

Proof :

Let f,g be the standard, continuous densities (with respect to Lebesgue measure in $\mathbb{R}^n$) of P,Q respectively. Then, if $A = C^{-1}$, $B = D^{-1}$,

$$f(x) = (2\pi)^{-n/2}(\det A)^{\frac{1}{2}} \exp - \frac{1}{2} A(x-p)\cdot(x-p)$$

$$g(x) = (2\pi)^{-n/2}(\det B)^{\frac{1}{2}} \exp - \frac{1}{2} B(x-q)\cdot(x-q)$$

where $x \in \mathbb{R}^n$ and $x \cdot y$ denotes the scalar product of x,y in $\mathbb{R}^n$. Thus (the integrals being on $\mathbb{R}^n$)

$$\int \left(\frac{dQ}{dP}\right)^{\beta} dP = \int g^{\beta}(x)\, f^{(1-\beta)}(x)\, dx$$

$$= (2\pi)^{-n/2}(\det A)^{(1-\beta)/2}(\det B)^{\beta/2} \int h(x)\, dx$$

where

$$h(x) = \exp - \frac{1}{2}\left\{(1-\beta)Ax\cdot x + \beta B(x-r)\cdot(x-r)\right\}$$

with $r = q-p$. To calculate $\int h$ we choose an orthogonal matrix V such that

$$V'\, A^{-\frac{1}{2}}\, B\, A^{-\frac{1}{2}}\, V = \Lambda$$

where Λ is the $n \times n$ diagonal matrix with $\lambda_1 \ldots \lambda_n$ in the main diagonal. The λ_j's are > 0 and are the eigen-values of the positive-definite matrix

$$A^{-\frac{1}{2}}\, B\, A^{-\frac{1}{2}} = C^{\frac{1}{2}}\, D^{-1}\, C^{\frac{1}{2}} .$$

Now, if we take $W = A^{-\frac{1}{2}}$, we see that

$$W'\, A\, W = I , \quad W'\, B\, W = \Lambda$$

where I is the $n \times n$ identity matrix. If we subsitute $x = Wy$ in $\int h(x)\, dx$ we have

$$\int h(x)\, dx = \int h(Wy)\, |\det W|\, dy$$

$$= \int \exp\left[-\frac{1}{2}\left\{(1-\beta)y \cdot y + \beta\Lambda(y-m)\cdot(y-m)\right\}\right] (\det A)^{-\frac{1}{2}}\, dy$$

(where m is such that $Wm = r$)

$$= (\det A)^{-\frac{1}{2}} \prod_{j=1}^{n} \int_{-\infty}^{\infty} k_j(t)\, dt$$

with

$$k_j(t) = \exp\left[-\frac{1}{2}\left\{(1-\beta)t^2 + \beta\lambda_j(t-m_j)^2\right\}\right], \quad m = (m_j) .$$

An easy calculation using the formula

$$\int_{-\infty}^{\infty} \exp\left[-(at^2+2bt+c)\right] dt = \left(\frac{\pi}{a}\right)^{\frac{1}{2}} \exp\left(\frac{b^2-ac}{a}\right), \quad a > 0$$

yelds

$$\int_{-\infty}^{\infty} k_j(t)dt = \left(\frac{2\pi}{\beta\lambda_j+1-\beta}\right)^{\frac{1}{2}} \exp\left[\frac{-\beta(1-\beta)}{2} \cdot \frac{\lambda_j m_j^2}{\beta\lambda_j+1-\beta}\right] .$$

From this and the relation

$$\frac{\det B}{\det A} = \det \Lambda = \lambda_1 \ldots \lambda_n$$

we obtain the formula given in lemma 5.

Lemma 6.

(i) If $b_j \geq 0$, $1 \leq j \leq N$, then

$$1 + \sum_{j=1}^{N} b_j \leq \prod_{j=1}^{N} (1+b_j) \leq \exp\left(\sum_{j=1}^{N} b_j\right) .$$

(ii) Let $f(x) = (1+x)/2\sqrt{x}$, $x > 0$; for any $M > 1$,

$$\{x \mid 1 \leq f(x) \leq M\} = [x_1, x_2] , \quad 0 < x_1 < x_2 < \infty$$

and there exist constants $d_1, d_2 (0 < d_1 < d_2 < \infty)$ such that for $x \in [x_1, x_2]$,

$$d_1(1-x)^2 \leq f(x)-1 \leq d_2(1-x)^2.$$

(iii) If $-1 < x_1 \leq x \leq x_2$ and $0 < \beta < 1$ then there exists a constant $d > 0$, depending on x_1, x_2 but independent of β, such that

$$\ln(1+\beta x) - \beta\ln(1+x) \leq d\beta x^2.$$

Proof.

(i) is well-known and simple.

(ii) The existence of x_1, x_2 follows from the facts that

$$\inf_{x>0} f(x) = f(1) = 1, \quad \lim_{x\to 0} f(x) = \lim_{x\to\infty} f(x) = \infty$$

and that f decreases in $]0,1[$ and increases in $]1,\infty[$. Also

$$f(x) - 1 = (1-x)^2/2\sqrt{x}(1+\sqrt{x})^2$$

which gives the affirmation concerning inequalities.

(iii) Note first that there are constants c_1, c_2, $0 < c_1 < c_2 < \infty$, such that

$$x - c_1 x^2 \leqslant \ell n(1+x) \leqslant x - c_2 x^2$$

for $-1 < x_1 \leqslant x \leqslant x_2 < \infty$; this can be shown by using the second order Taylor expansion for $\ell n(1+x)$. Without loss of generality we may take $x_1 < 0 < x_2$, for otherwise we just enlarge our interval. Then if $x \in [x_1, x_2]$ and $0 < b < 1$, βx is in $]x_1, x_2[$ whence

$$\begin{aligned} \ell n(1+\beta x) - \beta \ell n(1+x) &\leqslant \beta x - c_2 \beta^2 x^2 - \beta(x - c_1 x^2) \\ &\leqslant (\beta c_1 - c_2 \beta^2) x^2 \\ &\leqslant c_1 \, x^2 \end{aligned}$$

which proves this part (with $d = c_1$). This completes the proof of lemma 6.

<u>Proof of Theorem 3 :</u>

As noted in lemma 3, we need only consider the case where T is a denumerable set. Suppose that P and Q are not mutually singular. We shall establish that $P \sim Q$.

Now, according to lemma 2, we may suppose further, without loss of generality, that no ξ_t is the finite linear combination of 1 and the other ξ_s, $s \in T \setminus \{t\}$. In other words, writing $T = \{1,2,3,\ldots\}$, we may suppose that $(\xi_1,\ldots,\xi_n)$ has a Gaussian distribution in $\mathbb{R}^n$ which is absolutely continuous with respect to Lebesgue measure there — both under P as well as under Q. Let Σ_n be the sigma-algebra generated by $\xi_1,\ldots,\xi_n$ and let $P_n = P|\Sigma_n$, $Q_n = Q|\Sigma_n$. The hypothesis that P,Q are not mutually singular is now translated into

$$\inf_n \int \left(\frac{dQ_n}{dP_n}\right)^{\frac{1}{2}} dP_n > 0 \qquad \ldots \quad (1)$$

(cf. Theorem 1(b) (ii)).

Taking $\beta = 1/2$ in lemma 5, (1) is seen to be equivalent to

$$\sup_n \Pi_n \exp(S_n) < \infty \qquad \dots \quad (2)$$

where

$$\Pi_n = \prod_{j=1}^{n} \left\{ \frac{\lambda_j(n)+1}{2\lambda_j^{\frac{1}{2}}(n)} \right\}$$

$$S_n = \sum_{j=1}^{n} \frac{\lambda_j(n) m_j^2(n)}{\lambda_j(n)+1}$$

and $\lambda_j(n)$, $m_j(n)$, $1 \leqslant j \leqslant n$, correspond to the λ's and m's of lemma 5. Note that $\Pi_n \geqslant 1$ and $S_n \geqslant 0$; hence, (2) is equivalent to the following two conditions :

$$\sup_n \Pi_n < \infty \qquad \dots \quad (3)$$

and

$$\sup_n S_n < \infty \qquad \dots \quad (4)$$

According to lemma 6 (i), (3) is equivalent to

$$\sup_n \sum_{j=1}^{n} \left\{ \frac{\lambda_j(n)+1}{2\sqrt{\lambda_j(n)}} - 1 \right\} < \infty$$

which is equivalent to

$$\begin{cases} \sup_n \sum_{j=1}^{n} (\lambda_j(n) - 1)^2 < \infty & \dots \quad (5) \\ 0 < c_1 = \inf_{n,j} \lambda_j(n) \leqslant \sup_{n,j} \lambda_j(n) = c_2 < \infty \end{cases}$$

by virtue of lemma 6 (ii). So (3) and (4) together are equivalent to (5) and

$$\sup_n \sum_{j=1}^{n} m_j^2(n) < \infty \qquad \dots \quad (6)$$

Thus P and Q are not mutually singular iff (5) and (6) hold. To establish that $P \sim Q$ we shall now show that under (5) and (6)

$$\lim_{\beta \to 0} \inf_n \int \left(\frac{dQ_n}{dP_n} \right)^{\beta} dP_n = 1 \qquad \dots \quad (7)$$

In fact, according to theorem 1(b) (iii), (7) implies that $P \ll Q$; since the roles of

P and Q are symmetric, we also have $Q \ll P$ whence $P \sim Q$.

Again, in view of lemma 5, (7) is equivalent to

$$\lim_{\beta\to 0} \inf_n M_n(\beta) \exp\left[-A_n(\beta)\right] = 1 \qquad \ldots \quad (8)$$

with

$$M_n(\beta) = \prod_{j=1}^{n} \left[\frac{\lambda_j^\beta(n)}{\beta\lambda_j(n)+1-\beta}\right]^{\frac{1}{2}}$$

$$A_n(\beta) = \frac{\beta(1-\beta)}{2} \sum_{j=1}^{n} \frac{\lambda_j(n) m_j^2(n)}{\beta\lambda_j(n)+1-\beta}$$

where $\lambda_j(n)$, $m_j(n)$ are as before.

Note that $A_n(\beta) \geq 0$ and $0 < M_n(\beta) \leq 1$ by the arithmetic-geometric mean inequality. Certainly (8) will follow if we can show that

$$\lim_{\beta\to 0} \inf_n M_n(\beta) = 1$$

and

$$\lim_{\beta\to 0} \inf_n \exp\left[-A_n(\beta)\right] = 1.$$

This, in turn, means that it is enough to show that

$$\lim_{\beta\to 0} \sup_n \left[-\ln M_n(\beta)\right] = 0 \qquad \ldots \quad (10)$$

and

$$\lim_{\beta\to 0} \sup_n A_n(\beta) = 0 \qquad \ldots \quad (11)$$

According to (5) and lemma 6 (iii),

$$0 \leq \ln\{\beta\lambda_j(n) + 1-\beta\} - \beta \ln \lambda_j(n)$$

$$= \ln\{1 + \beta(\lambda_j(n) - 1)\} - \beta \ln\{1+(\lambda_j(n)-1)\}$$

$$\leq \text{const. } \beta(\lambda_j(n)-1)^2$$

which, in view of (5), proves (10).

Further, taking into account (5) again,

$$A_n(\beta) \leq \frac{1}{2}\, \beta \sum_{j=1} c_2 \cdot m_j^2(n)$$

which, in view of (6), proves (11).

This completes the proof of theorem 3.

§4. Notes and Remarks.

Many problems of applied probability and of pure mathematics lead to the question of deciding when two measures are mutually singular or equivalent. This has given rise to a vast literature. Even the Gaussian case, one of the most important for applications, has a substantial volume of papers devoted to it some of which are indicated in the survey article of Chatterji and Mandrekar in Probabilistic analysis and related topics I, p. 169-197, Ed. A.T. Bharucha-Reid, Academic Press (1978); the monograph of Neveu ("Processus alēatoires gaussiens" 1968) referred to there seems to have an almost complete bibliography upto 1967-68. The text of Gihman and Skorohod devotes a substantial chapter (chapter VII) to the question of Gaussian dichotomy as well as more general absolute continuity — singularity problems.

The theorem of Kakutani (theorem 2) was published in 1948. This can now be found in many texts of probability theory (e.g. Neveu). Originally, the theorem was stated for the case where the marginal measures p_n, q_n are equivalent. For this case, our proof using the 0 - 1 law becomes much shorter. For the 0 - 1 law itself, the reader may consult Doob [1953] (p. 102).

Theorem 3 (known as the Gaussian dichotomy theorem) is due independently to Feldman and to Hajek — the relevant papers appearing in 1958. (References to Feldman's paper and an English translation of Hajek's paper can be found in the above-mentioned survey article of Chatterji and Mandrekar. An important question which we have not treated here is that of the explicit calculation of dP/dQ in important cases of equivalence. References to some of these can be looked up in the bibliographies already cited.

Our approach via theorem 1 appears in detail in an article by Chatterji and Mandrekar in "Functional Analysis : Surveys and recent results" p. 247-257, Ed. K.D. Bierstedt and B. Fuchssteiner, North-Holland (1977). The present chapter follows this latter fairly closely except for a few technical points (e.g. lemmas 1 and 2 of §3) which have been made more explicit.

It has been our fond hope for several years to obtain theorem 3 via theorem 2 (Kakutani's theorem) and a general reasoning (like an appeal to some 0 - 1 law) avoiding all calculations. The reason for wishing to do so is the desire to extend the dichotomy theorem to other classes of measures like symmetric, stable measures. This remains an open problem for the time being. An important progress has been accomplished by Fernique ("Comparaison de mesures gaussiennes et de mesures produit" Ann. Inst. Henri Poincaré, 20 (1984), p. 165-175) who has proved amongst other things that a Gaussian measure and a product measure in $\mathbb{R}^\infty$ are either mutually equivalent or mutually singular provided that their 1-dimensional marginals are equivalent.

Functional analysts often discuss theorem 3 in the context of two Gaussian measures P and Q defined on some suitable sigma-algebra of a locally convex space E. Since Gaussianness is defined there by the property that for each $x' \in E'$, the random variable $x \mapsto x'(x)$ defined on E be Gaussian, their study amounts to studying the Gaussian process $\{\xi_{x'}\}_{x' \in E'}$, $\xi_{x'}(x) = x'(x)$, defined on the measurable space E. On the other hand, our approach may be looked at as the study of Gaussian measures in the special locally convex space $E = \mathbb{R}^T$. Thus, the two approaches are, in some sense, equivalent. However, the point made here (in lemma 3, for example) is that as far as the dichotomy itself is concerned, the index set T plays an insignificant role.

Chapter 7.

A subsequence principle.

§1. Statement of a theorem.

Theorem 1.

Let $\{f_n\}$ be a bounded sequence in L^2 over an arbitrary measure space (Ω,Σ,μ). Then there is a subsequence $\{f_{n_j}\}$ and a function α in L^2 such that $\sum_j c_j\{f_{n_j}-\alpha\}$ converges a.e. (μ) whenever $\sum_j |c_j|^2<\infty$, $c_j \in \mathbb{R}$.

Proof :

Since the theorem deals with a denumerable family of real-valued functions in L^2 we may suppose that μ is sigma-finite i.e. $\Omega = \bigcup_i \Omega_i$, $\mu(\Omega_i)<\infty$. If we can find a suitable subsequence for each Ω_i, then the diagonal procedure would assure a subsequence for all of Ω. Thus we may and do suppose that μ is finite or even that $\mu(\Omega) = 1$. By passing to a subsequence we may suppose also that $f_n \to \alpha$ weakly in L^2. Replacing f_n by $(f_n-\alpha)$ we may even suppose that $f_n \to 0$ weakly in L^2. If $\{g_n\}$ is a sequence of simple random variables such that $\sum_n \int|f_n-g_n|^2\, d\mu < \infty$ then it is easy to see that $g_n \to 0$ weakly in L^2 and $\sum_n |f_n-g_n| < \infty$ a.e. (μ). Thus, if we can prove the theorem for $\{g_n\}$ i.e. find a subsequence $\{g_{n_j}\}$ such that $\sum_j c_j\, g_{n_j}$ converges a.e. (μ) for any $(c_j) \in \ell^2$ then the corresponding subsequence $\{f_{n_j}\}$ will do to prove the affirmation contained in our theorem. In other words, we have reduced our problem to the case of a bounded sequence $\{g_n\}$ in L^2 over a probability space (Ω,Σ,μ) which is such that $g_n \to 0$ weakly in L^2 and where the g_n's are all simple.

Let us now define a subsequence $\{g_{n_j}\}$ which would do for theorem 1. Take $n_1 = 1$; if $n_1 < n_2 < \ldots < n_{j-1}$ have already been chosen, define $n_j > n_{j-1}$ so that

$$|\mathbb{E}\{g_n | g_{n_1},\ldots,g_{n_{j-1}}\}| < 2^{-j}$$

for $n \geq n_j$. This is possible since the conditional expectation in question takes the value $\int_A g_n d\mu/\mu(A)$ on a set $A \in \sigma\{g_{n_1},\ldots,g_{n_{j-1}}\}$, $\mu(A) > 0$ and this value goes to 0 as $n \to \infty$ because of the hypothesis that $g_n \to 0$ weakly in L^2. Further, the number of sets A involved is finite since the g_n's are simple. Thus the choice indicated for n_j is possible. If we write

$$a_j = \mathbb{E}\{g_{n_j} | g_{n_1}, \ldots, g_{n_{j-1}}\}$$

then $\{(g_{n_j} - a_j)\}_{j\geq 1}$ is a martingale difference sequence which is bounded in L^2 since

$$\int |g_{n_j} - a_j|^2 \, d\mu \leq 2 \int \left\{ |g_{n_j}|^2 + |a_j|^2 \right\} d\mu$$

$$\leq 4 \int |g_{n_j}|^2 \, d\mu \, .$$

According to a corollary of the martingale convergence theorem (cf. Chapter 1, §4) $\sum_j c_j\{g_{n_j} - a_j\}$ converges a.e. (μ) for any choice of $(c_j) \in \ell^2$. Since $|a_j| < 2^{-j}$ this implies that $\sum_j c_j \, g_{n_j}$ converges a.e. (μ) and concludes the proof of theorem 1.

Corollary 1 :

From any orthonormal sequence $\{f_n\}$ in L^2 over any measure space (Ω, Σ, μ) we can choose a subsequence $\{f_{n_j}\}$ such that the Fourier series of any function φ in L^2 with respect to the orthonormal subsystem $\{f_{n_j}\}$ converges a.e.

Proof :

This follows from theorem 1 because of the fact that $f_n \to 0$ weakly in L^2 for any orthonormal sequence $\{f_n\}$ and that $c_n = \int \varphi f_n d\mu$ is in ℓ^2 — both being consequences of Bessel's inequality.

Corollary 2 :

Let the $\{f_n\}$ be as in theorem 1. Then we can choose a subsequence $\{f_{n_j}\}$ and α in L^2 such that for any $\varepsilon > 0$,

$$\lim_{N\to\infty} N^{-(\varepsilon+\frac{1}{2})} \sum_{j=1}^{N} (f_{n_j} - \alpha) = 0 \qquad \text{a.e. } (\mu) \, .$$

The subsequence can be so chosen that the same is true for any further subsequence.

Proof :

Choose $\{f_{n_j}\}$, α as guaranted by theorem 1. Since $c_j = j^{-(\varepsilon+\frac{1}{2})}$ is such that $\sum_j |c_j|^2 < \infty$ we have that $\sum_j c_j\{f_{n_j} - \alpha\}$ converges a.e. (μ). By a well-known lemma (Kronecker's lemma

$$c_N \sum_{j=1}^{N} \{f_{n_j} - \alpha\} \to 0 \qquad \text{a.e. } (\mu) \, .$$

This establishes the affirmation of the corollary for a fixed $\varepsilon > 0$ and for $\{f_{n_j}\}$ and any of its subsequences since obviously any of the latter has the same a.e. convergence property with respect to any $(c_j) \in \ell^2$ as postulated in theorem 1. To have the affirmation for all $\varepsilon > 0$, we need only choose a sequence $\varepsilon_n \to 0$ and obtain, by a diagonal procedure, a subsequence $\{f_{n_j}\}$ which will do for all ε_n simultaneously. This proves corollary 2.

Remark : It is clear that we can replace $N^{-(\varepsilon+\frac{1}{2})}$ in corollary 2 by various other sequences such as $N^{-\frac{1}{2}}(\ell n\ N)^{-(\varepsilon+\frac{1}{2})}$. The proof remains the same. However, the central limit theorem of probability theory tells us that we cannot in general substitute $N^{-\frac{1}{2}}$ in place of $N^{-(\varepsilon+\frac{1}{2})}$ in corollary 2.

§2. Notes and Remarks.

The subsequence principle alluded to in the title of the present chapter can be stated heuristically (and partially) as follows. Let $f_n : \Omega \to \mathbb{R}$, $n \in \mathbb{N}$, be any sequence of $\mathbb{R}$-valued measurable functions on some measure space (Ω, Σ, μ) such that $\sup_n \int |f_n|^p\, d\mu < \infty$ for some p, $0 < p < \infty$. Then a subsequence of $\{f_n\}$ can be determined in such a way that it and all its further subsequences will behave, in many respects, in the same way as sequences of independent, identically distributed random variables with finite p-th moments. Thus, for p=1, a suitable subsequence (and all its further subsequences) will satisfay the strong law of large numbers. This is a theorem due to Komlós (dating from 1967). If p=2, suitable central limit theorems and laws of iterated logarithms hold for appropriate subsequences. These types of results were proved by Berkes, Gaposhkin and myself around 1970-74. For detailed bibliographical references, cf. the recent survey given in my article in Jahresbericht der Deutschen Mathematiker-Vereinigung 87 (1985) p. 91-107.

Aldous in 1977 (cf. my survey article mentioned above) has given a precise mathematical formulation to the principle, thus proving at one stroke a number of the theorems mentioned above.

Banach space valued generalizations of these theorems have been considered as well. For these to be valid, the Banach space concerned, obviously cannot be arbitrary. Thus for Komlós theorem to be valid in this context, a consideration of bounded constant sequences shows that the space must have the so-called Banach-Saks property viz. inside any bounded sequence in the Banach space, there is a subsequence which converges strongly in Césaro-mean. In particular, the space has to be reflexive; however, reflexivity only would not suffice. Further information on this topic may be obtained from the above-mentioned survey article of mine.

Theorem 1 is due to Rêvész (1965; cf. my survey article above for exact reference). Corollary 1 seems to be due originally to Marcinkiewicz (cf. item [20] (1937) in the bibliography of the papers of Marcinkiewicz in his collected papers).

References.

We recall that the list of references here has been restricted to books and monographs only where further relevant bibliography can be found. Works of various authors used or referred to in the text are mentioned at the appropriate places in the Notes and Remarks of different chapters.

Bourbaki, N.
Espaces vectoriels topologiques. Masson, Paris (1981)

Chatterji, S.D.
Les martingales et leurs applications analytiques. In Lecture Notes in Mathematics, vol. 307, Springer-Verlag, Berlin (1973)

Dellacherie, C. and Meyer, P.-A.
Probabilité et potentiel (Chapitres V à VIII). Hermann, Paris (1980)

Diestel, J. and Uhl, J.J.
Vector Measures. Math. Surveys, No. 15, American Mathematical Society, Providence Rhode Island (1977)

Doob, J.L.
Stochastic processes. John Wiley and Sons, N.Y. (1953)
Classical potential theory and its probabilistic counterpart. Springer-Verlag, Berlin (1984)

Dugundi, J. and Granas, A.
Fixed point theory (vol. I). Polish Scientific Publishers, Warsaw (1982)

Dunford, N. and Schwartz, J.T.
Linear operators I. Interscience Publishers, N.Y. (1958)

Egghe, L.
Stopping time techniques for analysts and probabilists. London Math. Soc. Lecture Note Series : 100, Cambridge Univ. Press, Cambridge (1984)

Gihman, I.I. and Skorohod, A.V.
The theory of stochastic processes I. Springer-Verlag, Berlin (1974)

Guzman, Miguel de
Differentiation of integrals in $\mathbb{R}^n$. Lecture Notes in Maths. vol. 481, Springer Verlag, Berlin (1975)
Real variable methods in Fourier analysis. North. Holland, Amsterdam (1981)

Gut, A. and Schmidt, K.D.
Amarts and set function processes. Lecture Notes in Maths. vol. 1042, Springer-Verlag, Berlin (1983)

Hayes, C.A. and Pauc, C.Y.
Derivation and martingales. Springer-Verlag, Berlin (1970)

Hunt, G.A.
Martingales et processus de Markov. Dunod, Paris (1966)

Neveu, J.
Martingales à temps discret. Masson, Paris (1972)

Riesz, F. and Sz.-Nagy, B.
Functional analysis. Frederick Ungar Publishing Co., N.Y. (1955)

Saks, S.
Theory of the integral. Dover Publications Inc, N.Y. (1964; reprint of first English edition of 1937).

Probabilistic Methods
in the
Geometry of Banach Spaces

by

Gilles Pisier

Université Paris 6
Equipe d'Analyse
Tour 4610, 4e-étage
Pl. Jussieu, 75230 Paris Cedex 05
FRANCE

and

Texas A&M University
College Station
TX 77843
USA

Table of Contents

Introduction

Probabilistic methods play an important rôle in the recent developments of the geometry of Banach spaces, especially in the "local theory" which concentrates on quantitative studies of finite dimensional spaces. The aim of these lectures is to present some of these methods. We indicate in the last chapter 8 several important topics which we have omitted. The general theme of the notes is Dvoretzky's theorem and the theory of type and cotype. Recently, Maurey and the author found a very simple proof of an inequality which is the main point in the proof of Dvoretzky's theorem, if one follows the approach of Milman (cf. [Mil] [F-L-M]). In our approach, Dvoretzky's theorem can be reformulated as a theorem involving Banach space valued Gaussian random variables. This may be of interest in probabilistic applications. Actually, our proof also yields a number of inequalities for vector valued functions which seem new (cf. in particular theorem 2.2, corollary 2.4, (2.5), (2.9), and (2.14)) and are not published elsewhere. This covers chapters 1 and 2. In chapter 3 we present the main features of the theory of type and cotype. In chapter 4, we follow the same strategy used to prove Dvoretzky's theorem to show that a Banach space B is of stable p ($1 \leq p < 2$) iff B does not contain ℓ_p^n's uniformly. Here we use p-stable variables instead of Gaussian ones and we employ a deviation inequality proved by a martingale method at the end of chapter 2. We are thus able to include complete proofs of all the results of the paper [M-P] concerning the type. The results for the cotype are included without proofs in chapter 3. At the end of chapter 3, we indicate the connection between the cotype and the dimensions of the ℓ_2^n-subspaces in Dvoretzky's theorem, as first proved in [F-L-M].

In chapter 5, we present the main results concerning the notion of K-convexity which is essential to understand the relation between the type of a space B and the cotype and its dual B . This also has close connections with Dvoretzky's theorem through the notion of "locally π-euclidean" spaces, which is discussed in chapter 5 together with Banach spaces containing uniformly complemented ℓ_2^n's.

In chapter 6, we briefly compare (without proofs) the notions of type and cotype with the corresponding notions for sums of martingale differences. The martingale notions lead to a smaller class of Banach spaces (and the notion of UMD spaces defines an even smaller class)

In chapter 7, we present a new proof of some recent results (mainly from [BMW]) concerning the notion of type for a general (non linear!) metric space.

Finally in chapter 8 we briefly survey several results which can be viewed as applications of Probability Theory to the Geometry of Banach spaces, but which we chose not to develop in the present notes.

Acknowledgement: I would like to thank Susan Trussell for her outstanding typing job.

Chapter 1
Dvoretzky's theorem by Gaussian methods

In this chapter, we will prove the famous theorem of Dvoretzky on the spherical sections of convex bodies. We follow the basic approach of the paper [F-L-M], but we do not use the isoperimetric inequality on the Euclidean sphere. We use instead a Gaussian analogue which is much easier to prove (and which is discussed in detail in the next chapter). This Gaussian presentation may be of some interest for applications in the theory of Gaussian random vectors or Gaussian processes.

We start by recalling some basic facts and terminology. Let E, F be Banach spaces and let $\lambda \geq 1$. The spaces E, F will be called λ-isomorphic if there is an isomorphism $T : E \to F$ such that $\|T\|\|T^{-1}\| \leq \lambda$. We will write this $E \overset{\lambda}{\sim} F$. The so-called Banach-Mazur "distance" between E and F is defined as

$$d(E,F) = \inf\{\lambda | E \overset{\lambda}{\sim} F\}.$$

Note that if G is another Banach space, we have

$$d(E,G) \leq d(E,F)d(F,G).$$

We denote by ℓ_p^n the space $\mathbf{R}^n$ equipped with the norm $\|x\| = (\sum_1^n |x_i|^p)^{1/p}$. The space ℓ_2^n is *the* Hilbert space of dimension n (or equivalently the n dimensional Euclidean space).

Let $1 \leq p \leq \infty$, $\lambda \geq 1$. We say that a Banach space B contains ℓ_p^n's λ-uniformly if there is a sequence of subspaces $B_n \subset B$ such that $B_n \overset{\lambda}{\sim} \ell_p^n$ for all n. (It is important to note that we do *not* require that B_n is included in B_{n+1}.)

We will study the set of p's for which the preceding property holds and we will relate it to certain inequalities satisfied by the norm of B. See chapter 3 for more information on this. The following fundamental theorem of Dvoretzky is of central importance in the theory.

Theorem 1.1: For any $\epsilon > 0$, any infinite dimensional Banach space B contains ℓ_2^n's $(1+\epsilon)$-uniformly.

The original proof appeared in [D], subsequent proofs appeared in [Mi1] and [Sz], and also in [T1] combined with [K]. Actually, there is a stronger version of this theorem which can be formulated as a finite dimensional statement. (We say a "local" statement in the modern terminology). We abbreviate "finite dimensional" into f.d.

Theorem 1.2: For each $\epsilon > 0$ there is a number $\eta(\epsilon) > 0$ with the following property. Let E be a f.d. Banach space of dimension N. Then E contains a subspace F of dimension $n = [\eta(\epsilon) \log N]$ such that $F \overset{1+\epsilon}{\sim} \ell_2^n$.

Clearly, theorem 1.2 implies theorem 1.1 since dim $B = \infty$ allows us to take $E \subset B$ with dimension N arbitrarily large, we then obtain n arbitrarily large, which is enough in theorem 1.1.

Remark: As stated above, theorem 1.2 appeared in [Mil]. The paper [D] contained a worse quantitative estimate of n as a function of N. The "logarithmic" estimate in theorem 1.2 is essentially best possible. Indeed, it can be shown that for $N > 1$, $E = \ell_\infty^N$ and $\epsilon > 0$ fixed, if $F \subset E$ and $F \overset{1+\epsilon}{\sim} \ell_2^n$, then necessarily $n \leq \tilde{\eta}(\epsilon) \log N$ for some function $\tilde{\eta}$ depending only on ϵ. This shows that, in general, $\log N$ cannot be replaced by any essentially larger function of N in theorem 1.2. However, we will see that if the spaces E are "far" from the ℓ_∞^N spaces in some sense then the logarithmic estimate can be considerably improved. We now turn to Banach space valued Gaussian random variables.

We will consider random variables defined on some probabaility space $(\Omega, \mathcal{A}, \mathbf{P})$ and with values in a Banach space B. We will always assume that B is separable, so that no measurability difficulty will arise. We will say that a random variable $X : \Omega \to B$ is Gaussian if, for any ξ in B^*, the real valued r.v. $\xi(X)$ is Gaussian (symmetric). We only consider symmetric Gaussian random variables and we allow degeneracy (i.e. the zero valued variable is Gaussian). It is known (cf. [F], [LS]) that any B-valued Gaussian variable X must satisfy

$$E\|X\|^p < \infty \quad \text{for all } p < \infty.$$

We will also consider the "weak" moments of X, and for that purpose we define

$$\sigma(X) = \sup\{(\mathbf{E}\xi(X)^2)^{1/2} \mid \xi \in B^*, \|\xi\| \leq 1\}.$$

Clearly we have $\sigma(X) \leq (\mathbf{E}\|X\|^2)^{1/2}$. We introduce the "dimension" of X simply as follows

$$d(X) = \frac{\mathbf{E}\|X\|^2}{\sigma(X)^2}. \tag{1.1}$$

Note that if $B = \ell_2^N$ and if the distribution of X is the canonical Gaussian measure on $\mathbf{R}^N$, then we find $\sigma(X) = 1$ and $\mathbf{E}\|X\|^2 = N$, so that $d(X) = N$ in that case. We will say that such a r.v. is a "canonical" Gaussian r.v. with values in ℓ_2^N. The "dimension" $d(X)$ is the square of the ratio between the strong and weak second moments of X. It is important to note that it depends very much on the Banach space B in which X takes its values; if we change the norm, the dimension of the "same" random vector may change. {Note that we always have $d(X) \leq \dim B$, this is rather easy to check.}

Let us recall a well known fact which will be proved in passing in the next chapter. There is a numerical constant $\alpha > 0$ such that any Banach space valued Gaussian r.v. X satisfies

$$\alpha(\mathbf{E}\|X\|^2)^{1/2} \leq \mathbf{E}\|X\|. \tag{1.2}$$

Therefore, we have

$$\alpha^2 d(X) \leq \left(\frac{\mathbf{E}\|X\|}{\sigma(X)}\right)^2 \leq d(X). \tag{1.3}$$

We can now give the "Gaussian formulation" of Dvoeretzky's theorem.

Theorem 1.3: For each $\epsilon > 0$ there is a number $\eta'(\epsilon) > 0$ with the following property. Let X be a Gaussian r.v. with values in a Banach space B. Then B contains a subspace F of dimension $n = [\eta'(\epsilon)d(X)]$ which is $(1+\epsilon)$-isomorphic to ℓ_2^n. Conversely, if B contains a subspace F with $F \overset{1+\epsilon}{\sim} \ell_2^n$, then there is a B-valued Gaussian r.v. X such that $d(X) \geq (1+\epsilon)^{-2}n$.

By this result, the problem to find ℓ_2^n-subspaces in B is reduced to produce B-valued Gaussian r.v.'s with dimension essentially as large as n.

The subspace F in theorem 1.3 will be obtained below in a very simple way: for n as above we will consider a sequence $X_1, \ldots, X_n$ of independent copies of the r.v. X defined on some probability space $(\Omega, \mathcal{A}, \mathbf{P})$. Let $M = \mathbf{E}\|X\|$. We will then show that there is a subset $\Omega_0 \subset \Omega$ with $\mathbf{P}(\Omega_0) > 0$ such that for all ω in Ω_0 we have $\forall (\alpha_i) \in \mathbf{R}^n$

$$(1+\epsilon)^{-1/2}\Big(\sum |\alpha_i|^2\Big)^{1/2} \leq M^{-1}\Big\|\sum_1^n \alpha_i X_i(\omega)\Big\| \leq (1+\epsilon)^{1/2}\Big(\sum |\alpha_i|^2\Big)^{1/2}.$$

We may then take for F the span of $\{X_1(\omega), \ldots, X_n(\omega)\}$ for ω in Ω_0.

Remark 1.4: The converse part of theorem 1.3 is very easy. Indeed, if $F \subset B$ and $F \overset{1+\epsilon}{\sim} \ell_2^n$ let $T : \ell_2^n \to F$ be such that $\|T\|\|T^{-1}\| \leq 1+\epsilon$. Let X_1 be a canonical Gaussian r.v. with values in ℓ_2^n, and let $X = T(X)$. Then $\sigma(X) \leq \|T\|$ and $\mathbf{E}\|X\|^2 \geq \|T^{-1}\|^{-2}n$ so that $d(X) \geq (1+\epsilon)^{-2}n$.

The main tool for the proof of theorem 1.2 is the following inequality which originates in the work of C. Borell [Bo]; Borell deduced a similar result from Paul Lévy's isoperimetric inequality on the Euclidean sphere, we give a direct proof (together with various related results) in the next chapter.

Theorem 1.5: Let X be a B-valued Gaussian r.v.

$$\forall t > 0 \quad \mathbf{P}\{|\,\|X\| - \mathbf{E}\|X\|\,| > t\mathbf{E}\|X\|\} \leq 2\exp -Kt^2 d(X) \tag{1.4}$$

where K is a numerical constant, $K = 2\pi^{-2}$.

Apart from this result, we will need two elementary (but a bit technical) lemmas which aim at "discretizing" the problem of finding an ℓ_2^n-subspace.

The first lemma shows that to prove that two norms are $(1+\epsilon)$-equivalent on $\mathbf{R}^n$, it is enough to check it on a δ-net in the unit sphere of one of the norms, with δ depending only on ϵ.

Lemma 1.6: For each $\epsilon > 0$, there is a $\delta = \delta(\epsilon) > 0$ with the following property. Let n be any integer and let $|||\ \ |||$ be a norm on $\mathbf{R}^n$. Let S be a δ-net in the unit sphere of $(\mathbf{R}^n, |||\ \ |||)$ and let $x_1, \ldots, x_n$ be elements of a Banach space B. If $\forall \alpha \in S$ $1-\delta \leq \|\sum_1^n \alpha_i x_i\| \leq 1+\delta$, then

$$\forall \alpha \in R^n \quad (1+\epsilon)^{-1/2}|||\alpha||| \leq \|\sum_1^n \alpha_i x_i\| \leq (1+\epsilon)^{1/2}|||\alpha|||. \tag{1.5}$$

In particular, if F is the subspace of B spanned by $x_1, \ldots, x_n$, then F is $(1+\epsilon)$-isomorphic to $(\mathbf{R}^n, |||\ \ |||)$.

Proof: Assume $|||\alpha||| = 1$. There is y^0, in S such that $|||\alpha - y^0||| \leq \delta$ hence $\alpha = y^0 + \lambda_1 \alpha'$ with $|\lambda_1| \leq \delta$ and $|||\alpha'||| = 1$. Continuing this process we obtain $\alpha = y^0 + \lambda_1 y^1 + \lambda_2 y^2 + \ldots$ with $y^j \in S$, and $|\lambda_j| \leq \delta^j$. It follows that

$$\|\sum \alpha_i x_i\| \leq \sum_{j \geq 0} \delta^j \|\sum_{i=1}^{n} y_i^j x_i\|$$
$$\leq (1+\delta)(1-\delta)^{-1}.$$

Similarly $\|\sum \alpha_i x_i\| \geq 1 - \delta - \delta(1+\delta)(1-\delta)^{-1} = (1-3\delta)(1-\delta)^{-1}$. Hence, if δ is chosen small enough so that $(1-3\delta)(1-\delta)^{-1} \geq (1+\epsilon)^{-1/2}$ and $(1+\delta)(1-\delta)^{-1} \leq (1+\epsilon)^{1/2}$, we obtain (by homogeneity) the announced result (1.5). Note that we can obtain a suitable $\delta > 0$ depending only on ϵ. The last part of lemma 1.6 is obvious.

The following lemma is a classical fact.

Lemma 1.7: Let $|||\ \ |||$ be any norm on $\mathbf{R}^n$. Then there is a δ-net S in $\{x \in \mathbf{R}^n |\ |||x||| = 1\}$ with cardinality

$$|S| \leq (1 + 2/\delta)^n.$$

Proof: Let $K = \{x \in \mathbf{R}^n |\ |||x||| \leq 1\}$. Let $(y_i)_{i<m}$ be a maximal subset of ∂K. The balls $y_i + 2^{-1}\delta K$ are mutually disjoint and are all contained in $(1+2/\delta)K$. By comparing the volumes, we obtain $m(\delta/2)^n \leq (1+2/\delta)^n$ which proves the lemma.

Proof of Theorem 1.3: Let $X_1, X_2, \ldots, X_n$ be a sequence of i.i.d copies of X. We will prove that with positive probability the vectors $(X_1(\omega), \ldots, X_n(\omega))$ span a subspace F which is $(1+\epsilon)$-isomorphic to ℓ_2^n. To prove this, we apply the two preceding lemmas in the case when $|||\alpha||| = \left(\sum |\alpha_i|^2\right)^{1/2}$ is the Euclidean norm on $\mathbf{R}^n$. Assume that $\sum |\alpha_i|^2 = 1$. Then the r.v. $\sum_1^n \alpha_i X_i$ has the same distribution as X. This is a basic property of Gaussian r.v.'s. Therefore $\mathbf{E}\|\sum \alpha_i X_i\| = \mathbf{E}\|X\|$ and by (1.14)

$$\mathbf{P}\{|\ \|\sum \alpha_i X_i\| = \mathbf{E}\|X\|\ | > \delta \mathbf{E}\|X\|\} \leq 2\exp - K\delta^2 d(X).$$

Let $M = \mathbf{E}\|X\|$ and let S be as in lemmas 1.6 and 1.7. The preceding inequality implies

$$\text{(1.6)} \qquad \mathbf{P}\{\exists \alpha \in S | M^{-1}\|\sum \alpha_i X_i\| - 1| > \delta\} \leq 2|S| \exp - K\delta^2 d(X)$$

Hence by lemma 1.7

$$\leq 2\exp\left\{\frac{2n}{\delta} - K\delta^2 d(X)\right\},$$

where δ is the function of ϵ given by lemma 1.6. This yields the desired estimate. Indeed, observe that if (1.7) $\frac{2n}{\delta} \leq \frac{1}{2}K\delta^2 d(X)$, the probability (1.6) is not greater than $2\exp - \frac{1}{2}K\delta^2 d(X)$; moreover, we can always assume that $d(X)$ is large enough (otherwise there is nothing to prove) so that

$$2\exp - \frac{1}{2}K\delta^2 d(X) < 1.$$

theorem 1.3. We then obtain that with positive probability we have:

$$\forall \alpha \in S \quad |M^{-1}\|\sum \alpha_i X_i(\omega)\| - 1| \leq \delta.$$

By Lemma 1.6, this implies that $x_i = X_i(\omega)$ satisfies (1.5). Hence if F_ω is the span of $(X_1(\omega), \ldots, X_n(\omega))$, we have proved that with positive probability we have $F_\omega \overset{1+\epsilon}{\sim} \ell_2^n$, as long as n satisfies (1.7). This allows to take n as large as $[4^{-1}K\delta^3 d(X)]$. This concludes the proof of theorem 1.3.

Remark: Note that (1.6) is valid for any $\delta > 0$, therefore, a simple computation shows that, for some numerical constant C_1, we have

$$\left(\mathbf{E} \sup_{\alpha \in S} |M^{-1}\|\sum \alpha_i X_i\| - 1|^2\right)^{1/2} \leq C_1 \delta^{-1}(n/d(X))^{1/2}$$

and if $n \leq \eta'(\epsilon) d(X)$

$$\leq C_1 \delta^{-1} \eta'(\epsilon)^{1/2}.$$

Hence, by lemma 1.6, we can certainly adjust the function $\eta'(\epsilon)$ so that if $n \leq \eta'(\epsilon) d(X)$ we have

$$\left(\mathbf{E} \sup\left\{\left\|\sum_1^n \alpha_i X_i\right\|^2 \,\Big|\, \sum |\alpha_i|^2 = 1\right\}\right)^{1/2} \leq (1+\epsilon)^{1/2} \mathbf{E}\|X\|.$$

A similar remark applies of course to the lower bound.

Remark: The preceding proof yields a function $\eta'(\epsilon)$ which is of order ϵ^3 when $\epsilon \to 0$. Actually, looking more carefully into the proof, one finds $\eta'(\epsilon)$ is like $\epsilon^2 |\log \epsilon|^{-1}$ when $\epsilon \to 0$ as in [FLM]. We should mention that recently Gordon [Go1] obtained $\eta'(\epsilon)$ of order ϵ^2, which cannot be improved. Gordon's proof uses a refinement of a classical lemma of Slepian on Gaussian processes. We refer to [Go1] and [Go2] for this interesting approach.

We now turn to the proof of theorem 1.2. For this all we need is the following lemma which essentially goes back to a well known paper of Dvoretzky-Rogers [DR].

Lemma 1.8: Let E be an N-dimensional space. Let $\overline{N} = N/2$ if N is even, $\overline{N} = \frac{N-1}{2}$ if N is odd. Then there are $\overline{N}$ elements $(x_i)_{i \leq \overline{N}}$ in E satisfying

$$\text{(1.8)} \qquad \forall (\alpha_i) \in \mathbf{R}^{\overline{N}} \quad \|\sum \alpha_i x_i\| \leq (\sum |\alpha_i|^2)^{1/2} \quad \text{and}$$

$$\text{(1.9)} \qquad \|x_i\| \geq 1/2 \quad \text{for all} \quad i \leq \overline{N}.$$

With the help of this lemma it is easy to deduce theorem 1.2 from theorem 1.3.

Proof of Theorem 1.2: Let (g_i) be a sequence of i.i.d. standard Gaussian r.v.'s as before. Let then $(x_i)_{i\leq N}$ be given by lemma 1.8. Let

$$X = \sum_{i\leq N} g_i x_i.$$

Then by (1.8) we have $\sigma(X) \leq 1$. On the other hand, we have

$$\mathbf{E} \sup_{i\leq N} \|g_i x_i\| \leq \mathbf{E}\|X\|. \tag{1.10}$$

Indeed, let (A_j) be a partition of our probability space such that

$$A_j \subset \{\omega \mid |g_j(\omega)| \, \|x_j\| = \sup_i \|g_i(\omega)\|\},$$

for all j. Then

$$\sup \|g_i x_i\| = \|\sum g_j 1_{A_j} x_j\|$$

so that

$$\begin{aligned}\mathbf{E}\sup \|g_i x_i\| &= \mathbf{E}\|\sum g_j 1_{A_j} x_j\| \\ &\leq \mathbf{E}\|\sum g_j x_j\|\end{aligned}$$

(the last line follows from the triangle inequality and the fact that, by symmetry, $\sum g_j(1-21_{A_j})x_j$ has the same distribution as X.) This establishes (1.10). Now (1.9) implies

$$\frac{1}{2}\mathbf{E} \sup_{i\leq N} |g_i| \leq \mathbf{E}\|X\|.$$

By an elementary and well known computation, there is a numerical constant $C > 0$ such that $\frac{1}{C}(\log n)^{1/2} \geq \mathbf{E} \sup_{i\leq n} |g_i| \geq C(\log n)^{1/2}$ for all n. Therefore we conclude that

$$d(X) \geq c'(\log N)^{1/2}$$

for some numerical constant c', so that we may apply theorem 1.3 to obtain theorem 1.2.

Proof of lemma 1.8: This is often proved using the "maximum volume ellipsoid" contained in the unit ball of E. The idea of this ellipsoid goes back to F. John. We will use the language of determinants rather than volume but the idea here is similar. We follow [L]. Given our N-dimensional space E, we claim that there is an operator $u : \ell_2^N \to E$ such that $\|u\| \leq 1$ and $\|u_{|F}\| \geq \frac{\dim F}{N}$ for all subspaces $F \subset \ell_2^N$. In particular $\|u\| = 1$. To prove this claim consider any determinant function (associated to any fixed matrix representation) $v \to \det(v)$. Let u be an operator such that $\det(u) = \max\{\det(v) | v : \ell_2^N \to E \quad \|v\| \leq 1\}$. Then, for $\epsilon > 0$, we have, for any $v : \ell_2^N \to E$

$$\det(u + \epsilon v) \leq \det(u)\|u + \epsilon v\|^N. \tag{1.11}$$

Now

$$\det(u+\epsilon v) = \det(u)\det(1+\epsilon u^{-1}v)$$
$$= \det(u)(1+\epsilon\, tr(u^{-1}v)+o(\epsilon))$$

hence substituting in (1.11) we get

$$1+\epsilon\, tr(u^{-1}v) \leq \|u+\epsilon v\|^N + o(\epsilon)$$
$$\leq (1+\epsilon\|v\|)^N + o(\epsilon)$$
$$\leq 1 + N\epsilon\|v\| + o(\epsilon)$$

so that

$$tr\ u^{-1}v \leq N\|v\|.$$

Now consider a subspace $F \subset \ell_2^N$ and let $P : \ell_2^N \to F$ be the orthogonal projection. Taking $v = uP$ we find

$$\dim F = tr\ P = tr(u^{-1}v) \leq N\|v\|$$

so that

$$\|uP\| \geq \dim F/N.$$

We can now easily complete the proof of lemma 1.8. Let u be as claimed above. Choose y_1 in ℓ_2^N such that $\|y_1\| = 1$ $\|u(y_1)\| \geq \|u\| = 1$. Then choose $y_2 \in y_1^\perp$ such that $\|y_2\| = 1$ and $\|uy_2\| \geq \|uP\| \geq (1-\frac{1}{N})$ where P is the projection onto $\{y_1\}^\perp$. Continuing in this way and using the preceding claim, one constructs a sequence such that $y_i \in \{y_1,\ldots,y_{i-1}\}^\perp$ $\|y_i\| = 1$ and $\|uy_i\| \geq (N-i+1)/N$. It is then easy to check that $x_i = u(y_i)$ $(i = 1,2,\ldots \overline{N})$ has the properties announced in lemma 1.8.

Chapter 2
Deviation inequalities for random vectors

In this chapter, we will show several upper bounds for the deviation of the norm — or a more general function — of a random vector. In the first part, we concentrate on the Gaussian case, while in the second part we treat sums of independent random vectors by a martingale method. Throughout the notes we abbreviate random variable by r.v..

Let us first reformulate theorem 1.5.

Theorem 2.1: Let X be a Gaussian r.v. with values in a Banach space B. Then for all $t > 0$

$$\mathbf{P}\{|\,\|X\| - \mathbf{E}\|X\|\,| > t\} \leq 2\ \exp -Kt^2\sigma(X)^{-2}$$

with $K = 2\pi^{-2}$.

A similar result has been known for some time through the work of C. Borell [B]. Borell proved an analogue of the isoperimetric inequality for Gaussian measures using the isoperimetric inequality on the euclidean sphere and a classical limiting argument. This approach leads to $K = 1/2$ (but with the median of $\|X\|$ instead of its mean) in the above inequality. For an interesting development of the "isoperimetric approach" in the Gaussian case, see [Eh]. On the other hand, Maurey found a proof of theorem 2.1 based on Ito's formula which yields $K = 1$ (see the remark p. below). We present a different approach, quite direct and elementary to prove theorem 2.1. Our approach has the advantage to apply to more general functions of X than the norm of X (and even vector valued functions) provided a suitable bound is known for the gradients of the functions. Moreover, $\pi/2$ is the best constant in (2.1) and (2.5) for $p = 1$

We first consider finite dimensional Banach spaces E, F and a locally lipschitzian function $f : E \to F$. The function f has in almost every point x a derivative $f'(x)$ which is a linear map from E into F. For y in E we denote by $f'(x) \cdot y$ the value of $f'(x)$ on y, so that

$$f'(x) \cdot y = \lim_{t \to 0} t^{-1}(f(x + ty) - f(x)).$$

The next result can be viewed as a Sobolev inequality for vector valued functions of a Gaussian variable. (Of course it is essential that the dimensions of E or F do not appear in (2.1)).

Theorem 2.2: Let X be an E-valued Gaussian random vector and let f be as above. Let Y be an independent copy of X. Then, for any measurable convex function $\Phi : F \to \mathbf{R}$ we have

$$\mathbf{E}\Phi(f(X) - \mathbf{E}f(X)) \leq \mathbf{E}\Phi(\frac{\pi}{2} f'(X) \cdot Y) \tag{2.1}$$

Proof: The proof below is a simplification, due to Maurey, of my original proof which used an expansion in Hermite polynomials (analogous to the proof of Lemma 7.3 below). Let

$X(\theta) = X\sin\theta + Y\cos\theta$ for $0 \leq \theta < 2\pi$. and $X'(\theta) = X\cos\theta - Y\sin\theta$. We first fix X and Y. Then clearly

$$f(X) - f(Y) = \int_0^{\pi/2} \frac{d}{d\theta}(f(X(\theta)))d\theta$$
$$= \int_0^{\pi/2} f'(X(\theta)) \cdot X'(\theta) d\theta$$

By the convexity of Φ, we have

$$\mathbf{E}\Phi(f(X) - f(Y)) \leq \frac{2}{\pi}\int_0^{\pi/2} \mathbf{E}\Phi(\frac{\pi}{2}f'(X(\theta)) \cdot X'(\theta))d\theta. \tag{2.2}$$

Now we observe that for each fixed θ the couple $(X(\theta), X'(\theta))$ has the same distribution as the couple (X, Y). This follows from the invariance of Gaussian measures under orthogonal transformations. In particular we have for all θ

$$\mathbf{E}\Phi(\frac{\pi}{2}f'(X(\theta)) \cdot X'(\theta)) = \mathbf{E}\Phi(\frac{\pi}{2}f'(X) \cdot Y),$$

therefore (2.1) follows from (2.2).

Remark: Note that the convexity of Φ implies

$$\mathbf{E}\Phi(f(X) - \mathbf{E}f(X)) \leq \mathbf{E}\Phi(f(X) - f(Y)), \tag{2.3}$$

indeed we may integrate with respect to Y only, inside Φ.

Remark: Since the dimension of E and F do not appear, it is easy to reformulate (2.1) in an infinite dimensional setting. We leave this to the reader.

Suppose now that E is the euclidean space $\mathbf{R}^N$. Let us denote by $|\ \ |$ the euclidean norm. Let f be a real valued Lipschitzian function on E. Let

$$\|f\|_{\text{Lip}} = \text{ess sup}\{|f'(x)|\ |x \in E\}$$
$$= \sup\left\{\frac{|f(x) - f(y)|}{|x - y|}\ |\ x, y \in E\right\}.$$

For each λ in $\mathbf{R}$, we consider the function

$$\forall t \in \mathbf{R} \quad \Phi_\lambda(t) = \exp \lambda t.$$

By theorem 2.2 and (2.3) we have

$$\mathbf{E}\Phi_\lambda(f(X) - \mathbf{E}f(X)) \leq \mathbf{E}\Phi_\lambda(\frac{\pi}{2}f'(X) \cdot Y)$$
$$= \mathbf{E}\exp\frac{1}{2}\left(\frac{\lambda\pi}{2}\right)^2 \|f\|^2_{\text{Lip}}.$$

Therefore, for $t > 0$, $\lambda > 0$

$$\mathbf{P}\{f(X) - \mathbf{E}f(X) > t\} \leq \exp\left(-\lambda t + \frac{1}{2}\left(\frac{\lambda\pi}{2}\right)^2 \|f\|^2_{\text{Lip}}\right).$$

By choosing the optimal λ, we find

$$\mathbf{P}\{f(X) - \mathbf{E}f(X) > t\} \leq \exp -Kt^2\|f\|^{-2}_{\text{Lip}}$$

with $K = 2\pi^{-2}$. Since this holds also for $-f$ instead of f, we have obtained the following

Corollary 2.3: Let $E = \mathbf{R}^N$ with the euclidean norm. let $f : E \to \mathbf{R}$ be lipschitzian. We have

$$\forall t > 0 \quad \mathbf{P}\{|f(X) - \mathbf{E}f(X)| > t\} \le \exp - Kt^2 \|f\|_{\mathrm{Lip}}^{-2}.$$

We can now explain how theorem 2.1 (and hence theorem 1.5) is derived.

Proof of Theorem 2.1: Let us first recall that any Gaussian r.v. X with values in a separable Banach space B admits an almost surely convergent expansion of the form $X(\omega) = \sum_{n=1}^{\infty} g_n(\omega) z_n$ with coefficients z_n in B, where (g_n) is an i.i.d. sequence of standard Gaussian r.v.'s. Here is a brief sketch of a proof of this well known fact: let H be the closure in L_2 of the variables of the form $\xi(X)$ with ξ in B^*. Then H is entirely formed of Gaussian variables. Let (g_n) be any orthonormal basis of H and let $\mathcal{A}_N$ be the σ-algebra generated by $(g_1, \dots g_N)$. It is easy to check that $\mathbf{E}(X|\mathcal{A}_N)$ is of the form $\sum_1^N g_i z_i$ with $z_i = \mathbf{E}(X g_i)$. Since we know that $X \in L^1(B)$, the martingale convergence theorem implies that $\sum_1^N g_i z_i$ converges a.s. and in $L^1(B)$ to X. This yields the announced representation of X as a series. This shows that it is enough to prove theorem 2.1 for X of the form $X = \sum_1^N g_i z_i$. But that case follows from corollary 2.3. Indeed, let

$$\forall \alpha \in \mathbf{R^N} \quad f(\alpha) = \|\sum_1^N \alpha_i z_i\|.$$

We have $\forall \alpha, \beta \in \mathbf{R}^N$

$$|f(\alpha) - f(\beta)| \le |\alpha - \beta| \sigma(X)$$

since

$$\sigma(X) = \sup\{(\sum |\xi(z_i)|^2)^{1/2} \mid \xi \in B^*,\ \|\xi\| \le 1\} = \sup\{\|\sum \alpha_i z_i\| \mid \sum \alpha_i^2 \le 1\}.$$

Therefore $f : \mathbf{R}^N \to \mathbf{R}$ satisfies $\|f\|_{\mathrm{Lip}} \le \sigma(X)$. We have $\|X\| = f(g_1, \dots g_N)$, therefore theorem 2.1 follows from Corollary 2.3 applied to the canonical Gaussian distribution on $\mathbf{R}^N$.

Let us state another application of theorem 2.2

Corollary 2.4: Let $f : \mathbf{R}^N \to F$ be a locally Lipschitzian function. Let γ_N be the canonical Gaussian probability measure on $\mathbf{R}^N$. Assume $\int \|f\| d\gamma_N < \infty$ and $\int f d\gamma_N = 0$. Then for all $p \ge 1$

$$\int \|f\|^p d\gamma_N \le \left(\frac{\pi}{2}\right)^p \int \|\sum_1^N \frac{\partial f}{\partial x_i}(x) y_i\|^p d\gamma_N(x) d\gamma_N(y). \tag{2.4}$$

This follows from (2.1) and (2.3) for the function $\Phi(z) = \|z\|^p$ for z in F.

Remark: If $F = \mathbf{R}$, the right side of (2.4) is equal to

$$\left(\frac{\pi}{2}\right)^p \mathbf{E}|g_1|^p \int \left(\sum \left|\frac{\partial f}{\partial x_i}(x)\right|^2\right)^{p/2} \gamma_N(dx)$$

so that (2.4) implies if $F = \mathbf{R}$

$$\|f\|_{L^p(d\gamma_N)} \le \frac{\pi}{2}\|g_1\|_p \left\| \left(\sum \left|\frac{\partial f}{\partial x_i}\right|^2\right)^{1/2} \right\|_{L^p(d\gamma_N)} \tag{2.5}$$

In passing, we can deduce from (2.5) a known inequality which is usually derived from the integrability theorem of [F] [LS].

Corollary 2.5: For $0 < q < p < \infty$, there is a constant $K(p,q)$ such that any Banach space valued Gaussian r.v. X satisfies

$$(\mathbf{E}\|X\|^p)^{1/p} \le K(p,q)(\mathbf{E}\|X\|^q)^{1/q}$$

and moreover $K(p,q)$ is $O(\sqrt{p})$ when q remains fixed and $p \to \infty$.

Proof: By the previous discussion, we may assume w.l.o.g. that $X = \sum_1^N g_i z_i$, with z_i in a space B. We consider the case $q = 1$ first. We have

$$\begin{aligned} \forall \xi \in B^* \quad \left(\sum |\xi(z_i)|^2\right)^{1/2} \left(\frac{2}{\pi}\right)^{1/2} &= \mathbf{E}|\xi(X)| \\ &\le \|\xi\|\mathbf{E}\|X\| \end{aligned}$$

since $\mathbf{E}|g_i| = (2/\pi)^{1/2}$. Hence

$$\sigma(X) \le \left(\frac{\pi}{2}\right)^{1/2} \mathbf{E}\|X\|. \tag{2.6}$$

For each $\alpha = (\alpha_i) \in \mathbf{R}^N$ let

$$f(\alpha) = \|\sum_1^N \alpha_i z_i\| - \int \|\sum \alpha_i z_i\| \gamma_N(d\alpha).$$

We have a.e.

$$\begin{aligned} \left(\sum \left|\frac{\partial f}{\partial \alpha_i}\right|^2\right)^{1/2} &\le \sup\{\|\sum \alpha_i z_i\| \mid \sum \alpha_i^2 \le 1\} \\ &= \sigma(X). \end{aligned}$$

Hence by (2.5) and (2.6)

$$\|f\|_{L^p(d\gamma_N)} \le \left(\frac{\pi}{2}\right)^{3/2} \|g_1\|_p \mathbf{E}\|X\|,$$

and by the triangle inequality

$$(\mathbf{E}\|X\|^p)^{1/p} \le (1 + \left(\frac{\pi}{2}\right)^{3/2} \|g_1\|_p)\mathbf{E}\|X\|. \tag{2.7}$$

Since $\|g_1\|_p$ is $O(\sqrt{p})$ when $p \to \infty$, this establishes corollary 2.5 when $q = 1$. If $0 < q < 1$, define θ by $\frac{1}{1} = \frac{\theta}{q} + \frac{1-\theta}{p}$, we have

$$\mathbf{E}\|X\| \le (\mathbf{E}\|X\|^q)^{\theta/q}(\mathbf{E}\|X\|^p)^{(1-\theta)/p}$$

hence substituting in (2.7) and dividing by $(\mathbf{E}\|X\|^p)^{(1-\theta)/p}$ we find

$$(\mathbf{E}\|X\|^p)^{1/p} \leq (1 + \left(\frac{\pi}{2}\right)^{3/2} \|g_1\|_p)^{1/\theta}(\mathbf{E}\|X\|^q)^{1/q}. \qquad q.e.d$$

Remark: It is possible to state a refinement of theorem 2.2 using the Ornstein-Uhlenbeck semi-group which we can introduce as follows. Let $0 \leq \epsilon \leq 1$. Let X be a Gaussian r.v. with values in a Banach space B and let Y be an independent copy of X, so that $\epsilon X + (1-\epsilon^2)^{1/2} Y \stackrel{d}{=} X$ (this sign means equality in distribution). Let then $f : B \to F$ be a measurable function with values in another Banach space F. We assume that $f(X)$ is in $L^1(F)$. We can then define

$$(T(\epsilon)f)(X) = \mathbf{E}_Y f(\epsilon X + (1-\epsilon^2)^{1/2} Y)$$

where we have denoted by $\mathbf{E}_Y$ the expectation with respect to Y (equivalently $\mathbf{E}_Y$ is the conditional expectation with respect to the σ-algebra generated by X). Let γ be the distribution of X on B. Clearly $T(\epsilon)$ is a bounded operator of norm 1 on $L^1(B,\gamma;F)$. It is easy to modify the proof of theorem 2.2 to obtain, for all Φ as in theorem 2.2, for $0 < \theta < \pi/2$

$$\mathbf{E}\Phi(f(X) - f(X(\theta))) \leq \mathbf{E}\Phi((\pi/2 - \theta) f'(X) \cdot Y),$$

hence integrating over Y inside the left hand side we find

$$\mathbf{E}\Phi(f(X) - (T(\sin\theta)f)(X)) \leq \mathbf{E}\Phi((\pi/2 - \theta) f'(X) \cdot Y),$$

and this yields immediately, for all $p \geq 1$ and all $0 < \epsilon < 1$.

$$\mathbf{E}\|f(X) - (T(\epsilon)f)(X)\|^p \leq (C(1-\epsilon)^{1/2})^p \mathbf{E}\|f'(X) \cdot Y\|^p \tag{2.8}$$

for some numerical constant C.

The Ornstein-Uhlenbeck semi-group T_t can be defined simply by setting

$$T_t = T(e^{-t/2}).$$

Then (2.8) yields the following interesting inequality (which is of interest when $t \to 0$):

$$\|f - T_t f\|_{L_p(B,\gamma;F)} \leq C\sqrt{t}\|f'(x) \cdot y\|_{L_p(\gamma\times\gamma;F)}. \tag{2.9}$$

Recall that F is an arbitrary Banach space in the above. In the case $F = \mathbf{R}$, some related stronger inequalities are proved in [Me] if $1 < p < \infty$.

Remark: B. Maurey found a proof of theorem 2.1 with the best constant $K = 1/2$. His proof uses stochastic integrals and apparently does not extend to the setting of theorem 2.2. Let us briefly indicate his argument. Let γ be the canonical Gaussian measure on $\mathbf{R}^N$. Let $(B_t)_{t>0}$ be the Brownian motion generated by γ so that γ is the distribution of B_1. Let $(P_t)_{t>0}$ be the associated Markovian semi-group. The argument is based on the following martingale inequality. Let $0 < t_0 < t_1 < t_2 < \ldots < t_n \leq 1$. Let $\mathcal{A}_k$ be the σ-algebra generated by $\{B_{t_0}, \ldots, B_{t_k}\}$. Let

(V_k) be a sequence of $\mathbf{R}^N$-valued r.v.'s such that V_k is $\mathcal{A}_{k-1}$-measurable for all k. Let us denote again by $|\ |$ the Euclidean norm on $\mathbf{R}^N$. Fix a positive number σ. Assume that $|V_k| \leq \sigma$ a.s. for all k. Let $S_n = \sum_{k=1}^n < V_k, B_{t_k} - B_{t_{k-1}} >$. A simple computation shows that, for any λ real

$$\begin{aligned}\mathbf{E}e^{\lambda S_n} &= \mathbf{E}(e^{\lambda S_{n-1}} \exp \frac{\lambda^2}{2}(t_n - t_{n-1})|V_n|^2) \\ &\leq \mathbf{E}e^{\lambda S_{n-1}} \exp \frac{\lambda^2}{2}(t_n - t_{n-1})\sigma^2\end{aligned}$$

Therefore

$$\mathbf{E}e^{\lambda S_n} \leq \exp \frac{\lambda^2}{2}\sigma^2. \tag{2.10}$$

Now if $f : \mathbf{R}^N \to \mathbf{R}$ is sufficiently smooth, one can check using Ito's formula that

$$f(B_1) - \mathbf{E}f(B_1) = \int_0^1 \operatorname{grad}(P_{1-t}f)(B_t) \cdot dB_t$$

Assume then $|f(x) - f(y)| \leq \sigma|x - y|$ for all x, y in $\mathbf{R}^N$. Then $P_{1-t}f$ satisfies the same inequality so that $|\operatorname{grad} P_{1-t}f| \leq \sigma$ a.e.. Then we can easily deduce from (2.10) the following which implies theorem 2.1 with $K = 1/2$

$$\mathbf{P}\{|f(B_1) - \mathbf{E}f(B_1)| > t\} \leq 2\exp -\frac{1}{2}\left(\frac{t}{\sigma}\right)^2.$$

Remark: Let $\phi : \mathbf{R}^N \to \mathbf{R}^N$ be a (not necessarily linear) map such that

$$\forall \alpha, \beta \in \mathbf{R}^N \quad |\phi(\alpha) - \phi(\beta)| \leq \ell|\alpha - \beta| . \tag{2.11}$$

Let γ_N be the canonical Gaussian probability on $\mathbf{R}^N$ and let $\lambda = \phi(\gamma_N)$ be the image measure. Let $f : \mathbf{R}^N \to F$ and let $\Phi : F \to \mathbf{R}$ be as in theorem 2.2. Assume $\int f d\lambda = 0$. Then applying (2.1) to $f \circ \phi$ instead of f we obtain

$$\int \Phi(f) d\lambda \leq \int\int \Phi\left(\frac{\pi}{2} f'(\phi(x)) \circ \phi'(x) \cdot y\right) \gamma_N(dx)\gamma_N(dy). \tag{2.12}$$

Now since $\phi'(x)$ is a linear operator of norm $\leq \ell$ on $\mathbf{R}^N$, it is easy to check that for all x

$$\int \Phi\left(\frac{\pi}{2} f'(\phi(x))\phi'(x) \cdot y\right) \gamma_N(dy) \leq \int \Phi\left(\frac{\pi}{2} \ell f'(\phi(x)) \cdot y\right) \gamma_N(dy).$$

(This follows from the fact that the extreme points in the set of linear operators $A : \mathbf{R}^N \to \mathbf{R}^N$ of norms ≤ 1 are the orthogonal transformations and they leave γ_N invariant). Hence (2.12) implies

$$\int \Phi(f) d\lambda \leq \int\int \Phi\left(\frac{\pi}{2} \ell f'(x) \cdot y\right) d\lambda(x) d\gamma_N(y).$$

This suggests that the class of measures λ of the form $\lambda = \phi(\gamma_N)$ for some contraction $\phi : \mathbf{R}^N \to \mathbf{R}^N$ is worthwhile to investigate. In particular, it would be interesting to characterize the symmetric probability measures on $\mathbf{R}^N$ which are of this form. Here is a simple but interesting example. Let $\phi : \mathbf{R}^N \to \mathbf{R}^N$ be defined by

$$\phi(t_1, \ldots, t_N) = (\psi(t_1), \ldots, \psi(t_N))$$

with ψ defined by

$$\psi(t) = (2\pi)^{-1/2} \int_{-\infty}^{t} e^{-u^2/2} du . \tag{2.13}$$

In other words,

$$\psi(t) = \gamma_1(]-\infty, t]),$$

It is clear from (2.13) that $|\psi'(t)| \leq (2\pi)^{-1/2}$, so that ϕ' is a diagonal matrix with $|\frac{\partial \psi_i}{\partial t_i}| \leq (2\pi)^{-1/2}$. Therefore ϕ satisfies (2.11) with $\ell = (2\pi)^{-1/2}$. Moreover, the image measure $\phi(\gamma_N)$ is nothing but the normalized Lebesgue measure m on the unit cube $[0,1]^N$. Hence we find if $\int f dm = 0$

$$\int \Phi(f) dm \leq \int \int \Phi\left(\frac{1}{2}\left(\frac{\pi}{2}\right)^{1/2} f'(x) \cdot y\right) dm(x) d\gamma_N(y). \tag{2.14}$$

We do not know if a similar inequality is satisfied when m is the normalized probability measure on $\{0,1\}^N$. This question is studied in chapter 7 below.

In the second part of this chapter, we will prove some deviation inequalities for Banach space valued r.v.'s of the form $X = \sum X_i$ with X_i independent. We use a simple but powerful idea due to Yourinski. His idea was developed and refined in [Ku], [A1]. It can be described as a general principle to estimate the distribution of $\|X\| - \mathbf{E}\|X\|$ when X is of the above form. Let us be more precise. Let (X_i) be a sequence of independent r.v.'s on a probability space $(\Omega, \mathcal{A}, \mathbf{P})$ with values in a space B. We assume that $\|X_i\|$ is integrable. Let $\mathcal{F}_i$ be the σ-algebra generated by $\{X_1, \ldots, X_i\}$, and let $\mathcal{F}_0$ be the trivial σ-algebra. Let $X = \sum_{i=1}^n X_i$. Then we can write $\mathbf{E}\|X\| = \mathbf{E}^{\mathcal{F}_0}\|X\|$ and

$$\|X\| - \mathbf{E}\|X\| = \sum_{i=1}^{n} d_i$$

with $d_i = \mathbf{E}^{\mathcal{F}_i}\|X\| - \mathbf{E}^{\mathcal{F}_{i-1}}\|X\|$. The crucial point is the following observation:

$$|d_i(\omega)| \leq \int \|X_i(\omega) - X_i(\omega')\| d\mathbf{P}(\omega'). \tag{2.15}$$

To prove this we may assume w.l.o.g. that X_j is a function of the j-th coordinate ω_j on the product probability space $(\Omega, \mathbf{P})^{\mathbf{N}}$. Then we have

$$\begin{aligned} d_i(\omega_1, \ldots, \omega_i) &= \int \|\sum_j X_j(\omega_j)\| d\mathbf{P}(\omega_{i+1}) \ldots d\mathbf{P}(\omega_n) \\ &\quad - \int \|\sum_{j<i} X_j(\omega_j) + X_i(\omega_i') + \sum_{j>i} X_j(\omega_j)\| d\mathbf{P}(\omega_i') d\mathbf{P}(\omega_{i+1}) \ldots d\mathbf{P}(\omega_n). \end{aligned} \tag{2.16}$$

By the triangle inequality,

$$\left| \|\sum X_j(\omega_j)\| - \|\sum_{j<i} X_j(\omega_j) + X_i(\omega_i') + \sum_{j>i} X_j(\omega_j)\| \right| \leq \|X_i(\omega_i) - X_i(\omega_i')\|$$

so that (2.15) follows from (2.16). Note that (2.15) implies immediately

$$|d_i(\omega)| \leq \|X_i(\omega)\| + \mathbf{E}\|X_i\|. \tag{2.17}$$

This method has numerous consequences, for instance (cf. [A1]) we have

$$\mathbf{E}|\ \|\sum X_i\| - \mathbf{E}\|\sum X_i\|\ |^2 \leq 4\sum \mathbf{E}\|X_i\|^2.$$

Indeed, since (d_i) is a sequence of martingale differences, we have

$$\begin{aligned} \mathbf{E}|\sum d_i|^2 &= \mathbf{E}\sum d_i^2 \quad \text{and by (2.17)} \\ &\leq 4\sum \mathbf{E}\|X_i\|^2. \end{aligned}$$

Similarly, by the Burkholder-Gundy inequalities (cf. [Bu]) if $1 \leq p < \infty$ there is a constant $C(p)$ such that

$$\|\ \|\sum X_i\| - \mathbf{E}\|\sum X_i\|\ \|_p \leq C(p)\|(\sum \|X_i\|^2)^{1/2}\|_p.$$

This principle can be used also when X_i is uniformly bounded. In that case let

$$\|X_i\|_\infty = \text{ess sup}\|X_i(\omega)\|.$$

Recall the following known inequality.

Lemma 2.6: Let d_i be a sequence of real valued martingale differences with $|d_i| \leq \alpha_i$ for each i. Then for all $t > 0$

$$\mathbf{P}\{|\sum d_i > t\} \leq 2\exp - \frac{t^2}{2\sum \alpha_i^2}.$$

Proof: (Sketch) Let λ be any real number. For all $-1 \leq x \leq 1$

$$\exp \lambda x \leq \text{ch } \lambda + x \text{ sh } \lambda.$$

(This is easy to check from the convexity of $x \to \exp \lambda x$ and $x = (1)\frac{1+x}{2} + (-1)\frac{1-x}{2}$.) Therefore any real valued r.v. ϕ such that $|\phi| \leq 1$ and $\mathbf{E}\phi = 0$ satisfies

$$\mathbf{E}\exp \lambda\phi \leq \text{ch}\lambda \leq \exp \lambda^2/2. \tag{2.18}$$

Now assume that (M_i) is a real valued martingale with respect to an increasing sequence of σ-algebras $(\mathcal{F}_i)$ and let $d_i = M_i - M_{i-1}$. Clearly, (2.18) implies for all i

$$\mathbf{E}^{\mathcal{F}_{i-1}} \exp \lambda M_i \leq \exp \lambda M_{i-1} \exp \frac{\lambda^2\|d_i\|_\infty^2}{2}$$

hence

$$\mathbf{E}\exp \lambda M_i \leq \mathbf{E}\exp \lambda M_{i-1} \exp \frac{\lambda^2}{2}\|d_i\|_\infty^2$$

which implies

$$\mathbf{E}\exp \lambda M_n \leq \exp \frac{\lambda^2}{2}\sum \|d_i\|_\infty^2.$$

It is then very easy to conclude. *q.e.d.*

With the same notation as before, (2.17) implies $\|d_i\|_\infty \le 2\|X_i\|_\infty$ hence Lemma 2.6 implies

$$\mathbf{P}\{|\ \|\sum X_i\| - \mathbf{E}\|\sum X_i\|\ | > t\} \le 2\exp - \frac{t^2}{8\sum \|X_i\|_\infty^2}. \tag{2.19}$$

On the other hand, we have trivially

$$\|\sum X_i\|_\infty \le \sum \|X_i\|_\infty. \tag{2.20}$$

We will now "interpolate" between (2.19) and (2.20) to obtain an inequality which will be needed in chapter 4.

Lemma 2.7: Let $1 < p < 2 < p' < \infty$ $\frac{1}{p} + \frac{1}{p'} = 1$. Let $X_1, \ldots, X_n$ be independent bounded B-valued r.v.'s. Then for all $t > 0$

$$\mathbf{P}\{|\ \|\sum X_i\| - \mathbf{E}\|\sum X_i\|\ | > t\} \le 2\exp - K_p \left(\frac{t}{\sup_{i\ge 1} i^{1/p}\|X_i\|_\infty}\right)^{p'}, \tag{2.21}$$

where K_p is a constant depending only on p.

Proof: By homogeneity we may assume that

$$\sup_i i^{1/p}\|X_i\|_\infty = 1.$$

Consider t of the form $t_m = 1 + 2^{-1/p} + \ldots + m^{-1/p} \sim p' m^{1/p'}$. By the triangle inequality, we have

$$\begin{aligned} |\ \|\sum X_i\| - \mathbf{E}\|\sum X_i\|\ | &\le |\ \|\sum_{i>m} X_i\| - \mathbf{E}\|\sum_{i>m} X_i\|\ | + \|\sum_{i\le m} X_i\| + \mathbf{E}\|\sum_{i\le m} X_i\| \\ &\le |\ \|\sum_{i>m} X_i\| - \mathbf{E}\|\sum_{i>m} X_i\|\ | + 2t_m. \end{aligned}$$

Let $X = \sum X_i$, and $R_m = \sum_{i>m} X_i$, then

$$\mathbf{P}\{|\ \|X\| - \mathbf{E}\|X\|\ | > 3t_m\} \le \mathbf{P}\{|\ \|R_m\| - \mathbf{E}\|R_m\|\ | > t_m\} \quad \text{hence by (2.19)}$$

$$\le 2\exp - t_m^2 8^{-1} \left(\sum_{i>m} i^{-2/p}\right)^{-1}$$

and since $\sum_{i>m} i^{-2/p} \le C m^{1-2/p}$ for some constant C we find the latter probability less than $2\exp - C'(t_m)^{p'}$ for some constant C'. This yields (2.21) for $t = t_m$ for some m; for t arbitrary we select m such that $t_m < t \le t_{m+1}$ and (2.21) follows.

Remark: In particular, if (Y_k) is a sequence of independent B-valued r.v.'s with $\|Y_k\|_\infty \le 1$ and if (β_k) satisfies $|\beta_k| \le k^{-1/p}$ for all k, then the variable $S = \sum \beta_k Y_k$ satisfies for all $t > 0$

$$\mathbf{P}\{|\ \|S\| - \mathbf{E}\|S\|\ | > t\} \le 2\exp - K_p t^{p'} \tag{2.22}$$

Chapter 3
Type and Cotype

In this chapter, we review the basic results of the theory of type and cotype.

We first introduce more notation. Let $D = \{-1,+1\}^{\mathbb{N}}$ and let μ be the uniform probability measure on D. We denote by $\epsilon_n : D \to \{-1,+1\}$ the n-th coordinate on D. Thus, the sequence (ϵ_n) is an i.i.d. sequence of symmetric $\{+1,-1\}$-valued random variables. Let B be a Banach space and let (Ω, m) be any measure space. When there is no ambiguity, we will denote simply by $\| \ \|_p$ the "norm" in the space $L_p(\Omega, m; B)$, for $0 < p \leq \infty$.

Definitions: i) Let $1 \leq p \leq 2$. A Banach space B is called of type p if there is a constant C such that, for all finite sequences (x_i) in B

$$(3.1) \qquad \|\sum \epsilon_i x_i\|_2 \leq C(\sum \|x_i\|^p)^{1/p}.$$

We denote by $T_p(B)$ the smallest constant C for which (3.1) holds.

ii) Let $2 \leq q \leq \infty$. A Banach space B is called of cotype q if there is a constant C such that for all finite sequences (x_i) in B

$$(3.2) \qquad \left(\sum \|x_i\|^q\right)^{1/q} \leq C\|\sum \epsilon_i x_i\|_2.$$

We denote by $C_q(B)$ the smallest constant C for which (3.2) holds. Clearly, if $p_1 \leq p_2$ then type $p_2 \Rightarrow$ type p_1 while cotype $p_1 \Rightarrow$ cotype p_2. Let us immediately observe that every Banach space is of type 1 and of cotype ∞ with constants equal to 1. In some cases this cannot be improved, for instance if $B = \ell_1$ it is easy to see that (3.1) holds for no $p > 1$. Similarly, if $B = \ell_\infty$ or c_0, then (3.2) holds for no $q < \infty$. We will clarify this question below. On the other hand, it is easy to see that if B is a Hilbert space then

$$\forall\, x_1, \ldots, x_n \in B \quad \|\sum \epsilon_i x_i\|_2 = \left(\sum \|x_i\|^2\right)^{1/2}.$$

Therefore a Hilbert space is of type 2 and cotype 2 (with constants 1). Since type and cotype are obviously isomorphic notions it follows that any space B which is isomorphic to a Hilbert space is of type 2 and cotype 2. It is a striking result of Kwapień [Kw] that the converse is true: if B is of type 2 and cotype 2, then B must be isomorphic to a Hilbert space.

Remark: Actually, the choice of the norm in $L_2(D, \mu, B)$ plays an inessential rôle in the above definitions. This follows from an inequality of Kahane [Ka] (for a simple proof see [P1]). For any $0 < r < \infty$, there are constants $A_r > 0$ and B_r such that any finite sequence (x_i) in any Banach space B satisfies

$$(3.3) \qquad A_r\|\sum \epsilon_i x_i\|_2 \leq \|\sum \epsilon_i x_i\|_r \leq B_r\|\sum \epsilon_i x_i\|_2.$$

In the case $B = \mathbf{R}$, the inequality (3.3) reduces to a classical inequality due to Khintchine. These inequalities make it very easy to analyze the type and cotype of the L_p-spaces: If $1 \le p \le 2$, every L_p-space is of type p and of cotype 2. If $2 \le p < \infty$, an L_p-space is of type 2 and of cotype p. These results are essentially best possible. The space L_∞ contains isometrically any separable Banach space, in particular ℓ_1 and c_0 which we mentioned earlier. Therefore, L_∞ is of type 1 and cotype ∞ and nothing more.

Using the above inequality (3.3), one can easily generalize the preceding observation. Let B be a Banach space of type p and of cotype q. Let (Ω, m) be any measure space and consider the space $L_r(\Omega, m; B)$. Then this space is of type $r \wedge p$ and of cotype $r \vee q$. Similar ideas lead to the following result which shows how to use type and cotype to study sums of independent random variables (cf. [H-J]).

Proposition 3.1: Let $(\Omega, \mathcal{A}, \mathbf{P})$ be a probability space. Let (Y_n) be a sequence of independent mean zero random variables with values in a Banach space B. Assume that B is of type p and cotype q, and that the series $\sum Y_n$ is a.s. convergent. Then for $0 < r < \infty$, we have

$$\alpha \mathbf{E}\left(\sum \|Y_n\|^q\right)^{r/q} \le \mathbf{E}\|\sum Y_n\|^r \le \beta \mathbf{E}\left(\sum \|Y_n\|^p\right)^{r/p}$$

where $\alpha > 0$ and β are constants depending only on B, p, q, and r.

Proof: Assume first that each Y_n is symmetric. Consider the sequence $(\varepsilon_n Y_n)_{n\ge 1}$ defined on $(D \times \Omega, \mu \times \mathbf{P})$. This sequence has the same distribution as $(Y_n)_{n\ge1}$. It is therefore easy to deduce proposition 3.1 in that case from (3.1), (3.2) and (3.3). The general case follows by an easy symmetrization argument. *q.e.d.*

In particular, taking $r = p$ (resp. $r = q$) in proposition 3.1 we find

(3.4) $$\mathbf{E}\|\sum Y_n\|^p \le \beta \sum \mathbf{E}\|Y_n\|^p$$

[resp. (3.5) $$\alpha \sum \mathbf{E}\|Y_n\|^q \le \mathbf{E}\|\sum Y_n\|^q.]$$

Let (g_n) be an i.i.d. sequence of standard Gaussian random variables. We may apply proposition 3.1 with Y_n of the form $Y_n = g_n x_n$ with x_n in B. We thus find that if B is of type p (resp. cotype q) then there is a constant $C' \le T_p(B)$ (resp. $C' \le C_q(B)$) such that

(3.1)' $$\|\sum g_i x_i\|_2 \le C' \left(\sum \|x_i\|^p\right)^{1/p}$$

[resp. (3.2)' $$(\sum \|x_i\|^q)^{1/q} \le C' \|\sum g_i x_i\|_2).]$$

Conversely, it can be shown that if a space B satisfies (3.1)' (resp. (3.2)') then B is of type p (resp. cotype q). For (3.1)' this is an immediate consequence of the next proposition. For (3.2)' the proof is a bit more delicate, we refer the interested reader to Corollary 1.3 in [MP].

Proposition 3.2: Let $r \geq 1$. Let (φ_n) be a sequence of i.i.d. symmetric real valued r.v.'s and let x_n be a sequence in a Banach space B.

(i) If B is arbitrary, we have

$$\|\sum_1^n \epsilon_i x_i\|_r \leq (\mathbf{E}|\phi_1|)^{-1}\|\sum_1^n \phi_i x_i\|_r.$$

(ii) On the other hand, if B is of cotype q_0 for some $q_0 < \infty$ and if $s > \max\{r, q_0\}$, then we have

$$\|\sum \phi_i x_i\|_r \leq C\|\phi_1\|_s\|\sum \epsilon_i x_i\|_r$$

where C is a constant depending only on $\{r, q_0, s\}$ and on B.

Proof: (i) Just observe that the sequence $(\epsilon_i|\phi_i|)$ has the same distribution as (ϕ_i) (assuming that (ϵ_i) and (ϕ_i) are independent of each other). Then we find that $\sum \epsilon_i x_i \mathbf{E}|\phi_i|$ can be obtained from $\sum \varphi_i x_i$ simply by integrating over $|\phi_i|$. Hence, by Jensen's inequality

$$\|\sum \epsilon_i x_i \mathbf{E}|\phi_i|\ \|_r \leq \|\sum \epsilon_i|\phi_i|x_i\|_r = \|\sum \phi_i x_i\|_r,$$

which concludes the proof of (i) since $\mathbf{E}|\phi_i| = \mathbf{E}|\phi_1|$.

We now turn to (ii). By the preceding discussion, we know that $L_r(B)$ is of cotype q for $q = \max\{r, q_0\}$. We will first show a preliminary result. Let $\{A^1, \ldots, A^m\}$ be a partition of some probability space into sets of probability $1/m$. Let $\{1_{A_i^1}, \ldots, 1_{A_i^m}\}$ be a sequence of independent copies of the m-tuple $\{1_{A^1}, \ldots, 1_{A^m}\}$. Let us assume that $\{1_{A_i^j}\}$ is independent of the sequence $\{\epsilon_i\}$. Let $\psi_i = \epsilon_i 1_{A_i^1}$.

We have then for some constant K

$$(*) \qquad \|\sum \psi_i x_i\|_r \leq Km^{-1/q}\|\sum \epsilon_i x_i\|_r .$$

Indeed, to prove $(*)$ let us introduce another copy $\{\epsilon'_i\}$ of the sequence $\{\epsilon_i\}$ and independent of all the other sequences. Since $L_r(B)$ is of cotype q (with constant K say) we have

$$\left(\sum_1^m \|\sum_i \epsilon_i 1_{A_i^j} x_i\|_r^q\right)^{1/q} < K\left\|\sum_{j=1}^m \epsilon'_j \left(\sum_i \epsilon_i 1_{A_i^j} x_i\right)\right\|_r .$$

the left hand side of the preceding is equal to $m^{1/q}\|\sum \psi_i x_i\|_r$ while the right hand side is majorized (since $|\sum \epsilon'_j 1_{A_i^j}| \leq 1$) by $K\|\sum \epsilon_i x_i\|_r$. Therefore this inequality implies $(*)$.

Note that we may drop in $(*)$ the assumption that $\mathbf{P}(A^1)$ is the inverse of an integer by changing the constant K. We can now conclude rather easily. We assume as earlier that $\{|\phi_i|\}$ and $\{\epsilon_i\}$ are independent of each other. Observe that

$$(**) \qquad |\phi_i| = \int_0^\infty 1_{\{|\phi_i| > t\}}\, dt .$$

Therefore, applying $(*)$ in the particular case $A' = \{|\phi_1| > t\}$ we obtain, for each $t > 0$

$$\|\sum \epsilon_i 1_{\{|\phi_i|>t\}} x_i\|_r \leq K(\mathbf{P}\{|\phi_1| > t\})^{1/q} \|\sum \epsilon_i x_i\|_r \,.$$

Hence, using $(**)$ and Jensen's inequality, we find

$$\|\sum \epsilon_i |\phi_i| x_i\|_r \leq K \int_0^\infty (\mathbf{P}\{|\phi_1| > t\})^{1/q} dt \|\sum \epsilon_i x_i\|_r$$

Finally, since (ϕ_i) and $(\epsilon_i|\phi_i|)$ have the same distribution and since (easy calculation) for each $s > q$ there is a constant $C(s,q)$ such that

$$\int_0^\infty (\mathbf{P}\{|\phi_1| > t\})^{1/q} dt \leq C(s,q) \|\phi_1\|_s \,,$$

we obtain the conclusion of part (ii). *q.e.d*

Remark: Proposition 3.2 ii) was first obtained in [MP]. The preceding simple and direct argument was shown to me some years ago by S. Kwapień.

Remark: The preceding proof remains valid assuming merely that (ϕ_i) is a sign invariant sequence (i.e. $(\epsilon_i \phi_i)$ and (ϕ_i) have the same distribution for every choice of signs $\epsilon_i = \pm 1$), and replacing $\mathbf{E}|\phi_1|$ by $\inf \mathbf{E}|\phi_i|$.

The notions of type and cotype have appeared in various problems involving the analysis of vector valued functions or random variables. One of the great advantages of the classification of Banach spaces in terms of type and cotype is the existence of a rather satisfactory "geometric" characterization of these notions.

We first explain the characterization of spaces which have a non trivial type or a non trivial cotype.

Theorem 3.3: [MP] Let B be a Banach space.
i) B is of type p for some $p > 1$ iff B does not contain ℓ_1^n's uniformly.
ii) B is of cotype q for some $q < \infty$ iff B does not contain ℓ_∞^n's uniformly.

Remark 3.4: In such results, the "only if" part is trivial. Indeed, assume B contains ℓ_p^n's λ-uniformly for some $\lambda > 1$. This means $\forall n\ \exists x_1, \ldots, x_n \in B$ such that

$$\forall\ (\alpha_i) \in \mathbf{R}^n \quad \left(\sum |\alpha_i|^p\right)^{1/p} \leq \|\sum_1^n \alpha_i x_i\| \leq \lambda \left(\sum |\alpha_i|^p\right)^{1/p}.$$

Hence we have

$$n^{1/p} \leq \|\sum \epsilon_i x_i\|_2 \leq \lambda n^{1/p}.$$

and

$$n^{1/r} \leq \left(\sum \|x_i\|^r\right)^{1/r} \leq \lambda n^{1/r}.$$

Therefore B cannot be of type $r > p$ or of cotype $r < p$. In particular if $p = 1$ (resp. $p = \infty$) B cannot have a nontrivial type (resp. cotype).

Actually theorem 3.3 can be extended as follows: Let $1 \le p_0 < 2 < q_0 < \infty$. A space B is of type p for some $p > p_0$ iff B does not contain $\ell^n_{p_0}$'s uniformly. We will state a similar result using the type and cotype indices which are defined as follows.

(3.6) $$p(B) = \sup\{p \mid B \text{ is of type } p\}$$

(3.7) $$q(B) = \inf\{q \mid B \text{ is of cotype } q\}.$$

Then the main theorem relating the type and cotype of B to the geometry of B is

Theorem 3.5: [MP] [K] Let B be an infinite dimensional Banach space. Then for each $\epsilon > 0$ B contains ℓ^n_p's $(1+\epsilon)$-uniformly both for $p = p(B)$ and for $p = q(B)$.

Remark 3.6: Krivine proved [K] that if a Banach space B contains ℓ^n_p's $(1+\epsilon)$-uniformly for *some* $\epsilon > 0$ then it also contains them $(1+\epsilon)$-uniformly for *all* $\epsilon > 0$. Therefore, from now on we simply say in that case that B contains ℓ^n_p's uniformly. For $p = 1$ or $p = \infty$, this result goes back to James and for $p = 2$ to Dvoretzky. For a proof in the case $1 \le p < 2$ see the next chapter. For $1 < p < \infty$, it is unknown whether any space isomorphic to ℓ_p contains a subspace $(1+\epsilon)$-isomorphic to ℓ_p for each $\epsilon > 0$. This is known as the "distortion problem". This would be an infinite dimensional analogue of Krivine's result.

By theorem 3.5 and remark 3.4, we have

(3.8) $$p(B) = \inf\{p \mid B \text{ contains } \ell^n_p\text{'s uniformly}\}$$

(3.9) $$q(B) = \sup\{p \mid B \text{ contains } \ell^n_p\text{'s uniformly}\}.$$

Therefore the set of p's for which B is of type p (or of cotype p) is closely related to the set of p's for which B contains ℓ^n_p's uniformly. The properties of the stable laws (see the next chapter) show that if $1 \le p_1 \le p_2 \le 2$, any space containing $\ell^n_{p_1}$'s uniformly must also contain $\ell^n_{p_2}$'s uniformly. Therefore, in general the set of p's for which B contains ℓ^n_p's uniformly contains the interval [p(B), 2] and the point $q(B)$; nothing more can be added in general.

We will prove theorems 3.3 and 3.5, but only the part concerning the "type", in the next chapter.

Remark: The duality between the type of B and the cotype of B^* will be discussed in chapter 5. In a *different* direction, the reader should note that $p(B) > 1$ implies $q(B) < \infty$. Moreover, $p(B) > 1$ iff $p(B^*) > 1$. These statements follow easily from theorem 3.3. Indeed, if we note that ℓ_1^n embeds isometrically (in the real case) into $\ell_\infty^{2^n}$, we immediately see that B contains ℓ_1^n's uniformly as soon as it contains ℓ_∞^n's uniformly. This shows that $p(B) > 1$ implies $q(B) < \infty$. Similarly, it is easy to see that B contains ℓ_1^n's uniformly iff its dual B^* also does. We leave this as an exercise to the reader (use the fact that it is the same to embed ℓ_1^n in a quotient of B^* or in B^* itself.) Moreover, it is rather easy to show that B is of type p (resp. cotype q) iff its bidual B^{**} has the same property. For various quantitative results related to the preceding remarks see [KT] and [T2].

It should be pointed out that theorems 3.3 and 3.5 have a weak point, they characterize only the index $p(B)$ (or $q(B)$) but they do not distinguish when the supremum (or the infimum) is attained in (3.6) (or (3.7)). For instance, a space B can satisfy $p(B) = q(B) = 2$ without being isomorphic to a Hilbert space. This problem does not arise for (3.8) (or (3.9)) since, by Krivine's theorem (see remark 3.6) the set of p's for which B contains ℓ_p^n's uniformly is a closed subset of $\mathbf{R}$.

For the "general" spaces, theorem 3.3 and 3.5 seem to be difficult to improve. But, for more classical concrete spaces, the type and cotype has been completely elucidated.

For instance (cf. [TJ]) the Schatten classes C_p or the more general non-commutative L_p-spaces (cf. [F]) have been treated. Their type and cotype is the same as for the usual L_p spaces (see above). Also, J. Bourgain [B1] proved that the space L_1/H^1 or the space $(H^\infty)^*$ is of cotype 2 (we mean here the spaces relative to the one dimensional torus).

In another direction, the case of Banach lattices is also completely elucidated, cf. [M1]. Here are the main results in that case (which includes Orlicz spaces, Lorentz spaces, etc.). Let us consider a Banach lattice B which is a sublattice of the lattice of all measurable functions on a measure space (Ω, m). Then if $x_1, \ldots, x_n$ are elements of B and if $0 < p < \infty$, the function $(\sum |x_i|^p)^{1/p}$ is well defined as a measurable function and is also in B (by the lattice property).

Maurey proved a Banach lattice generalization of Khintchine's inequality which reduces the study of type and cotype for lattices to some very simple "deterministic" inequalities.

Theorem 3.7 [M1]: Let B be a Banach lattice as above. Assume $q(B) < \infty$. Then there is a constant β depending only on B such that for all $x_1, \ldots x_n$ in B we have

$$\frac{1}{\sqrt{2}}\|(\sum |x_i|^2)^{1/2}\| \leq \|\sum \epsilon_i x_i\|_2 \leq \beta\|(\sum |x_i|^2)^{1/2}\|. \tag{3.10}$$

Note: The left side of (3.10) holds in any Banach lattice B; it follows from Khintchine's inequality for which the best constant $1/\sqrt{2}$ was found by Szarek.

It follows immediately that B (as above) is of type p (resp. cotype q) iff there is a constant C such that any finite sequence (x_i) in B satisfies

$$\|(\sum |x_i|^2)^{1/2}\| \leq C(\sum \|x_i\|^p)^{1/p}$$

$$\text{resp.}\quad \left(\sum \|x_i\|^q\right)^{1/q} \le C\left\|\left(\sum |x_i|^2\right)^{1/2}\right\|.\ )$$

In the case $p < 2$ (or $q > 2$), one can even obtain a much simpler result as shown by the following result of Maurey [M1].

Theorem 3.8: Let B be a Banach lattice as above.

(i) Let $2 < q < \infty$. Then B is of cotype q iff there is a constant C such that any sequence (x_i) of *disjointly supported* elements of B satisfies

$$\left(\sum \|x_i\|^q\right)^{1/q} \le C\left\|\sum x_i\right\|.$$

(ii) Assume $q(B) < \infty$. Let $1 < p < 2$. Then B is of type p iff there is a constant C such that any sequence (x_i) of disjointly supported elements satisfies

$$\left\|\sum x_i\right\| \le C\left(\sum \|x_i\|^p\right)^{1/p}.$$

Remark: For $q = 2$ (or $p = 2$) the preceding statement is false, the Lorentz spaces $L^{2,1}$ (or $L^{2,q}$ for $2 < q < \infty$) provide counterexamples.

Note that for a disjointly supported sequence (x_i) we have

$$\left\|\sum |x_i|\,\right\| = \|\sup |x_i|\,\| = \left\|\sum x_i\right\|.$$

Remark: In the particular case of Banach lattices, type and cotype are closely connected to the moduli of uniform smoothness or uniform convexity. This is investigated in great detail in the paper [F].

We should mention that there are several relatively natural spaces for which the type or cotype is not well understood. For instance, by [TJ] the projective tensor product $\ell_2\hat{\otimes}\ell_2$ is of cotype 2, but is unknown whether $\ell_2\hat{\otimes}\ell_2\hat{\otimes}\ell_2$ is also of cotype 2. Similarly the fact that L^1/H^1 is of cotype 2 (cf. [B1]) is open in dimensions greater than 1. It is also unknown whether the dual J^* of the classical James space J is of cotype 2 or any finite cotype. Also unknown is the cotype of the space $(C^1(\mathbf{T}^2))^*$, which is the dual of the space of continuously differentiable functions on the two-dimensional torus $\mathbf{T}^2$.

The notions of type and cotype are also closely related to the dimensions of the spherical sections of convex bodies, or equivalently the dimensions of the ℓ_2^n-subspaces of a given Banach space. This was discovered in the paper [FLM].

The main connection is given in the next result, which can be viewed as a refinement of theorems 1.1 and 1.2.

Theorem 3.9: [FLM] Let $2 \le q < \infty$. Let B be a Banach space of cotype q. For each $\epsilon > 0$ there is a number $\eta_B(\epsilon) > 0$ depending only on B, q and $\epsilon > 0$ with the following property:
Let E be any N-dimensional subspace of B. Then E contains a subspace F of dimension $n = [\eta_B(\epsilon)N^{2/q}]$ such that $F \overset{1+\epsilon}{\sim} \ell_2^n$.

Remark: Fix $\epsilon > 0$. For any Banach space B, let $\phi_\epsilon^B(N)$ be the greatest number n such that any subspace $E \subset B$ of dimension N contains a subspace $F \subset E$ of dimension n such that $F \overset{1+\epsilon}{\sim} \ell_2^n$. The preceding result combined with theorem 3.3 ii) has the following rather surprising consequence: if B is infinite dimensional, then either $\phi_\epsilon^B(N)$ is $O(\log N)$ when $N \to \infty$, or there is a $\delta > 0$ such that

$$\liminf_{N\to\infty} \phi_\epsilon^B(N) N^{-\delta} > 0.$$

In other words, either $\phi_\epsilon^B(N)$ grows like $\log N$ or it grows faster than a positive power of N. To prove this note that if the second case fails then (by theorems 3.9 and 3.3) B must contain ℓ_∞^n's uniformly, and as we have already mentioned, $\log N$ is the best possible estimate when $E = \ell_\infty^N$.

Remark 3.10: Actually the preceding remark can be pushed further. One can prove that for any fixed $\epsilon > 0$, we have

$$\frac{2}{q(B)} = \lim_{N\to\infty} \frac{\log \phi_\epsilon^B(N)}{\log N}. \tag{3.11}$$

Indeed, by theorem 3.9 we have $\phi_\epsilon^B(N) \geq [\eta_B(\epsilon) N^{2/q}]$ whenever B is of cotype q. Hence

$$\liminf_{N\to\infty} \frac{\log \phi_\epsilon^B(N)}{\log N} \geq \frac{2}{q(B)}.$$

For the converse, note that by theorem 3.5, B contains ℓ_q^n's uniformly for $q = q(B)$; moreover it is known (cf. [FLM] for more details) that there is a constant C'_ϵ such that

$$\phi_\epsilon^{\ell_q^N}(N) \leq C'_\epsilon N^{2/q}$$

for all N. It follows that

$$\phi_\epsilon^B(N) \leq C'_\epsilon N^{2/q}$$

so that

$$\limsup_{N\to\infty} \frac{\log \phi_\epsilon^B(N)}{\log N} \leq \frac{2}{q}$$

Thus we have checked (3.11).

Proof of theorem 3.9: By lemma 1.8, there is a sequence $\{x_i \mid i \leq \overline{N}\}$ in E such that if

$$X = \sum_1^{\overline{N}} g_i x_i$$

we have $\sigma(X) \leq 1$ and $\|x_i\| \geq 1/2$ for all $i \leq \overline{N}$. By the definition of cotype q spaces, there is a constant C such that

$$\left(\sum \|x_i\|^q\right)^{1/q} \leq C(\mathbf{E}\|X\|^2)^{1/2}$$

(we can take $C = C_q(B)$ here). Hence

$$(\mathbf{E}\|X\|^2)^{1/2} \geq (2C)^{-1}\overline{N}^{1/q}$$

so that

$$d(X) \geq (2C)^{-2}\overline{N}^{2/q}.$$

By theorem 1.3, we immediately obtain the conclusion of theorem 3.9.

Chapter 4
Stable Type

In this chapter, we discuss the notion of stable type p which is closely related to the usual notion of type p. This will enable us to prove the statements concerning the type in theorems 3.3 and 3.5. Let $0 < p \leq 2$. A real valued r.v. θ will be called p-stable if for some $\sigma \geq 0$ its Fourier transform satisfies

$$(4.1) \qquad \forall t \in \mathbf{R} \quad \mathbf{E}\exp(it\theta) = \exp -\frac{\sigma^p|t|^p}{2}.$$

The number σ will be called the parameter of θ. We will call θ standard if $\sigma = 1$. Of course if $p = 2$ we recover Gaussian variables, and σ^2 is the variance of θ. However, the case $p < 2$ is quite different from $p = 2$ in many respects. For instance, if $p < 2$ then $\mathbf{E}|\theta|^p = \infty$ (unless $\sigma = 0$, which means $\theta \equiv 0$) and the "best" that one can say is that $\mathbf{P}\{|\theta| > t\}$ is $O(t^{-p})$ when $t \to \infty$, which implies $\mathbf{E}|\theta|^r < \infty$ for all $r < p$. Actually, it is known that

$$(4.2) \qquad \lim_{t\to\infty} t^p \mathbf{P}\{|\theta| > t\} = c_p\sigma^p$$

where c_p is a positive constant depending only on p. The main property of such variables is their "stability": Let (θ_n) be an i.i.d. sequence of real valued p-stable r.v.'s then for any finite sequence of real numbers (α_i) we have

$$\sum_1^n \alpha_i\theta_i \stackrel{d}{=} (\sum_1^n |\alpha_i|^p)^{1/p}\, \theta_1$$

where we have denoted by $\stackrel{d}{=}$ equality in distribution. This implies that if $\sum_1^\infty |\alpha_n|^p < \infty$ then $\sum_1^\infty \alpha_n\theta_n$ converges a.s. and has the same distribution as $(\sum |\alpha_n|^p)^{1/p}\theta_1$.

Now let B be a separable Banach space over $\mathbf{R}$. A B-valued r.v. X will be called p-stable if, for all ξ in B^*, the real valued r.v. $\xi(X)$ is p-stable. The "stability" property still holds: if $X_1, \ldots, X_n, \ldots$ are i.i.d. copies of X, then we have

$$\sum_1^n \alpha_i X_i \stackrel{d}{=} (\sum_1^n |\alpha_i|^p)^{1/p}\, X_1$$

and similarly, for infinite series. For example, let (θ_n) be as above, let (x_n) be a sequence in B such that the series $X = \sum_1^\infty \theta_n x_n$ is a.s. convergent. Then X is an example of a p-stable r.v. with values in B. In the Gaussian case, we have seen that any 2-stable r.v. can be represented in this way, but this is no longer true for p-stable r.v.'s when $p < 2$. We will use a different series representation which is specific to the case $p < 2$.

First, we should mention that for any p-stable B-valued r.v. X, there is a measure Q on B such that $\int \|x\|^p Q(dx) < \infty$ and such that the Fourier transform of X admits the following representation

$$\forall \xi \in B^* \quad \mathbf{E}\exp i\xi(X) = \exp - \int |\xi(x)|^p Q(dx).$$

It is sometimes convenient to replace the measure $\|x\|^p Q(dx)$ by its image Q_1 by the map $x \to \frac{x}{\|x\|}$. We then obtain a measure Q_1 concentrated on the unit sphere of B with total mass

$$Q_1(B) = \int \|x\|^p Q(dx),$$

and satisfying

$$\forall \xi \in B' \quad \mathbf{E} \exp i\xi(X) = \exp - \int |\xi(x)|^p Q_1(dx).$$

We will say that Q (or Q_1) is a spectral measure for X. This representation has been rediscovered by many authors, we refer for instance to [AG], th. 3.6.16, p. 149.

In the case when X is an a.s. convergent series of the form $X = \sum_1^\infty \theta_n x_n$, then we can take $Q = \sum \delta_{x_n}$. Indeed, we have necessarily $\sup_n |\theta_n| \, \|x_n\| < \infty$ a.s. and then an easy computation using (4.2) leads to $\sum \|x_n\|^p < \infty$. We have then $Q_1 = \sum \|x_n\|^p \, \delta_{x_n \|x_n\|^{-1}}$. In that case, we have a discrete spectral measure. Note that by symmetrization, we may always assume that Q or Q_1 are symmetric.

The following representation of p-stable variables will be very useful. It goes back to the work of P. Lévy and was revived recently by R. LePage (see the paper [LWZ]). We first need more notation.

Let (A_k) be an i.i.d. sequence of exponential r.v.'s i.e. such that $\mathbf{P}(A_k > t) = e^{-t}$ for $t > 0$. Let $\Gamma_j = \sum_1^j A_k$. The sequence $(\Gamma_j)_{j \geq 1}$ is nothing else but the successive times of the jumps of a standard Poisson process. The representation is then as follows.

Theorem 4.1: Let B be a separable Banach space. Let Q be a symmetric probability measure on B such that $\int \|x\|^p Q(dx) < \infty$. Let (Y_j) be an i.i.d. sequence of B-valued r.v.'s with distribution equal to Q. Then if the series

$$X = \sum_1^\infty \Gamma_j^{-1/p} Y_j \tag{4.3}$$

converges a.s., it defines a p-stable B-valued r.v.. Moreover, we have for all ξ in B^*

$$\lim_{t \to \infty} t^p \mathbf{P}\{|\xi(X)| > t\} = \mathbf{E}|\xi(Y_1)|^p = \int |\xi(x)|^p Q(dx). \tag{4.4}$$

Therefore the measure $\frac{1}{2c_p} Q$ is a spectral measure for X, where c_p is the constant in (4.2).

Remark: The preceding result gives a representation of any p-stable r.v. X_1 with values in B. Indeed, by a suitable normalization, we may assume w.l.o.g. that X_1 admits a symmetric probability Q concentrated on the unit sphere of B as its spectral measure. Let X be as in (4.3). Then Q is a common spectral measure for both X_1 and $(2c_p)^{1/p} X$, which means equivalently that $X_1 \stackrel{d}{=} (2c_p)^{1/p} X$. The fact that the series defining X converges a.s. follows from a well known result of Ito-Nisio (cf. [IN], or [HJ2]).

For the proof we will need

Lemma 4.2: Let $(\Gamma_{j1})_{j\geq 1}$ and $(\Gamma_{j2})_{j\geq 1}$ be two independent copies of the sequence $(\Gamma_j)_{j\geq 1}$. Let $a_1 > 0$ $a_2 > 0$. Consider the non decreasing rearrangement of the countable set $\{\frac{1}{a_1}\Gamma_{j1} \mid j \geq 1\} \cup \{\frac{1}{a_2}\Gamma_{j2} \mid j \geq 1\}$, and denote it by $(\lambda_j^*)_{j\geq 1}$. Then the sequence $(\lambda_j^*)_{j\geq 1}$ has the same distribution as the sequence $\{\frac{1}{a_1+a_2}\Gamma_j \mid j \geq 1\}$.

Proof: Let $N^i(t) = \sum_{j\geq 1} 1_{\{\Gamma_{ji} < t a_i\}}$ $i = 1,2$. Then N^i is a Poisson process with parameter a_i. Since N^1 and N^2 are independent, it is well known that $N = N^1 + N^2$ is a Poisson process with parameter $a_1 + a_2$. Clearly $(\lambda_j^*)_{j\geq 1}$ is the sequence of the successive times of the jumps of the process $N^1 + N^2$, therefore it has the same distribution as the same sequence for N which is like $\left(\frac{1}{a_1+a_2}\Gamma_j\right)_{j\geq 1}$ in distribution. *q.e.d.*

Proof of Theorem 4.1: We first observe that theorem 4.1 reduces immediately to the one dimensional case by replacing X by $\xi(X)$ with ξ in B'. Thus we consider the case $B = \mathbf{R}$. To show that X is p-stable we use the following classical criterion: Let X^1, X^2 be independent copies of X, assume that for all real numbers α_1, α_2 we have

$$(\alpha_1 X_1 + \alpha_2 X_2) \overset{d}{=} (|\alpha_1|^p + |\alpha_2|^p)^{1/p} X$$

where we have written $\overset{d}{=}$ for equality in distribution. Then X is p-stable. (Recall that we deal here only with symmetric variables). Let then $\{(\Gamma_{ji})_{j\geq 1}, (Y_{ji})_{j\geq 1}\}$ (for $i = 1, 2.$) be two independent copies of $\{(\Gamma_j)_{j\geq 1}, (Y_j)_{j\geq 1}\}$. Let $X_i = \sum_{j\geq 1}(\Gamma_{ji})^{-1/p} Y_{ji}$. Then X_1, X_2 are independent copies of X. Consider α_1, α_2 in $\mathbf{R}$. Let $a_i = |\alpha_i|^p$. We apply lemma 4.2. Let $(\lambda_j^*)_{j\geq 1}$ be as in lemma 4.2, assume that $(\lambda_j^*)_{j\geq 1}$ and $(\Gamma_j)_{j\geq 1}$ are independent. Clearly

$$\begin{aligned}
\alpha_1 X_1 + \alpha_2 X_2 &\overset{d}{=} \sum (\lambda_j^*)^{-1/p} Y_j, \quad \text{therefore by lemma 4.2} \\
&\overset{d}{=} (|\alpha_1|^p + |\alpha_2|^p)^{1/p} \sum \Gamma_j^{-1/p} Y_j \\
&\overset{d}{=} (|\alpha_1|^p + |\alpha_2|^p)^{1/p} X.
\end{aligned}$$

Hence X is p-stable.

Using elementary computations, the reader will check that

$$Z = X - Y_1 \Gamma_1^{-1/p} = \sum_{j\geq 2} Y_j \Gamma_j^{-1/p}$$

satisfies (hint : this is easy for Y_1 bounded, the general case follows)

$$\lim_{t\to\infty} t^p \mathbf{P}\{|Z| > t\} = 0$$

so that

$$\begin{aligned}
\lim_{t\to\infty} t^p \mathbf{P}\{|X| > t\} &= \lim_{t\to\infty} t^p \mathbf{P}\{|\Gamma_1^{-1/p} Y_1| > t\} \\
&= \lim t^p \mathbf{P}\{\Gamma_1 < t^{-p}|Y_1|^p\} \\
&= \lim t^p \mathbf{E}\{1 - \exp -(t^{-p}|Y_1|^p)\} \\
&= \mathbf{E}|Y_1|^p.
\end{aligned}$$

This establishes (4.4) in case $B = \mathbf{R}$, and hence also in the general case. Using (4.1) and (4.2) we find using (4.4)

$$\mathbf{E}\exp i\xi(X) = \exp -(2c_p)^{-1}\int |\xi(x)|^p Q(dx)$$

which shows that $(2c_p)^{-1}Q$ is a spectral measure for X. *q.e.d.*

Another approach to p-stable random variables goes through stochastic integrals. It can be described as follows. Let (S, Σ, m) be any measure space. Let us consider a p-stable random measure M based on (S, Σ, m). By this we mean a collection of real valued p-stable r.v.'s $(M(A))_{A\in\Sigma}$ indexed by the sets in Σ with the following properties:

i) $M(A)$ is p-stable with parameter equal to $m(A)^{1/p}$.

ii) For any mutually disjoint sequence $(A_n)_n$, the sequence $(M(A_n))_n$ is independent.

For a step function of the form $f = \sum 1_{A_i}\alpha_i$ with $\alpha_i \in \mathbf{R}$ and $A_i \in \Sigma$ mutually disjoint, one defines $\int f dM = \sum \alpha_i M(A_i)$. Now, if $f \in L_p(S, \Sigma, m)$, it is easy to define by a density argument the stochastic integral $\int f dM$. This will be a real valued stable r.v. with parameter equal to $(\int |f|^p dm)^{1/p}$. This is the p-stable analogue of a classical construction of Kakutani.

In the Banach space valued case, if $f \in L_p(S, \Sigma, m; B)$ the integral $\int f dM$ cannot in general be defined. This is one of the motivations behind the following.

Definition: Let $1 \le p \le 2$. Let (θ_n) be an i.i.d. sequence of standard p-stable r.v.'s. A Banach space B is called of stable type p if there is a constant C such that, for any finite sequence (x_i) in B, we have

$$\mathbf{E}\left\|\sum \theta_i x_i\right\| \le C\left(\sum \|x_i\|^p\right)^{1/p} \quad \text{if } p > 1,$$

$$\sup_{t>0} t\mathbf{P}\{\|\sum \theta_i x_i\|\} \le C\sum \|x_i\| \quad \text{if } p = 1.$$

We will denote by $ST_p(B)$ the smallest constant C for which this holds.

Remarks: i) It follows from results of [HJ1] and [A2] that, for each $r < p$, there is constant $C(p,r)$ such that any Banach space valued p-stable r.v. X satisfies

$$\left(\sup_{t>0} t^p\mathbf{P}\{\|X\| > t\}\right)^{1/p} \le C(p,r)(\mathbf{E}\|X\|^r)^{1/r} < \infty\,. \tag{4.5}$$

This shows in particular that all the moments

$$\left(\mathbf{E}\left\|\sum \theta_i x_i\right\|^r\right)^{1/r}$$

are equivalent when $0 < r < p$, so that we could have used other moments in the preceding definition.

ii) It is then easy to see that B is of stable type p iff for any sequence (x_n) in B, the convergence of $\sum \|x_n\|^p$ implies the a.s. convergence of $\sum \theta_n x_n$.

Proposition 4.3: B is of stable type p iff for any measure space (S, Σ, m) and any p-stable random measure M as above, the stochastic integral $\int f dM$ defines a continuous linear operator from $L_p(S, \Sigma, m; B)$ into $L_0(B)$. In that case for f in $L_p(m; B)$, $\int f dM$ is a B-valued p-stable r.v.

Proof: Assume that B is of stable type p. Let f be a step function of the form $f = \sum 1_{A_i} x_i$ with $A_i \subset S$, $A_i \in \Sigma$ mutually disjoint. Then

$$\int f dM = \sum x_i M(A_i)$$
$$\stackrel{d}{=} \sum x_i m(A_i)^{1/p}\theta_i.$$

Hence we have if $p > 1$

$$\mathbf{E}\|\int f dM\| \le ST_p(B)\Big(\sum \|x_i\|^p m(A_i)\Big)^{1/p}$$
$$\le ST_p(B)\Big(\int \|f\|^p dm\Big)^{1/p}.$$

This shows that $f \to \int f dM$ can be extended to a bounded operator from $L_p(m;B)$ into $L_1(B)$, and a fortiori into $L_0(B)$. For $p = 1$, a similar argument works.

To prove the converse all we need is *one* sufficiently rich measure space (S,Σ,m) for which the stochastic integral can be defined. (For instance $S = [0,1]$ equipped with the Lebesgue measure suffices.) Let then A_n be a disjoint sequence with $m(A_n) > 0$. For any sequence (x_n) in B such that $\sum \|x_n\|^p < \infty$ the function $f = \sum 1_{A_n} m(A_n)^{-1/p} x_n$ is in $L_p(m;B)$. If $\int f dM$ makes sense, then the series $\sum M(A_n) m(A_n)^{-1/p} x_n$ must also make sense, which implies that $\sum \theta_n x_n$ converges a.s. Hence B must be of stable type p by the above remark. *q.e.d.*

The relation between type and stable type is described by the next result. Recall that for $p = 2$, we have seen (see the remarks after proposition 3.1) that stable type 2 and type 2 are the same notion.

Proposition 4.4: Let B be a Banach space. If $1 \le p_1 < p \le 2$ we have:

a) If B is of stable type p, then B is of type p.

b) If B is of type p, then B is of stable type p_1.

Proof: The first part follows immediately from proposition 3.2. For the second part, we use the following fact which we prove below: there is a constant $C(p,p_1)$ such that if (θ_n) is an i.i.d. sequence of p_1-stable r.v.'s and if (α_n) is any sequence of real numbers, we have

(4.6) $$\sup_{t>0} t^{p_1} \mathbf{P}\{(\sum |\alpha_n\theta_n|^p)^{1/p} > t\} \le C(p,p_1)\sum |\alpha_n|^{p_1}$$

Clearly (4.6) implies that for $r < p_1$ there is a constant $C(p,p_1,r)$ such that

(4.7) $$\|(\sum |\alpha_n\theta_n|^p)^{1/p}\|_r \le C(p,p_1,r)(\sum |\alpha_n|^{p_1})^{1/p_1}.$$

Now assume B of type p. By proposition 3.1 we have for any finite sequence (x_i) in B

$$\|\sum \theta_i x_i\|_r \le C\|(\sum (|\theta_i|\,\|x_i\|)^p)^{1/p}\|_r \quad \text{hence by (4.7)}$$
$$\le CC(p,p_1,r)(\sum \|x_n\|^{p_1})^{1/p_1}$$

and this shows that B is of stable type p_1 (using (4.5) if necessary).

Remark: One can also prove part b) above using theorem 4.1 and the elementary fact that

$$\mathbf{E}\Big(\sum \Gamma_j^{-p/p_1}\Big)^{r/p} < \infty.$$

Remark: Let us briefly prove (4.6). Let $t > 0$, $\Omega_1 = \{\sup |\alpha_n\theta_n| > t\}$ and $\Omega_2 = \{(\sum |\alpha_n\theta_n|^p)^{1/p} > t\}$. Clearly $\mathbf{P}(\Omega_1) \le \sum \mathbf{P}\{|\theta_n| > t|\alpha_n|^{-1}\} \le K \sum |\alpha_n|^{p_1} t^{-p_1}$ for some constant K. On the other hand,

$$\begin{aligned}
\mathbf{P}(\Omega_2) &\le \mathbf{P}(\Omega_1) + \mathbf{P}(\Omega_2 - \Omega_1) \\
&\le \mathbf{P}(\Omega_1) + \mathbf{P}\Big\{\sum |\alpha_n\theta_n|^p \, 1_{\{|\alpha_n\theta_n| \le t\}} > t^p\Big\} \\
&\le \mathbf{P}(\Omega_1) + t^{-p} \sum \mathbf{E}(|\alpha_n\theta_n|^p \, 1_{\{|\alpha_n\theta_n| \le t\}}) \\
&\le \mathbf{P}(\Omega_1) + t^{-p}\Big(K' \sum |\alpha_n|^{p_1} t^{-p_1+p}\Big)
\end{aligned}$$

for some constant K'. Hence we conclude $\mathbf{P}(\Omega_2) \le (K + K')t^{-p_1} \sum |\alpha_n|^{p_1}$ which establishes (4.6).

The spaces of stable type p are completely characterized for $p < 2$ by the following result from [MP]. The equivalence (ii) $\Leftrightarrow$ (iii) is due to Krivine [K].

Theorem 4.5: Let $1 \le p < 2$. The following properties of a Banach space B are equivalent.

i) B is of stable type p.
ii) For each $\epsilon > 0$, B does not contain ℓ_p^n's $(1+\epsilon)$-uniformly.
iii) For some $\epsilon > 0$, B does not contain ℓ_p^n's $(1+\epsilon)$-uniformly.

Note in particular that ℓ_p or L_p is not of stable type p.

For the proof, we follow [P2] and first state a quantitative version of theorem 4.5.

Lemma 4.6: Let $1 < p < 2$, let $\frac{1}{p} + \frac{1}{p'} = 1$. For each $\epsilon > 0$, there is a number $\eta_p(\epsilon) > 0$ with the following property: any Banach space E of stable type p contains a subspace F of dimension $n = [\eta_p(\epsilon) ST_p(E)^{p'}]$ such that $F \overset{1+\epsilon}{\sim} \ell_p^n$.

Remark: Of course the preceding is void if $n = 1$, it is interesting only if the stable type constant $ST_p(E)$ is large.

Before proving lemma 4.6, we first derive its main consequences, in particular it implies theorem 4.5.

Proof of Theorem 4.5: We first assume $1 < p < 2$. We start by proving (iii) $\Rightarrow$ (i). If B is not of stable type p, then $ST_p(B) = \infty$, therefore we can find f.d. subspaces $E \subset B$ with $ST_p(E)$ arbitrarily large. Then lemma 4.6 implies that B contains subspaces $F \overset{1+\epsilon}{\sim} \ell_p^n$ with n arbitrarily large. This means that B contains ℓ_p^n's $(1+\epsilon)$-uniformly, for each $\epsilon > 0$, so that (iii) fails. This shows that (iii) $\Rightarrow$ (i). To prove (i) $\Rightarrow$ (ii), assume that for *some* $\epsilon > 0$ B contains ℓ_p^n's $(1+\epsilon)$-uniformly, we will show that B is not of stable type p. Indeed, for each n there are $x_1, \ldots, x_n$ in B such that

$$\forall (\alpha_i) \in \mathbf{R}^n \quad (\sum |\alpha_i|^p)^{1/p} \le \|\sum \alpha_i x_i\| \le (1+\epsilon)(\sum |\alpha_i|^p)^{1/p}.$$

Hence

$$(\sum \|x_i\|^p)^{1/p} \leq (1+\epsilon)n^{1/p}$$

and

$$\mathbf{E}\|\sum \theta_i x_i\| \leq \mathbf{E}(\sum_1^n |\theta_i|^p)^{1/p}.$$

But it is easy to check (using the fact that $\mathbf{E}|\theta_1|^p = \infty$) that $\alpha_n = \mathbf{E}(\frac{1}{n}\sum_1^n |\theta_i|^p)^{1/p} \to \infty$ when $n \to \infty$. Therefore B cannot be of stable type p, so that (i) $\Rightarrow$ (ii).

{Here is a quick proof that α_n is unbounded. Let

$$\alpha_n^t = \mathbf{E}(\frac{1}{n}\sum_1^n |\theta_i|^p \, 1_{\{|\theta_i|<t\}})^{1/p},$$

we have

$$\forall t > 0 \quad \alpha_n^t \leq \alpha_n$$

but when $n \to \infty$

$$\alpha_n^t \to (\mathbf{E}|\theta_1|^p \, 1_{\{|\theta_1|<t\}})^{1/p}$$

by the law of large numbers and this is unbounded in t since $\mathbf{E}|\theta_1|^p = \infty$.}

Finally (ii) $\Rightarrow$ (iii) is trivial, so that the proof of theorem 4.5 is complete in the case $p > 1$. Let us now consider the case $p = 1$. The proof that (i) $\Rightarrow$ (ii) $\Rightarrow$ (iii) is the same. To prove that (iii) $\Rightarrow$ (i), observe that if B does not contain ℓ_1^n's uniformly, then there is a $p > 1$ such that B does not contain ℓ_p^n's uniformly (this follows from the equivalence (ii) $\Leftrightarrow$ (iii) established above for $1 < p < 2$). Therefore, B must be of stable type p for some $p > 1$, hence of stable type 1 by proposition 4.4. *q.e.d.*

Among the consequences of theorem 4.5, we have

Corollary 4.7: Let $1 \leq p < 2$. If a space B is of stable type p then it is also of stable type p_1 (and hence of type p_1) for some $p_1 > p$.

Proof: The equivalence (ii) $\Leftrightarrow$ (iii) in theorem 4.5 implies immediately that the set of p's of [1,2] for which B contains ℓ_p^n's uniformly is a closed subset of [1,2]. Therefore, its complement must be open in [1,2] and this implies corollary 4.7.

Corollary 4.8: Any infinite dimensional Banach space B contains ℓ_p^n's uniformly for all p such that $p(B) \leq p \leq 2$.

Proof: Indeed, if $p(B) < p \leq 2$, then by proposition 4.4 B cannot be of stable type p, hence by theorem 4.5 B contains ℓ_p^n's uniformly. Similarly if $p = p(B)$ by corollary 4.7. Moreover if $p(B) = 2$, then we obtain the same conclusion from Dvoretzky's theorem.

We come now to the main point which is lemma 1.6. To prove it, we will need the following elementary fact which will allow us to replace Γ_j by j in several estimates.

Lemma 4.9: $\mathbf{E}\sum_{j\geq 1}|\Gamma_j^{-1/p}-j^{-1/p}|<\infty$.

Proof: This is easy to prove using

$$\forall t>0 \quad \mathbf{P}\{\Gamma_j<t\}=\int_0^t \frac{x^{j-1}}{(j-1)!}e^{-x}dx$$

hence

$$\sum_{j\geq 1}\mathbf{E}|\Gamma_j^{-1/p}-j^{-1/p}|=\int_0^\infty \sum_{j\geq 1}|x^{-1/p}-j^{-1/p}|\frac{x^{j-1}}{(j-1)!}e^{-x}dx.$$

Elementary computations (using Stirling's formula) show that this integral converges.

Proof of Lemma 4.6: Let $C=\frac{1}{2}ST_p(E)$. By the definition of $ST_p(E)$, there is a B-valued p-stable r.v. of the form $S=\sum_1^N \theta_i x_i$ such that $\mathbf{E}\|S\|>C$ and $\sum\|x_i\|^p=1$. Let $Q=\frac{1}{2N}\sum_1^N\|x_i\|^p(\delta_{x_i\|x_i\|^{-1}}+\delta_{-x_i\|x_i\|^{-1}})$. Clearly Q is a spectral measure for X. Let (Y_j) be an i.i.d. sequence of B-valued r.v.'s each with distribution Q. Then by the comment after theorem 4.1, we have

$$\sum_{j\geq 1}\Gamma_j^{-1/p}Y_j \stackrel{d}{=} a_p S$$

where $a_p=(2c_p)^{-1/p}$. (The a.s. convergence of $\sum_1^\infty \Gamma_j^{-1/p}Y_j$ is elementary here).

Let $X=\sum\Gamma_j^{-1/p}Y_j$. We have

$$\mathbf{E}\|X\|=a_p\mathbf{E}\|S\|>a_pC\ .$$

Using lemma 4.9 we will compare X with the variable $\tilde{X}=\sum_{j\geq 1}j^{-1/p}Y_j$.

We will need to define i.i.d. copies of X and $\tilde{X}$ as follows. Let $(\Omega,\mathcal{A},\mathbf{P})$ be our underlying probability space. We define i.i.d. sequences (X_n) and $(\tilde{X}_n)$ on the product space $(\Omega,\mathcal{A},\mathbf{P})^{\mathbf{N}}$ in the usual way:

$$\forall\omega=(\omega_k)_k\in\Omega^{\mathbf{N}}$$
$$X_k(\omega)=X(\omega_k) \quad \text{and}$$
$$\tilde{X}_k(\omega)=\tilde{X}(\omega_k).$$

Recall that Y_j takes its values in the unit sphere of B. This allows us to write for all $k\geq 1$

$$\mathbf{E}\|X_k-\tilde{X}_k\|\leq M_p=\mathbf{E}\sum_{j\geq 1}|\Gamma_j^{-1/p}-j^{-1/p}|. \tag{4.8}$$

We will show that with positive probability $\{\tilde{X}_1,\ldots,\tilde{X}_n\}$ spans a subspace $(1+\epsilon)$-isomorphic to ℓ_p^n with n as specified in lemma 4.6. For that purpose, we will use the deviation inequality (2.22) from chapter 2. This inequality implies, if $\sum_1^n|\alpha_i|^p=1$, the following

$$\mathbf{P}\{|\,\|\sum_1^n\alpha_i\tilde{X}_i\|-\mathbf{E}\|\sum_1^n\alpha_i\tilde{X}_i\|\,|>t\}\leq 2\exp -K_pt^{p'}, \tag{4.9}$$

for all $t > 0$. To prove (4.9) let us denote by $(\beta_k^*)_{k\geq 1}$ a non-increasing rearrangement of the collection $\{|\alpha_i|j^{-1/p} \mid i \leq n,\ j \geq 1\}$. Since the Y_j's are i.i.d. and symmetric, we have

$$\sum_1^n \alpha_i \tilde{X}_i \stackrel{d}{=} \sum_{k\geq 1} \beta_k^* Y_k. \tag{4.10}$$

But for all $k \geq 1$ we have

$$\text{card}\,\{(i,j) \mid |\alpha_i|j^{-1/p} > k^{-1/p}\} < \sum_i |\alpha_i|^p k = k$$

hence

$$\beta_k^* \leq k^{-1/p}.$$

Therefore, (4.9) immediately follows from (4.10) and (2.22).

To continue, the basic idea is the same as in the proof of theorem 1.3, but here the variables $\tilde{X}_i$ are no longer p-stable so that we have to use instead the following inequality

$$|\mathbf{E}\|\sum \alpha_i X_i\| - \mathbf{E}\|\sum \alpha_i \tilde{X}_i\|\,| \leq M_p \sum |\alpha_i|. \tag{4.11}$$

This follows from (4.8) and the triangle inequality. Moreover, since (X_i) are i.i.d. and p-stable

$$\sum \alpha_i X_i \stackrel{d}{=} \left(\sum |\alpha_i|^p\right)^{1/p} X$$

hence

$$\mathbf{E}\|\sum \alpha_i X_i\| = \left(\sum |\alpha_i|^p\right)^{1/p} \mathbf{E}\|X\|.$$

Let us now assume

$$M_p n^{1/p'} \leq (\delta/2)(a_p C) \tag{4.12}$$

and recall that $a_p C \leq \mathbf{E}\|X\|$. Let $A = \mathbf{E}\|X\|$ and assume $\sum |\alpha_i|^p = 1$. Then (4.11) and (4.12) imply

$$|\mathbf{E}\|\sum_1^n \alpha_i \tilde{X}_i\| - A| < \delta A/2$$

hence (4.9) implies

$$\begin{aligned}\mathbf{P}\{|\,\|\sum \alpha_i \tilde{X}_i\| - A| > \delta A\} &\leq 2\exp -K_p\left(\frac{\delta A}{2}\right)^{p'} \\ &\leq 2\exp -K_p\left(\frac{\delta a_p C}{2}\right)^{p'}.\end{aligned}$$

We can now complete the proof by exactly the same reasoning as earlier for theorem 1.3. We use lemmas 1.6 and 1.7. Let S be a δ-net in the unit sphere of ℓ_p^n, with $|S| \leq (1+2/\delta)^n$. If

$$2(1+2/\delta)^n \exp -K_p\left(\frac{\delta a_p C}{2}\right)^{p'} < 1 \tag{4.13}$$

then we find with positive probability

$$\forall \alpha \in S \quad |A^{-1}\|\sum_1^n \alpha_i \tilde{X}_i\| - 1| \leq \delta.$$

By lemma 1.6, if $\delta = \delta(\epsilon)$ is suitably chosen, the sequence $\{A^{-1}\tilde{X}_1, \ldots, A^{-1}\tilde{X}_n\}$ spans with positive probability a subspace F such that $F \overset{1+\epsilon}{\sim} \ell_p^n$.

It remains to recapitulate the restrictions on n that this reasoning imposes; first (4.12) requires $n \leq ((\delta a_p)(2M_p)^{-1}C)^{p'}$, while for (4.13) it suffices that

$$\frac{2n}{\delta} \leq \frac{1}{2}K_p(\delta C/2)^{p'}$$

with

$$2\exp -\frac{n}{\delta} < 1.$$

It is then easy to check that we can obtain n of the form announced in lemma 4.6. *q.e.d.*

In some cases, the dimension given by lemma 4.6 is remarkably large. This is the case for instance when $E = \ell_1^N$ as shown by the following result discovered in [JS].

Theorem 4.10: Let $1 < p < 2$. For each $\epsilon > 0$, there is a number $\delta_p(\epsilon) > 0$ such that, for any $N \geq 1$, the space ℓ_1^N contains a subspace $F \overset{1+\epsilon}{\sim} \ell_p^n$ of dimension $n = [\delta_p(\epsilon)N]$.

Proof: Clearly $\mathbf{E}\sum_1^N |\theta_i| = N\mathbf{E}|\theta_1|$ if (θ_i) are i.i.d. standard p-stable r.v.'s. Therefore $ST_p(\ell_1^N) \geq \mathbf{E}|\theta_1|N^{1/p'}$, so that theorem 4.10 follows from lemma 4.6 with $\delta_p(\epsilon) = (\mathbf{E}|\theta_1|)^{p'}\eta_p(\epsilon)$.

Remark: It is natural to ask whether theorem 4.10 is valid with ℓ_r^n $(1 < r < p)$ in the place of ℓ_1^N. This can be shown by known factorization arguments but only for *some* $\epsilon > 0$ depending on p and r. It seems open for ϵ arbitrarily small (of course $\delta_p(\epsilon)$ is then allowed to depend on r).

To conclude this chapter we give several equivalent reformulations of stable type p, for $p < 2$.

We first observe that for a space B to contain ℓ_p^n's uniformly it suffices that much less is true: If there is a constant C such that for all n there are $x_1, \ldots, x_n$ in B satisfying.

$$\text{(4.14)} \qquad \begin{cases} & \|x_i\| \leq C \\ \text{and } \forall(\alpha_i) \in \mathbf{R}^n & \left(\sum |\alpha_i|^p\right)^{1/p} \leq \|\sum \alpha_i x_i\|, \end{cases}$$

then B contains ℓ_p^n's uniformly. Indeed, the property (4.14) is enough to contradict stable type p (cf. the proof of theorem 4.5).

Let (x_n) be a sequence in B. Let

$$\|(x_n)\|_{p,\infty} = \left(\sup_{t>0} t^p \sum 1_{\{\|x_n\|>t\}}\right)^{1/p}.$$

Equivalently if (β_n^*) is a non-increasing rearrangement of $(\|x_n\|)_n$ we have

$$\|(x_n)\|_{p,\infty} = \sup n^{1/p}\beta_n^*.$$

This is the weak-ℓ_p norm of the sequence $(\|x_n\|)$. This notion allows us to give a formulation of stable type p analogous to the definition of type p. A space B is of stable type p iff

$$\exists C\ \forall n\ \forall x_1, \dots x_n \in B \qquad (4.15)$$

$$\|\sum \epsilon_i x_i\|_2 \leq C\|(x_n)\|_{p,\infty}.$$

Indeed, if B is of stable type p then by corollary 4.7 it is of type p_1 for some $p_1 > p$, a fortiori it satisfies (4.15). Conversely, it is easy to see that (4.15) cannot hold if B contains ℓ_p^n's uniformly.

Note however that (4.15) is of limited interest since we know from corollary 4.7 and proposition 4.4 that B is of stable type p iff it is of type p_1 for some $p_1 > p$.

Let us now consider sums of independent random variables with values in a stable type p space. We will use the following lemma (of independent interest) which comes from [MaP]. The simple proof below is due to Joel Zinn.

Lemma 4.11: Let (Z_i) be a sequence of independent positive r.v.'s. Then for $0 < p < \infty$

$$\sup_{t>0} t^p \mathbf{P}\{\sup_{u>0} u^p \sum_i 1_{\{Z_i > u\}} > t\} \leq 2e \sup_{t>0} t^p \sum_i \mathbf{P}\{Z_i > t\}. \qquad (4.16)$$

Proof: This clearly reduces to the case $p = 1$ (replace Z_i by $Z_i^{1/p}$). Let us denote by (Z_i^*) the non-increasing rearrangement of the sequence Z_i. Note that $Z_n^* > t$ iff $\sum 1_{\{Z_i > t\}} \geq n$, and also

$$\sup_n nZ_n^* = \sup_{t>0} t \sum 1_{\{Z_i > t\}}. \qquad (4.17)$$

Hence we can write for all $c > 0$

$$\begin{aligned} \mathbf{P}(Z_n^* > t) = \mathbf{P}\{\sum 1_{Z_i > t} \geq n\} &\leq e^{-cn}\mathbf{E}\exp c(\sum 1_{Z_i > t}), \\ &\leq e^{-cn} \prod_i (1 + (e^c - 1)\mathbf{P}(Z_i > t)) \\ &\leq \exp[-cn + (e^c - 1)\sum \mathbf{P}(Z_i > t)]. \end{aligned}$$

Let us assume (by homogeneity) that $\sum \mathbf{P}(Z_i > t) \leq 1/t$. Then the above computation yields

$$\begin{aligned} \mathbf{P}(Z_n^* > t) &\leq \exp[-cn + (e^c - 1)\frac{1}{t}] \quad \text{hence taking } c = \log nt \\ &\leq \exp n(-\log(nt) + 1 - \frac{1}{tn}), \end{aligned}$$

Therefore

$$\begin{aligned} \mathbf{P}(\sup nZ_n^* > t) &\leq \sum \mathbf{P}\{Z_n^* > t/n\} \\ &\leq \sum_n \exp n(-\log t + 1 - \frac{1}{t}), \quad \text{and if } t > 2e, \\ &\leq \sum_n (\frac{e}{t})^n \\ &\leq \frac{e}{t}\left(1 - \frac{e}{t}\right)^{-1}, \quad \text{so that} \end{aligned}$$

$$\mathbf{P}(\sup nZ_n^* > t) \leq \frac{2e}{t}.$$

Finally we observe that for $t \leq 2e$ the preceding inequality is trivial, so that (4.16) follows from this and (4.17).

We close this chapter by a slight refinement of a result appearing in [R].

Theorem 4.12: For any B-valued r.v. X let

$$\Lambda_p(X) = \left(\sup_{t>0} t^p \mathbf{P}\{\|X\| > t\}\right)^{1/p}.$$

Let $1 \leq p < 2$. Then if B is of stable type p, there is a constant C such that for any sequence (X_i) of independent symmetric B-valued r.v.'s we have

$$\Lambda_p\left(\sum_1^n X_i\right) \leq C\left(\sup_{t>0} t^p \sum_1^n \mathbf{P}\{\|X_i\| > t\}\right)^{1/p} \tag{4.18}$$

$$\text{a fortiori} \quad < C\left(\sum \Lambda_p(X_i)^p\right)^{1/p}. \tag{4.19}$$

Remarks: (i) Note that (4.19) implies conversely that B is of stable type p by letting simply $X_i = \theta_i x_i$.

(ii) If $p > 1$, an easy symmetrization procedure shows that the preceding statement extends to independent mean zero r.v.'s.

Proof: Let (ϵ_i) be as before a sequence of signs independent of the sequence (X_i). Let us denote by $\mathbf{E}_\epsilon$ (resp. $\mathbf{E}_X$) the expectation signs with respect to (ϵ_i) (resp. (X_i)). For each ω fixed, we have (see the end of chapter 2)

$$\left\|\sum_1^n \epsilon_i X_i(\omega)\right\| = \mathbf{E}_\epsilon \left\|\sum_1^n \epsilon_i X_i(\omega)\right\| + Z(\epsilon, \omega) \tag{4.20}$$

where $Z(\epsilon, \omega) = \sum_1^n d_i(\epsilon, \omega)$ is a sum of martingale differences with respect to the filtration induced by $(\epsilon_1, \ldots, \epsilon_n)$, and moreover

$$|d_i| \leq \|X_i(\omega)\|.$$

by (2.15).

By a classical result of Burkholder-Gundy [Bu] and an interpolation argument we have

$$\Lambda_p(Z) \leq C_p \Lambda_p\left(\left(\sum d_i^2\right)^{1/2}\right)$$

for some constant C_p, hence

$$\Lambda_p(Z) \leq C_p \Lambda_p\left(\left(\sum \|X_i\|^2\right)^{1/2}\right). \tag{4.21}$$

On the other hand, if B is of stable type p then it is of type p_1 for some $p_1 > p$ by corollary 4.7, hence

$$\Lambda_p\left(\mathbf{E}_\epsilon \left\| \sum \epsilon_i X_i \right\|\right) \leq C'_p \Lambda_p\left(\left(\sum \|X_i\|^{p_1}\right)^{1/p_1}\right) \tag{4.22}$$

for some constant C'_p.

From Lemma 4.11, we deduce immediately that if $p_1 > p$

$$\Lambda_p\left(\sum \|X_i\|^{p_1}\right)^{1/p_1} \leq C(p, p_1)\left(\sup_{t>0} t^p \sum \mathbf{P}\{\|X_i\| > t\}\right)^{1/p} \tag{4.23}$$

for some constant $C(p, p_1)$.

Finally, combining (4.20) with (4.21), (4.22) and (4.23) we obtain the announced result (4.18).

Chapter 5
Duality and K-Convexity

In this chapter, we study the notion of K-convexity which was introduced at the end of [MP]. This notion appears now as the key to understand the duality between type and cotype. More precisely, let B be a Banach space. We will see below (proposition 5.2) that if B is of type p, then B^* is of cotype p' with $\frac{1}{p}+\frac{1}{p'}=1$, the converse fails in general, but it is true if B is a K-convex space. The real meaning of K-convexity was elucidated in [P5], where it is proved that a Banach space B is K-convex if (and only if) B does not contain ℓ_1^n's uniformly. Spaces which do not contain ℓ_1^n's uniformly are sometimes called B-convex; so that with this terminology B- and K-convexity are equivalent properties.

This geometric characterization of K-convexity also has an important application to the "spherical sections of convex bodies", that is to say to the ℓ_2^n subspaces of Banach spaces, as in Dvoretzky's theorem. Indeed, we will show below that if B does not contain ℓ_1^n's uniformly, then B contains uniformly complemented ℓ_2^n's. This means that there is a constant C such that for each $\epsilon>0$ there is a subspace $B_n\subset B$ with $B_n \overset{1+\epsilon}{\sim} \ell_2^n$ and a projection $P_n : B\to B_n$ with $\|P_n\|\leq C$. Moreover, one can roughly say that these complemented ℓ_2^n subspaces are present in every suitably large subspace of B. This is stated more precisely below in theorem 5.10, where we show that B is K-convex if and only if B is "locally π-euclidean".

We now define K-convexity. We need some notation. We denote by I_B the identity operator on a Banach space B. We will often write simply $L_2(B)$ instead of $L_2(D,\mu;B)$. Let us denote by R_1 the orthogonal projection from $L_2(D,\mu)$ onto the closed span of the sequence $\{\epsilon_n|n\in\mathbf{N}\}$. A Banach space B is called K-convex if the operator $R_1\otimes I_B$ (defined a priori only on $L_2(D,\mu)\otimes B$) extends to a bounded operator from $L_2(D,\mu;B)$ into itself. We will denote by $K(B)$ the norm of $R_1\otimes I_B$ considered as an operator acting on $L_2(D,\mu;B)$. Clearly $R_1\otimes I_B$ is bounded on $L_2(B)$ iff $R_1\otimes I_{B^*}$ is bounded on $L_2(B^*)$. Let us first treat a simple example, the case when $B=\ell_1^k$ with $k=2^n$. Then, we may isometrically identify B with $L_1(D_n)$ where $D_n=\{-1,+1\}^n$, equipped with its normalized Haar measure. Let us denote by b_i the i-th coordinate on $\{-1,+1\}^n$ considered as an element of $L_1(D_n)$. Consider then the B-valued function $F: D\to B$ defined by $F(\omega)=\prod_{i=1}^n(1+\epsilon_i(\omega)b_i)$. We have $\|F(\omega)\|_B=1$ hence $\|F\|_{L_2(B)}=1$. But on the other hand, we have clearly

$$((R_1\otimes I_B)F)(\omega)=\sum_{i=1}^n\epsilon_i(\omega)b_i,$$

so that

$$\begin{aligned}\|(R_1\otimes I_B)(F)\|_{L_2(B)}&=\mathbf{E}\Big|\sum_1^n\epsilon_i\Big|\\&\geq A_1n^{1/2}\end{aligned}$$

for some positive numerical constant A_1. Returning the definition of $K(B)$, we find

$$K(\ell_1^{2^n})\geq A_1n^{1/2}\,.$$

In particular, $K(\ell_1^n)$ is unbounded when $n \to \infty$. From this (and the observation that if S is a closed subspace of B than $K(S) \leq K(B)$) we deduce immediately.

Proposition 5.1: A K-convex Banach space cannot contain ℓ_1^n's uniformly.

We now turn to the duality between type and cotype. We first state some simple observations.

Proposition 5.2: Let B be a Banach space. Let $1 \leq p \leq 2 \leq p' \leq \infty$ be such that $\frac{1}{p} + \frac{1}{p'} = 1$.
(i) If B is of type p, then B^* is of cotype p'.
(ii) If B is K-convex, and if B^* is of cotype p' then B is of type p.

Remark: Part (i) is due to Hoffman-Jørgensen [HJ1] and Maurey independently. Part (ii) comes from [MP].

To clarify the proof we state the following

Lemma 5.3: Consider $x_1, \dots x_n$ in an arbitrary Banach space B. Define

$$|||(x_i)||| = \sup\Big\{\sum_1^n < x_i, x_i^* > \Big| x_i^* \in B^* \, \|\sum_1^n \epsilon_i x_i^*\|_{L_2(B^*)} \leq 1\Big\} \tag{5.1}$$

Then

$$|||(x_i)||| = \inf\Big\{\|\sum_1^n \epsilon_i x_i + \Phi\|_{L_2(B)} \tag{5.2}$$

where the infimum is over all Φ in $L_2(B)$ such that $\mathbf{E}(\epsilon_i \Phi) = 0$ for all $i = 1, 2, \dots, n$ (or equivalently over all Φ in $L_2 \otimes B$ such that $R_1 \otimes I_B(\Phi) = 0$).

Proof of Lemma 5.3: We consider the natural duality between $L_2(B)$ and $L_2(B^*)$. Let $S \subset L_2(B^*)$ be the subspace

$$S = \Big\{\sum_1^n \epsilon_i x_i^* \; x_i^* \in B^*\Big\}.$$

The norm which appears on the right side of (5.2) is the norm of the space $\mathcal{X} = L_2(B)/S^{\perp}$. Clearly $\mathcal{X}^* = S^{\perp\perp} = S$. Therefore, the identity (5.2) is nothing but the familiar equality

$$\forall z \in \mathcal{X} \qquad \sup\{< z, z^* > | z^* \in \mathcal{X}^*, \|z^*\| \leq 1\} = \|z\|.$$

Proof of Proposition 5.2: We leave part (i) as an exercise for the reader. Let us prove (ii). Assume B^* of cotype p' so that $\exists C \; \forall n \; \forall x_i^* \in B^*$

$$\Big(\sum \|x_i^*\|^{p'}\Big)^{1/p'} \leq C \|\sum \epsilon_i x_i^*\|_{L_2(B^*)}$$

This implies for all x_i in B

$$|||(x_i)||| \leq C\Big(\sum \|x_i\|^p\Big)^{1/p}.$$

Assume $\sum \|x_i\|^p < 1$. By (5.2) there is a Φ in $L_2(B)$ such that $\mathbf{E}(\epsilon_i \Phi) = 0$ for all i and such that

$$\|\sum \epsilon_i x_i + \Phi\|_{L_2(B)} < C .$$

We have

$$\sum \epsilon_i x_i = (R_1 \otimes I_B)(\sum \epsilon_i x_i + \Phi)$$

hence

$$\|\sum \epsilon_i x_i\|_{L_2(B)} \leq K(B)C .$$

By homogeneity, this proves that B is of type p with constant not more than $K(B)C$.

We come now to the main result of this chapter which is the converse of proposition 5.1.

Theorem 5.4: A Banach space B is K-convex if (and only if) it does not contain ℓ_1^n's uniformly.

The projection R_1 can be replaced by all kinds of projections which behave similarly in the preceding statement. For instance, let (g_n) be an i.i.d. sequence of normal Gaussian r.v.'s on some probability space $(\Omega, \mathcal{A}, \mathbf{P})$, and let G_1 be the orthogonal projection from $L_2(\Omega, \mathcal{A}, \mathbf{P})$ onto the closed span of $\{g_n | n \in \mathbf{N}\}$. Then (see [P5]) a space B is K-convex iff $G_1 \otimes I_B$ is a bounded operator from $L_2(\Omega, \mathcal{A}, \mathbf{P}; B)$ into itself. This allows us to reproduce the proof of proposition 5.2 in a Gaussian setting, replacing (3.1) and (3.2) by (3.1)' and (3.2)', if we wish.

We can proceed similarly with (3.4) or (3.5), by introducing a projection Q_1 as follows.

Let $(\Omega, \mathcal{A}, \mathbf{P})$ be a probability space. We write simply L_2 for $L_2(\Omega, \mathcal{A}, \mathbf{P})$. Let $(\mathcal{A}_n)_{n \geq 1}$ be a sequence of *independent* σ-subalgebras of $\mathcal{A}$. Let S_0 be the (one dimensional) subspace of L_2 formed by the constant functions. Let S_1 be the subspace formed by all the functions of the form

$$\sum_{n=1}^{\infty} y_n$$

with $y_n \in L_2(\mathcal{A}_n)$ for all n,

$$\mathbf{E} y_n = 0$$

and

$$\sum \mathbf{E}|y_n|^2 < \infty .$$

We denote by Q_1 the orthogonal projection from L_2 onto S_1. One can then show (see theorem 5.5 below) that if B is K-convex then $Q_1 \otimes I_B$ is bounded on $L_2(B)$. Note that, in the case $(\Omega, \mathbf{P}) = (D, \mu)$, if we take for $\mathcal{A}_n$ the σ-algebra generated by ϵ_n then Q_1 coincides with R_1.

Let us return to our probability space $(\Omega, \mathcal{A}, \mathbf{P})$. We may as well assume that $\bigcup_n \mathcal{A}_n$ generates the σ-algebra $\mathcal{A}$. Actually we can define a sequence of projections $(Q_k)_{k \geq 0}$ as follows. Let us denote by F_k the closed subspace of L_2 spanned by all the functions f for which there are $n_1 < n_2 < \ldots < n_k$ such that f is measurable with respect to the σ-algebra generated by $\mathcal{A}_{n_1} \cup \ldots \cup \mathcal{A}_{n_k}$.

{ Consider the following special case: let (θ_n) be a sequence of independent r.v.'s and let $\mathcal{A}_n$ be the σ-algebra generated by θ_n. Then F_k is the subspace of all the functions in L_2 which depend on at most k of the functions $\{\theta_n | n \geq 1\}$. }

Note that $F_k \subset F_{k+1}$ and $\cup F_k$ is dense in L_2. Let then $S_k = F_k \cap F_{k-1}^{\perp}$, and let Q_k be the orthogonal projection from L_2 onto S_k.

{ Note: In the special case considered above, let us denote by λ_n the law of θ_n. Then S_k is the subspace spanned by all the functions of the form $F(\theta_{n_1}, \dots, \theta_{n_k})$ such that

$$\int F(x_1, \dots, x_i, \dots, x_k) d\lambda_{n_i}(x_i) = 0$$

for all $i = 1, 2, \dots, k$. }

We can now formulate a strengthening of theorem 5.4.

Theorem 5.5: Let $(Q_k)_{k \geq 0}$ be as above. If a Banach space B does not contain ℓ_1^n's uniformly then $Q_k \otimes I_B$ defines a bounded operator on $L_p(\Omega, \mathcal{A}, \mathbf{P}; B)$ for $1 < p < \infty$ and any $k \geq 0$. Moreover there is a constant $C = C(p, B)$ such that the norm of $Q_k \otimes I_B$ on $L_p(B)$, which we denote by $\|Q_k \otimes I_B\|_p$ satisfies

$$\|Q_k \otimes I_B\|_p \leq C^k \quad \text{for all } k \geq 0 \,.$$

Clearly theorem 5.4 is a consequence of theorem 5.5. The proofs of these results are intimately connected with the theory of *holomorphic semi-groups*. Let us describe the semi-group which is naturally associated to $(Q_k)_{k \geq 0}$.

Let us denote by π_m the conditional expectation operator on $L_2(\Omega, \mathbf{P})$ with respect to the σ-subalgebra generated by $\bigcup_{n \neq m} \mathcal{A}_n$. Let us identify (without loss of generality) $\mathcal{A}$ with $\mathcal{A}_1 \otimes \mathcal{A}_2 \otimes \dots$. For any $0 \leq \epsilon \leq 1$, we consider the operator $T(\epsilon)$ defined formally as follows

$$T(\epsilon) = \prod_{m=1}^{\infty} [\pi_m + \epsilon(1 - \pi_m)] \,. \tag{5.3}$$

Note that if a function f in L_2 "depends" only on $\mathcal{A}_1, \mathcal{A}_2, \dots, \mathcal{A}_N$ then clearly $\pi_m f = f$ for all $m > N$ so that

$$T(\epsilon) f = \prod_{m=1}^{N} [\pi_m + \epsilon(1 - \pi_m)] f \tag{5.4}$$

is unambiguously defined. Therefore $T(\epsilon)$ is unambiguously defined and is clearly a contraction on the subspace $\mathcal{F} \subset L_2$ of all the functions which depend only on finitely many σ-algebras among the sequence $(\mathcal{A}_n)_{n \geq 1}$. Thus we may extend $T(\epsilon)$ as a linear contraction on L_2. We let

$$T_t = T(e^{-t}) \quad \text{for all } t > 0 \,.$$

Note that since the π_m's are mutually commuting operators, $(T_t)_{t \geq 0}$ is a contractive semi-group on L_2. It is easy to see that it is a strongly continuous semi-group on L_p for all $1 \leq p < \infty$.

Now let B be a Banach space. Since the π_m's are conditional expectation operators, the operators $\epsilon 1 + (1-\epsilon)\pi_m$ are contractions on $L_p(B)$ for $1 \leq p \leq \infty$. Therefore (5.4) makes sense also if f is in $L_p(B)$ and "depends" only on $\mathcal{A}_1, \ldots, \mathcal{A}_N$. This shows that $T_t \otimes I_B$ can be defined on B-valued functions using the same formulas (5.4) and (5.3) and by simple density arguments one sees that $S_t = T_t \otimes I_B$ is a strongly continuous semi-group on the space $Y = L_p(B)$ for $1 < p < \infty$.

We will show that if B does not contain ℓ_1^n's uniformly then S_t is a *holomorphic* semi-group on Y. This means that, for any y in Y, the function $t \to S_t y$ admits a holomorphic extension on an open sector of the complex plane containing the positive real axis. For that purpose, we will use the following criterion

Theorem 5.6: (Beurling-Katô [Be] [K]) Let $(S_t)_{t\geq 0}$ be a strongly continuous contractive semi-group on a complex Banach space Y. Assume that $\sup_{t\geq 0} \|1 - S_t\| = \rho < 2$. Then there exist constants $\phi = \phi(\rho) > 0$ and $\Delta = \Delta(\rho)$ (depending only on ρ) for which the semi-group admits a holomorphic extension $\varsigma \to S_\varsigma$ defined in the sector

$$V_\phi = \{\varsigma \in \mathbf{C} \mid |\operatorname{Arg} \varsigma| < \phi\}$$

and satisfying moreover

$$\sup_{\varsigma \in V_\phi} \|S_\varsigma\| \leq \Delta .$$

We will use the following simple variant.

Corollary 5.7: Let $(S_t)_{t\geq 0}$ be a strongly continuous semi-group on a complex Banach space Y. Assume that there are constants $\alpha < 1$ and M such that for any finite subset $A \subset \mathbf{R}_+$ and for any $s \geq 0$ we have

$$\Big\| \prod_{t\in A} (1 - S_t) S_s \Big\| \leq M 2^{\alpha|A|}. \tag{5.5}$$

Then there are constants $\phi = \phi(\alpha, M)$ and $\Delta = \Delta(\alpha, M)$ such that the conclusion of theorem 5.6 holds.

Proof: We introduce the following equivalent norm on Y. For all y in Y let

$$|y| = \sup\{2^{-\alpha n} \Big\| \prod_{t\in A} (1 - S_t) S_s y \Big\|\}$$

where the sup runs over all $n \geq 0$, all subsets $A \subset \mathbf{R}_+$ of cardinality n and all $s \geq 0$. Clearly by our assumptions

$$\|y\| \leq |y| \leq M\|y\| .$$

Moreover

$$|S_t y| \leq |y| \quad \text{and} \quad |(1 - S_t) y| \leq 2^\alpha |y| \quad \text{for all } t \geq 0 ,$$

so that we can apply theorem 5.6 to $(Y, |\ |)$ to obtain corollary 5.7 with $\phi = \phi(2^\alpha)$ and $\Delta = M\Delta(2^\alpha)$. *q.e.d.*

The proof of theorem 5.5 is based on corollary 5.7 and the following

Lemma 5.8: Let B be a Banach space and let $1 < p < \infty$. Consider the semi-group $S_t = T_t \otimes I_B$ defined as above on $L_p(B)$. If B does not contain ℓ_1^n's uniformly, there are constants $\alpha < 1$ and M such that (5.5) holds.

Proof: Let $Y = L_p(B)$. We know by theorem 3.3 i) and by the remarks preceding proposition 3.1 that Y also does not contain ℓ_1^n's uniformly. We claim that this implies the following property for Y.

Claim: This is an integer n and a number $\alpha < 1$ such that for any n-tuple $P_1, \dots, P_n$ of mutually commuting contractive projections on Y we have

$$\left\|\prod_{j=1}^{n}(1-P_j)\right\| \leq 2^{\alpha n} .$$

Indeed, suppose that this fails. Then for any n and any $\delta > 0$ there are $P_1, \dots, P_n$ as above and y in the unit ball of Y such that

$$\left\|\prod_{j=1}^{n}(1-P_j)y\right\| > 2^n - \delta . \tag{5.6}$$

For any $A \subset \{1,\dots,n\}$, let $P_A = \prod_{j\in A} P_j$. Let $A^c = \{1,\dots,n\} - A$ be the complement of A. Clearly (5.6) implies by the triangle inequality

$$\sum_{\beta\subset A^c}\left\|P_\beta \prod_{j\in A}(1-P_j)y\right\| > 2^n - \delta \tag{5.7}$$

and since $\|P_\beta \prod_{j\in A}(1-P_j)y\| \leq 2^{|A|}$, (5.7) implies

$$\begin{aligned}\left\|P_{A^c}\prod_{j\in A}(1-P_j)y\right\| &> 2^N - (2^{|A^c|}-1)2^{|A|} - \delta \\ &> 2^{|A|} - \delta .\end{aligned} \tag{5.8}$$

Consider now any choice of signs $(\epsilon_1, \dots, \epsilon_n)$ in $\{-1,+1\}^n$. We have then

$$\left\|\prod(1+\epsilon_j P_j)y\right\| > 2^n - 2^n\delta. \tag{5.9}$$

Indeed, if $A = \{j \mid \epsilon_j = -1\}$ we have

$$\begin{aligned}\left\|\prod(1+\epsilon_j P_j)y\right\| &\geq \left\|P_{A^c}\prod(1+\epsilon_j P_j)y\right\| \\ &= 2^{|A^c|}\left\|P_{A^c}\prod_{j\in A}(1-P_j)y\right\|\end{aligned}$$

hence by (5.8)

$$\geq 2^n - 2^{|A'|}\delta .$$

This establishes (5.9). Using the triangle inequality again, we deduce from (5.9)

$$(2^n - n) + \|\sum \epsilon_j P_j y\| > 2^n - 2^n\delta$$

hence

$$\|\sum \epsilon_j P_j y\| > n - 2^n\delta \tag{5.10}$$

for all choices of signs.

Consider now real coefficients (α_j) such that $\sum|\alpha_j| = 1$. Let $\epsilon_j = \text{sign}(\alpha_j)$. We have

$$\begin{aligned}\|\sum \alpha_j P_j y\| &= \|\sum \epsilon_j |\alpha_j| P_j y\| \\ &\geq \|\sum \epsilon_j P_j y\| - \|\sum \epsilon_j(1 - |\alpha_j|)P_j y\| \\ &\geq n - 2^n\delta - \sum |1 - |\alpha_j|| \\ &\geq n - 2^n\delta - (n-1) = 1 - 2^n\delta .\end{aligned}$$

By homogeneity, we have proved that (5.10) implies

$$\forall(\alpha_j) \in \mathbf{R}^n \quad \|\sum \alpha_j P_j y\| \geq (1 - 2^n\delta)\sum|\alpha_j| .$$

Since

$$\|\sum \alpha_j P_j y\| \leq \sum |\alpha_j| ,$$

this shows that the subspace of B spanned by $(P_1 y, \ldots, P_n y)$ is $(1 - 2^n\delta)^{-1}$-isomorphic to ℓ_1^n. Since $\delta > 0$ and n are arbitrary, we conclude (assuming that the above claim is not true) that Y contains ℓ_1^n's uniformly. This contradiction concludes the proof of the above claim.

{ Note: Actually in the preceding proof the subspace spanned by $(y, P_1 y, \ldots, P_n y)$ is uniformly isomorphic to ℓ_1^{n+1}. }

We can now complete the proof of lemma 5.8. We will use (5.3) to show that $S_t = T(e^{-t})\otimes I_B$ is an average of conditional expectations. Indeed let λ_t be the probability on $\{0,1\}^{\mathbf{N}}$ which is the infinite product of $(1 - e^{-t})\delta_0 + e^{-t}\delta_1$. For any $\xi = (\xi_m)$ in $\{0,1\}^{\mathbf{N}}$, we define

$$P(\xi) = \prod_{m=1}^{\infty} [\pi_m + \xi_m(1 - \pi_m)] \tag{5.11}$$

As before, this formula makes sense on $\mathcal{F}$ hence on L_2; actually $P(\xi)$ is the conditional expectation with respect to the σ-algebra generated by $\bigcup_{\xi_n = 1} \mathcal{A}_n$.

Comparing (5.11) and (5.3), we see that

$$T_t = \int P(\xi)d\lambda_t(\xi) . \tag{5.12}$$

Let us denote by $\tilde{P}(\xi)$ the (conditional expectation) operator $P(\xi)$ considered as acting on $Y = L_p(B)$. Then (5.12) implies

$$S_t = \int \tilde{P}(\xi) d\lambda_t(\xi). \tag{5.13}$$

Note that all the projections $\tilde{P}(\xi)$ are mutually commuting, so that the above claim implies

$$\|(1 - \tilde{P}(\xi^1)) \dots (1 - \tilde{P}(\xi^n))\| \leq M 2^{\alpha n} \tag{5.14}$$

for all $\xi^1, \dots, \xi^n$ in $\{0,1\}^{\mathbf{N}}$.

Consider now $t_1, \dots, t_n$ in $\mathbf{R}_+$. Averaging (5.14) with respect to $\lambda_{t_1}(d\xi^1) \dots \lambda_{t_n}(d\xi^n)$, we obtain

$$\|(1 - S_{t_1}) \dots (1 - S_{t_n})\| \leq M 2^{\alpha n}$$

and this concludes the proof of corollary 5.7 (since $\|S_s\| \leq 1$ for all $s \geq 0$).

Remark: If a real Banach space does not contain ℓ_1^n's uniformly, then neither does any reasonable complexification of it. Indeed, the complexifications are **R**-isomorphic to $B \oplus B$ and if B is of type p, then clearly $B \oplus B$ must be of type p also (recall theorem 3.3). This allows us to assume (without further notice) that B is a complex Banach space in a situation such as in theorem 5.5.

Proof of theorem 5.5: We will exploit the conclusion of corollary 5.7. In order to do that it is important to note the following identity

$$T(\epsilon) = \sum_{k \geq 0} \epsilon^k Q_k , \tag{5.15}$$

where the right hand side should be interpreted as defined on $\mathcal{F}$ (and then extended by density to the whole of L_2). This identity (5.15) follows from (5.3) after the coefficients have been identified (this is left to the reader). We find

$$Q_1 = \sum_{m=1}^{\infty} (1 - \pi_m) \prod_{j \neq m} \pi_j$$

$$Q_2 = \sum_{1 \leq m_1 < m_2 < \infty} (1 - \pi_{m_1})(1 - \pi_{m_2}) \prod_{\substack{j \neq m_1 \\ j \neq m_2}} \pi_j \quad \text{etc.} \dots$$

Note that for all f in L_2 we have

$$\mathbf{E}f = \prod_{j=1}^{\infty} \pi_j f \quad \text{and} \quad \mathbf{E}^{A_m} f = \prod_{j \neq m} \pi_j f$$

so that

$$Q_1 f = \sum_{m=1}^{\infty} \mathbf{E}^{A_m} f - \mathbf{E}f .$$

Let $S = (\mathcal{F} \otimes B) \cap L_p(B)$. Clearly S is a dense subspace of $L_p(B)$. Let $\tilde{Q}_k = Q_k \otimes I_B$ considered as an operator on $S \subset L_p(B)$. By (5.15), we have

$$S_t = \sum_{k \geq 0} e^{-kt} \tilde{Q}_k. \tag{5.16}$$

By the conclusion of corollary 5.7, $t \to S_t$ admits a holomorphic extension to a sector $V = V_\phi$ with $\sup_{\varsigma \in V} \|S_\varsigma\| \leq M$. By (5.16) and the unicity of analytic continuation we have necessarily

$$S_\varsigma f = \sum_{k \geq 0} e^{-k\varsigma} \tilde{Q}_k f \tag{5.17}$$

for all ς in V and all f in S.

We may of course assume $0 < \phi < \pi/2$, hence we may define $a = \pi/tg\phi$ so that for all b in $[-\pi, \pi]$, the point $a + ib$ belongs to V. It is then easy to check from (5.17) that

$$\tilde{Q}_k = \frac{e^{ka}}{2\pi} \int_{-\pi}^{\pi} S_{a+ib} e^{ikb} db$$

at least on S. This implies that the norm of $\tilde{Q}_k$ as an operator from S equipped with the norm of $L_p(B)$ into itself is majorized as follows

$$\|\tilde{Q}_k\| \leq e^{ka} \sup_{|b| \leq \pi} \|S_{a+ib}\| \leq M e^{ka} .$$

This shows that $\tilde{Q}_k$ is bounded, hence can be uniformly extended to an operator on $L_p(B)$. Moreover, we conclude as announced that

$$\limsup_{k \to \infty} \|Q_k \otimes I_B\|^{1/k} \leq e^a < \infty .$$

This ends the proof of theorem 5.5.

Proof of theorem 5.4: This is now obvious. By theorem 5.5 if B does not contain ℓ_1^n's uniformly, $R_1 \otimes I_B$ is bounded on $L_p(B)$ for $1 < p < \infty$, hence B is K-convex. The converse was already proved in proposition 5.1. *q.e.d.*

For a shortcut in the proof of theorem 5.4, see [F2].

Remark 5.9: One can treat similarly several other projections than the already mentioned R_1, G_1 or $(Q_k)_{k \geq 0}$.

(i) For instance, consider the Euclidean sphere S_n in $\mathbf{R}^{n+1}$ equipped with its normalized canonical measure λ_n. Let us denote by L_n the span of the $n+1$ coordinate functions in $L_2(S_n, \lambda_n)$, and by ρ_n the orthogonal projection from $L_2(S_n, \lambda_n)$ onto L_n. Let $1 < p < \infty$. Using similar ideas as above, it is easy to show that a Banach space B is K-convex iff $\{\rho_n \otimes I_B\}$ is a *uniformly* bounded sequence of operators on $L_p(S_n, \lambda_n)$.

One could also consider spherical harmonics of higher degree $k > 1$ in analogy with Q_k.

(ii) In the Gaussian case, let us denote by H_k the k-th Wiener chaos (i.e. the closed span of all the Hermite polynomials of degree exactly k in the infinitely many variables $g_1, g_2, \ldots, g_n, \ldots$). Then if B is K-convex the orthogonal projection G_k onto H_k defines a bounded operator $G_k \otimes I_B$ on $L_p(\Omega, \mathbf{P}; B)$ for all $k \geq 0$ and $1 < p < \infty$, and moreover $\|G_k \otimes I_B\| \leq C^k$ for some constant $C = C(p, B)$. See [P5] for more details.

(iii) Actually it is proved in [P5] that if B does not contain ℓ_1^n's uniformly, then any Markovian convolution semi-group relative to any compact group G is holomorphic on $L_p(G; B)$, if $1 < p < \infty$.

We now turn to the connection between the notion of K-convexity and the ℓ_2^n-subspaces of a Banach space (or the spherical sections of convex bodies in Dvoretzky's terminology). We need more definitions. Let B be a Banach space. We will say that B contains *uniformly complemented* ℓ_2^n's if there is a constant C such that for each $\epsilon > 0$ and for each n, there is a subspace $F_n \subset B$ and a projection $P_n : B \to F_n$ such that $F_n \overset{1+\epsilon}{\sim} \ell_2^n$ and $\|P_n\| \leq C$. Note that by Dvoretzky's theorem (cf. theorem 1.2), if this holds for *some* $\epsilon > 0$, then it automatically holds for *all* $\epsilon > 0$. We will see below (following [FT]) that K-convex spaces possess the preceding property.

Let us say that a subspace $F \subset B$ is C-complemented in B if there is a linear projection $P : B \to F$ with $\|P\| \leq C$.

Actually K-convexity is equivalent to a strengthened form of the preceding property where we require that the spaces B_n can be found roughly everywhere in B. More precisely, we will say that B is *locally π-euclidean* if there is a constant C such that for each $\epsilon > 0$ and each integer n, there is an integer $N(n, \epsilon)$ such that every subspace $E \subset B$ with $\dim E \geq N$ contains an n-dimensional subspace $F \subset E$ such that $F \overset{1+\epsilon}{\sim} \ell_2^n$ and F is C-complemented in B. We will prove below the following

Theorem 5.10: A Banach space B is locally π-euclidean iff B is K-convex.

Note: The "if" part was proved in [FT] and the converse in [P5].

To prove the "if" part, we will proceed as in chapter 1, i.e. we first state a theorem connecting B-valued Gaussian r.v.'s and complemented ℓ_2^n's (in analogy with theorem 1.3).

Theorem 5.11: Let B be a real Banach space. Let X (resp. X^*) be a B-valued (resp. B^*-valued) Gaussian r.v.. Let us denote simply by $\|X\|_2$ (resp. $\|X^*\|_2$) the norm of X (resp. X^*) in $L_2(B)$ (resp. $L_2(B^*)$). We assume that for some constant C

$$\|X\|_2 \|X^*\|_2 \leq C\mathbf{E} < X^*, X > . \tag{5.18}$$

Then for each $\epsilon > 0$ there is a number $\eta_1(\epsilon) > 0$ such that B contains a subspace F of dimension $n = [\eta_1(\epsilon) C^{-1} \min(d(X), d(X^*))]$ such that $F \overset{1+\epsilon}{\sim} \ell_2^n$ and F is $(8C)$-complemented in B. Moreover, F is included in the span of the range of X.

For the proof, we will use the following technical but elementary lemma.

Lemma 5.12: Consider operators $\alpha : \ell_2^k \to B$ and $\beta : B \to \ell_2^k$ satisfying $\|\alpha\|\|\beta\| \leq 2$ and $tr(\beta\alpha) \geq C^{-1}k$. Then there is a subspace F of the range of α of dimension $n = [(4C)^{-1}k]$ which is $4C$-complemented in B and 2-isomorphic to ℓ_2^n.

Proof: The idea is to show that by suitably restricting $\beta\alpha$ we can obtain a factorization of the identity of ℓ_2^n through B with $n \geq (4C)^{-1}k$. Here are the routine details. First note that $tr|\beta\alpha| \geq tr\ \beta\alpha$. By the polar decomposition $\beta\alpha = U|\beta\alpha|$, we may assume (replacing β by $U^*\beta$) that $\beta\alpha$ is hermitian and positive. After a change of orthonormal basis in ℓ_2^k, we may as well assume that $\beta\alpha$ is a diagonal operator relative to an orthonormal basis (e_i) with positive coefficients $\lambda_1 \geq \ldots \geq \lambda_k$ such that

$$(5.19) \qquad \sum_1^k \lambda_i \geq C^{-1}k \quad \text{and} \quad |\lambda_i| \leq 2\,.$$

Let $k' = [k/4C]$. Clearly (5.19) implies $2k' + k\lambda_{k'} \geq C^{-1}k$ hence

$$(5.20) \qquad \lambda_{k'} \geq (2C)^{-1}\,.$$

Let now $F_1 \subset \ell_2^k$ be the span of $\{e_1, \ldots, e_{k'}\}$ and let $\gamma : \ell_2^k \to \ell_2^k$ be the operator defined by $\gamma e_i = \lambda_i^{-1} e_i$ if $i \leq k'$ and $\gamma e_i = 0$ if $i > k'$. By (5.20) we have $\|\gamma\| \leq 2C$. Is is then easy to check that $P = \alpha\gamma\beta : B \to B$ is a projection onto $F = \alpha(F_1)$ and moreover $\|P\| \leq \|\alpha\|\|\beta\|\|\gamma\| \leq 4C$. On the other hand $\alpha\gamma$ restricted to F_1 is an isomorphism between F_1 and $\alpha(F_1)$ so that $\dim F = \dim F_1 = k'$ and clearly F is 2-isomorphic with F_1. *q.e.d.*

The preceding lemma reduces the existence of complemented ℓ_2^n-subspaces to the factorization of certain "thick" operators through B. As we will immediately show, the latter factorization is easy to obtain in the situation of theorem 5.11.

Proof of theorem 5.11: Let (X_i, X_i^*) be an i.i.d. sequence of $B \times B^*$-valued r.v.'s, each with the same distribution as (X, X^*). Let (e_i) be the canonical basis of ℓ_2^k. We introduce the random operators

$$\alpha_\omega : \ell_2^k \to B \quad \text{and} \quad \beta_\omega : B \to \ell_2^k$$

defined by $\alpha_\omega e_i = X_i(\omega)$ and $\beta_\omega^* e_i = X_i^*(\omega)$. By the remark following the proof of theorem 1.3, there is a function $\eta'(\epsilon)$ such that if

$$k \leq \eta'(\epsilon)\min\{d(X), d(X^*)\}$$

we have

$$(\mathbf{E}\|\alpha_\omega\|^2)^{1/2} \leq (1+\epsilon)^{1/2}\mathbf{E}\|X\| \leq (1+\epsilon)^{1/2}\|X\|_2$$

and

$$(\mathbf{E}\|\beta_\omega\|^2)^{1/2} \leq (1+\epsilon)^{1/2}\mathbf{E}\|X^*\| \leq (1+\epsilon)^{1/2}\|X^*\|_2\,.$$

Assume that $\epsilon = 1$. We have then

$$\mathbf{E}\|\alpha_\omega\|\|\beta_\omega\| \leq (\mathbf{E}\|\alpha_\omega\|^2\mathbf{E}\|\beta_\omega\|^2)^{1/2} \leq 2\|X\|_2\|X^*\|_2 ,$$

hence by (5.18)

(5.21) $$\mathbf{E}\|\alpha_\omega\|\|\beta_\omega\| \leq 2C\mathbf{E} < X^*, X > .$$

On the other hand, a simple computation shows that the matrix of $\beta_\omega\alpha_\omega$ is nothing but $(< X_j^*, X_i >)_{ij}$ so that $tr\ \beta_\omega\alpha_\omega = \sum_1^k < X_i^*, X_i >$ hence

$$\mathbf{E}\ tr\ \beta_\omega\alpha_\omega = k\mathbf{E} < X^*, X > .$$

Therefore (5.21) implies

$$\mathbf{E}\|\alpha_\omega\|\|\beta_\omega\| \leq 2Ck^{-1}\mathbf{E}\ tr(\beta_\omega\alpha_\omega) ,$$

so that there is necessarily a choice of ω such that

$$\|\alpha_\omega\|\|\beta_\omega\| \leq 2Ck^{-1}tr(\beta_\omega\alpha_\omega) .$$

By homogeneity, we may obtain exactly the assumption of Lemma 5.12, and this yields theorem 5.11 in the particular case $\epsilon = 1$. To obtain the general case, we can simply apply theorem 1.3 to obtain a further subspace (which will still be $(8C)$-complemented) $(1+\epsilon)$-isomorphic to a Euclidean space. { Note: One can also obtain this by suitably modifying the preceding proof. } q.e.d

We now give an application of theorem 5.11 which first appeared in [FT].

Theorem 5.13: Let B be a K-convex Banach space. Assume that B^* is of cotype q_*. There are positive constants K_1, K_2 with the following property. Let $E \subset B$ be a subspace 2-isomorphic to ℓ_2^N. There is a subspace $F \subset E$ of dimension $n = [K_1 N^{2/q_*}]$ which is 2-isomorphic to ℓ_2^n and K_2-complemented in B.

As in [FT] we rely on the following

Lemma 5.14: Let Y be a Banach space of cotype q. Let $v : \ell_2^N \to Y$ be an operator, let

$$\lambda = \left\|\sum_1^N g_i v e_i\right\|_2$$

Let $m = N - [N/2]$. Then there is a subspace $S \subset \ell_2^N$ of dimension m with a basis $(f_1, \ldots, f_m)$ such that

(5.22) $$\|v_{|S}\| \leq \lambda C_q(Y)(2/N)^{1/q}$$

and

$$\Big\|\sum_{i\le m} g_i v f_i\Big\|_2 \le \lambda\,. \tag{5.22'}$$

Proof: Let $\overline{N} = [N/2]$. Proceeding as in the proof of lemma (1.8), we construct by induction an orthonormal basis $\{y_1,\dots,y_N\}$ of ℓ_2^n such that

$$\|vy_1\| = \|v\|,\quad \|vy_2\| = \|v_{|\{y_1\}^\perp}\|,\dots$$
$$\|vy_{k+1}\| = \|v_{|\{y_1,\dots,y_k\}^\perp}\|\,.$$

Note that by the rotational invariance of Gaussian measures

$$\sum g_i v y_i \stackrel{d}{=} \sum g_i v e_i \tag{5.23}$$

so that we have

$$\Big(\sum \|vy_i\|^q\Big)^{1/q} \le C_q(Y)\Big\|\sum g_i v y_i\Big\|_2 = C_q(Y)\lambda.$$

Since $\{\|vy_i\|\}$ is non-increasing, this implies

$$\|vy_{\overline{N}+1}\| \le \lambda C_q(Y)(2/N)^{1/q}\,.$$

We let then $S = [y_1,\dots,y_{\overline{N}}]^\perp$, so that (5.22) holds. Let $f_i = y_{\overline{N}+i}$ with $\overline{N} = [N/2]$. We have then obviously

$$\Big\|\sum_{i\le m} g_i v f_i\Big\|_2 \le \Big\|\sum_{i\le N} g_i v y_i\Big\|_2$$

hence by (5.23)

$$\le \lambda\,,$$

so that (5.22)' also holds. This concludes the proof of lemma 5.14.

Proof of theorem 5.13: By theorem 5.11, the proof reduces to producing suitable Gaussian r.v.'s X (E-valued) and X^* (B^*-valued). Here is how one can proceed. Let $\beta : E \to \ell_2^N$ be an isomorphism such that $\|\beta\| = 1$ and $\|\beta^{-1}\| \le 2$. Note that E is a quotient of B^*. Let us denote by $Q : B^* \to E^* = B^*/E^\perp$ the quotient map. We have clearly

$$\Big\|\sum \epsilon_i \beta^* e_i\Big\|_2 \le \|\beta\|\Big\|\sum \epsilon_i e_i\Big\|_2 = \sqrt{N}\,. \tag{5.24}$$

(Here again (e_i) is the canonical basis of ℓ_2^N). Let $x_i^* = \beta^* e_i$. By an obvious lifting argument, for each fixed choice of sign $\{\epsilon_1,\dots,\epsilon_N\}$ we can find an element $F(\epsilon_1,\dots,\epsilon_N)$ in B^* such that

$$QF(\epsilon_1,\dots,\epsilon_N) = \sum \epsilon_i x_i^* \tag{5.25}$$

and

$$\|F(\epsilon_1,\dots,\epsilon_N)\| \le 2\Big\|\sum \epsilon_i x_i^*\Big\|\,. \tag{5.26}$$

Considering now F as an element of $L_2(D,\mu;B^*)$, we will apply to F the projection $R_1 \otimes I_{B^*}$ and use the fact that this projection is bounded on $L_2(B^*)$ with norm equal to $K(B^*) = K(B)$. Let z_i be the elements of B^* defined by the identity $(R_1 \otimes I_{B^*})(F) = \sum_1^N \epsilon_i z_i$. It is easy to check that

$$(R_1 \otimes I_{B^*})QF = Q(R_1 \otimes I_{B^*})(F)$$

so that (5.25) implies $Q(\sum \epsilon_i z_i) = \sum \epsilon_i x_i^*$ or equivalently $Qz_i = x_i^*$ for all $i = 1, \ldots, N$. By (5.26) and (5.24) we have

$$\begin{aligned}\|\sum \epsilon_i z_i\|_2 &\leq K(B)\|F\|_2 \leq 2K(B)\|\sum \epsilon_i x_i^*\|_2 \\ &\leq 2K(B)\sqrt{N}\end{aligned}$$

Now since B^* is of cotype q_*, by proposition 3.2 ii) there is a constant C_1 such that

$$\|\sum g_i z_i\|_2 \leq C_1\|\sum \epsilon_i z_i\|_2 \; . \tag{5.27}$$

Therefore, we find a constant C_2 such that

$$\|\sum g_i z_i\|_2 \leq C_2 N^{1/2} \; . \tag{5.28}$$

Let $v : \ell_2^N \to B^*$ be the operator defined by $ve_i = z_i$. Note that $Qv = \beta^*$. By lemma 5.14 there is a subspace $S \subset \ell_2^N$ of dimension $m = N - \overline{N} \geq [N/2]$ such that

$$\|v_{|S}\| \leq C_3 N^{\frac{1}{2} - \frac{1}{q_*}} \tag{5.29}$$

for some constant C_3. In the sequel C_4, C_5, C_6 will be constants independent of N. With the notation of lemma 5.14, we set

$$X^* = \sum_{i \leq m} g_i v f_i \quad \text{and} \quad X = \sum_{i \leq m} g_i \beta^{-1} f_i \; .$$

We claim $d(X^*) \geq C_4 N^{2/q_*}$. Indeed, we have $\sigma(X^*) \leq \|v_{|S}\|$ and

$$\begin{aligned}\|X^*\|_2 &\geq \|\sum_{i \leq m} g_i Q v f_i\|_2 = \|\sum g_i \beta^* f_i\|_2 \\ &\geq \|\beta^{-1}\|^{-1}\|\sum g_i f_i\|_2 \geq 2^{-1} m^{1/2} \; .\end{aligned}$$

For the r.v. X, we can write

$$\|X\|_2 \geq \frac{1}{\|\beta\|}\Big\|\sum_{i \leq m} g_i f_i\Big\|_2 = m^{1/2}$$

while $\sigma(X) \leq \|\beta^{-1}\| \leq 2$, so that $d(X) \geq C_5 N$. Finally, we have

$$\begin{aligned}E < X^*, X > &= \sum_{i \leq m} < v f_i, \beta^{-1} f_i > \\ &= \sum < v f_{i|_E}, \beta^{-1} f_i > \\ &= \sum < Q v f_i, \beta^{-1} f_i >\end{aligned}$$

and since $Qv = \beta^*$

$$\sum_{i \leq m} \langle \beta^* f_i, \beta^{-1} f_i \rangle = m$$

Moreover $\|X\|_2 \leq \|\beta^{-1}\| \|\sum g_i f_i\|_2 \leq 2m^{1/2}$ and by (5.22)' and (5.28) we have

$$\|X^*\|_2 \leq C_2 N^{1/2} \ .$$

In conclusion, we do find

$$\|X\|_2 \|X^*\|_2 \leq C_6 \mathbf{E} < X^*.X > ,$$

so that an application of theorem 5.11 now yields theorem 5.13. *q.e.d.*

Remark: It is known that the order of magnitude of n in theorem 5.13 cannot be improved asymptotically (for instance if $B = L_p$, $\frac{1}{p} + \frac{1}{q_*} = 1$, $q_* \geq 2$).

Remark: We refer the reader to [BG] for a different exposition of the results of [FT] using random matrices. In particular, the following result (implicit in [FT]) is proved explicitly in [BG]: Let B be a K-convex Banach space. Assume B of cotype q and B^* of cotype q^*. Then *every* N-dimensional subspace $E \subset B$ contains a subspace $F \subset E$ of dimension $n = [K_1 N^\alpha]$ with $\alpha = \min\{\frac{2}{q}, \frac{2}{q_*}\}$ which is 2-isomorphic to ℓ_2^n and C-complemented in B with $C = 4K(B)$. (Here again K_1 is a positive constant independent of N.).

Proof of theorem 5.10: Assume that B is K-convex, then by proposition 5.1, theorem 3.3 and proposition 5.2 B is of cotype q and B^* is of cotype q_* for some q and q_*. Therefore theorem 5.13 together with Dvoretzky's theorem (theorem 1.2 or theorem 3.9) immediately gives that B is locally π-euclidean.

Conversely, if B is locally π-euclidean, then B does not contain ℓ_1^n's uniformly (see the remark below) so that B is K-convex by theorem 5.4. *q.e.d.*

Remark: It is known that there is a numerical constant c such that for any factorization of the form

$$\begin{array}{ccc} & \ell_1^m & \\ \alpha \nearrow & & \searrow \beta \\ \ell_2^n & & \ell_2^n \end{array}$$

the Hilbert-Schmidt norm $\|\beta\alpha\|_{HS}$ satisfies $\|\beta\alpha\|_{HS} \leq C\|\alpha\|\|\beta\|$. This implies that if a subspace of ℓ_1^m is 2-isomorphic to ℓ_2^n, it can be λ-complemented in ℓ_1^m only if $\lambda \geq (2C)^{-1}\sqrt{n}$. We leave this as an exercise to the reader. It implies clearly that if B is locally π-euclidean then B does not contain ℓ_1^n's uniformly.

Chapter 6
Martingale Type and Cotype

We have seen in chapter 3 (cf. proposition 3.1) that type and cotype are equivalent to certain inequalities for sums of independent mean zero r.v.'s. It is natural to consider similar inequalities for sums of Banach space valued martingale differences. This was investigated in [P3]. We will say that a space B is of martingale type p (in short M-type p) if there is a constant C such that, for all martingales $(M_n)_n$ with values in B, we have

$$\sup_n \mathbf{E}\|M_n\|^p \leq C \sum_{n\geq 0} \mathbf{E}\|M_n - M_{n-1}\|^p$$

with the convention $M_{-1} \equiv 0$.

Similarly, we will say that B is of martingale cotype q (in short M-type q) if there is a C such that, for all B-valued martingales (M_n) we have

$$\sum_{n\geq 0} \mathbf{E}\|M_n - M_{n-1}\|^q \leq C \sup_n \mathbf{E}\|M_n\|^q.$$

Using classical methods from martingale theory (cf. e.g. [Bu]), one can show easily the following: If B is of M-type p (resp. M-cotype q) then for all $1 \leq r < \infty$ there is a constant C' such that for all B-valued martingales (M_n) we have

$$\mathbf{E}\sup_n \|M_n\|^r \leq C'\mathbf{E}\Big(\sum_{n\geq 0} \|M_n - M_{n-1}\|^p\Big)^{r/p}$$

$$\Big(\text{resp. } \mathbf{E}\Big(\sum_{n\geq 1} \|M_n - M_{n-1}\|^q\Big)^{r/q} \leq C'\mathbf{E}\sup \|M_n\|^r\Big).$$

We should recall that by Doob's inequality, if $1 < r < \infty$

$$(\mathbf{E}\sup \|M_n\|^r)^{1/r} \leq \frac{r}{r-1} \sup_n (\mathbf{E}\|M_n\|^r)^{1/r},$$

and

$$\sup_{t>0} t\mathbf{P}\{\sup \|M_n\| > t\} \leq \sup_n \mathbf{E}\|M_n\|.$$

These notions are only superficially similar to the usual type and cotype.

Clearly M-type p implies type p and M-cotype q implies type q. Obviously, every Banach space is of M-type 1 and of M-cotype ∞. Also, it is easy to show that B is of M-type p iff B^* is of M-cotype p' with $\frac{1}{p} + \frac{1}{p'} = 1$. These notions have a geometric characterization in terms of uniform smoothness and uniform convexity.

To each space B, we associate the moduli of smoothness and convexity which are defined as follows.

$$\rho_B(t) = \sup\{\frac{\|x + ty\| + \|x - ty\|}{2} - 1 \mid x, y \in B \quad \|x\| = \|y\| = 1\}$$

$$\delta_B(\epsilon) = \inf\{1 - \|\frac{x+y}{2}\| \mid x, y \in B \quad \|x\| = \|y\| = 1 \quad \|x - y\| \geq \epsilon\}$$

B is called uniformly smooth (resp. uniformly convex) if $\rho_B(t)/t \to 0$ when $t \to 0$ (resp. $\delta_B(\epsilon) > 0$ for all $0 < \epsilon \leq 2$).

We will say here that B is p-smooth (resp. q-convex) if it admits an equivalent norm for which the modulus of smoothness $\rho(t)$ (resp. convexity $(\delta(\epsilon))$ satisfies $\rho(t) \leq Kt^p$ for all $t > 0$ (resp. $\delta(\epsilon) \geq K\epsilon^q$ for all $0 < \epsilon \leq 2$) for some constant K. The notions of M-type p and M-cotype q are then completely elucidated by the following

Theorem 6.1: Let $1 \leq p \leq 2 \leq q < \infty$. A Banach space B is of M-type p (resp. M-cotype q) iff it is p-smooth (resp. q-convex).

Theorem 6.2: A space B is of M-type p for some $p > 1$ (resp. M-cotype q for some $q < \infty$) iff B has an equivalent uniformly smooth (resp. convex) norm.

For the proofs of these results, we refer to [P3]. We should mention that the class of spaces appearing in theorem 6.2 coincides with the class of super-reflexive spaces studied by James and Enflo, see [P3] and the references there for more details. See also [Ga].

By theorem 6.2, if B is of M-type $p > 1$ (or M-cotype $q < \infty$) then B is reflexive since uniform smoothness (resp. convexity) implies reflexivity. This shows how different these notions are from the usual type and cotype, since we have

Theorem 6.3: [J] There is a Banach space of type 2 (hence of cotype q for some $q < \infty$) which is not reflexive, therefore this space is of M-type p for *no* $p > 1$ and of M-cotype q for *no* $q < \infty$.

Remark: It is possible to find a uniformly convex space B for which the index of type $p(B)$ differs from the corresponding index for the M-type. Similarly for the cotype. See [P4] for details.

Finally, let us discuss the connection of these notions with that of UMD spaces which is studied in the lectures of D. Burkholder. It is very easy to see that for UMD spaces these complications do not appear. Indeed, if a UMD space B is of type p then clearly B is of M-type p. Similarly for the cotype.

Moreover, a UMD space cannot contain ℓ_1^n's or ℓ_∞^n's uniformly (simply because L_1 and L_∞ are not UMD). Therefore, by theorem 3.3 and the above remark, a UMD space must be of M-type p and M-cotype q for some $1 < p \leq q < \infty$. In particular, a UMD space has an equivalent uniformly convex norm. However, the converse is not true as shown by the counterexample in [P4], which is uniformly convex but *not* UMD. For a similar example in the Banach lattice situation see [B2]. This shows that even among Banach lattices (or rearrangement invariant spaces) the class of UMD spaces is smaller than the class of uniformly convex spaces.

Chapter 7
Type for Metric Spaces

In this chapter we study the notion of type for metric spaces which was introduced in [BMW]. We present some of the results of [BMW] as well as a new inequality (lemma 7.3) which relates the usual type of a Banach space with its "metric type". This gives a simpler proof of the corresponding theorem of [BMW]. We start with the observation that certain inequalities such as the parallelogram inequality

$$(\dagger) \qquad \|x+y\|^2 + \|x-y\|^2 \le 2\|x\|^2 + 2\|y\|^2$$

(which characterizes Hilbert spaces) are actually purely metric. Indeed, if d is the distance associated to the norm, $(\dagger)$ is equivalent to: for any collection of four points $(x_{\epsilon_1\epsilon_2})$ indexed by $\{-1,\ +1\}^2$ we have

$$(*) \qquad \sum_{\epsilon_1,\epsilon_2} d(x_{\epsilon_1\epsilon_2},\ x_{-\epsilon_1-\epsilon_2})^2 \le \sum_{\epsilon_1,\epsilon_2} d(x_{\epsilon_1\epsilon_2},\ x_{-\epsilon_1\epsilon_2})^2 + d(x_{\epsilon_1\epsilon_2},\ x_{\epsilon_1-\epsilon_2})^2.$$

The left (resp. right) side can be interpreted as the sum of the squared lengths of the main diagonals (resp. edges) of the "2 dimensional cube" with vertices $\{x_{\epsilon_1\epsilon_2} \mid \epsilon_1 = \pm 1,\ \epsilon_2 = \pm 1\}$. For four points in the plane, our terminology corresponds to the following diagram:

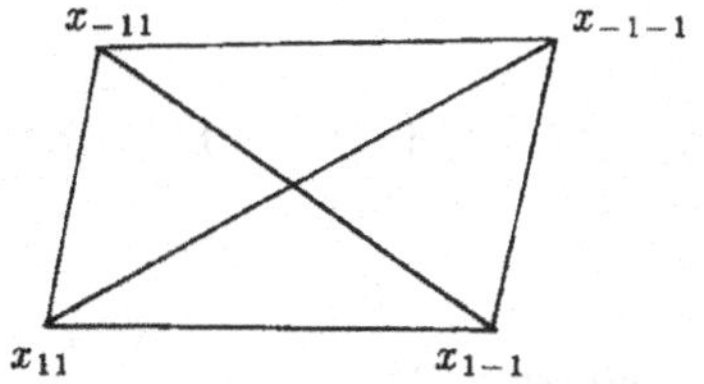

Various considerations lead to study $(*)$ with an exponent p replacing 2 and also to generalize $(*)$ to "n-dimensional cubes" instead of 2-dimensional ones.

Let (T,d) be a metric space. By an "n-dimensional cube" in T, we mean a function

$$f : \{-1,+1\}^n \to T\ .$$

Recall the notation $D = \{-1,+1\}^{\mathbb{N}}$, we again denote by ϵ_i the i-th coordinate on $\{-1,+1\}^{\mathbb{N}}$ or sometimes on $\{-1,+1\}^n$ and we denote by μ the normalized uniform measure on D. For each $\epsilon = (\epsilon_1,\ldots,\epsilon_n)$ in $\{-1,+1\}^n$, we denote $-\epsilon = (-\epsilon_1,-\epsilon_2,\ldots,-\epsilon_n)$. For any $f : \{-1,+1\}^n \to T$ and for any $i = 1,2,\ldots,n$, we denote

$$\Delta_i f = d(f(\epsilon_1,\ldots,\epsilon_i,\ldots,\epsilon_n),\ f(\epsilon_1,\ldots,-\epsilon_i,\ldots,\epsilon_n)).$$

(only the i-th sign is changed in the preceding line).

Definition: A metric space (T,d) will be called of metric type p $(1 \le p \le 2)$ if there is a constant C such that for all n and for all $f : \{-1,+1\}^n \to T$ we have

$$\left(\int d(f(\epsilon),f(-\epsilon))^2 d\mu\right)^{1/2} \le C n^{\frac{1}{p}-\frac{1}{2}} \left(\sum_1^n \int (\Delta_i f)^2 d\mu\right)^{1/2}. \tag{7.1}$$

Clearly, if $1 \le p_1 \le p_2 \le 2$, metric type p_2 implies metric type p_1. Moreover, it is rather easy to see that every metric space is of type 1 (with constant 1). Indeed, let $f_0 = f$ and

$$f_i(\epsilon) = f(-\epsilon_1,\dots,-\epsilon_i,\epsilon_{i+1},\dots,\epsilon_n).$$

Then by the triangle inequality we have

$$\begin{aligned} d(f(\epsilon),f(-\epsilon)) &\le \sum_{i=1}^n d(f_{i-1},f_i) \\ &\le \sqrt{n}\left(\sum d(f_{i-1},f_i)^2\right)^{1/2} \end{aligned} \tag{7.2}$$

and integrating the square of the last inequality, we find that T satisfies (7.1) with $p = C = 1$. (Note that $d(f_{i-1},f_i)$ and $\Delta_i f$ have the same distribution). In analogy with the linear theory, one can characterize the metric spaces which have a nontrivial type.

Let us denote by C_p^n the set $\{-1,+1\}^n$ equipped with the metric d_p induced by the norm of ℓ_p^n, i.e. $\forall \epsilon,\epsilon' \in \{-1,+1\}^n$

$$d_p(\epsilon,\epsilon') = \left(\sum |\epsilon_i - \epsilon_i'|^p\right)^{1/p} = 2 \text{ card } \{i \mid \epsilon_i \ne \epsilon_i'\}^{1/p}.$$

Let ϕ be a map between metric spaces, we will denote

$$\|\phi\|_{\mathrm{Lip}} = \sup_{s \ne t} \frac{d(\phi(s),\phi(t))}{d(s,t)}.$$

We say that a metric space T contains C_p^n's $(1+\epsilon)$-uniformly if for each n there is a subset $T_n \subset T$ and a bijection $\phi_n : C_p^n \to T_n$ such that

$$\|\phi_n\|_{\mathrm{Lip}} \|\phi_n^{-1}\|_{\mathrm{Lip}} \le 1+\epsilon.$$

We can now state

Theorem 7.1 [BMW] The following properties of a metric space T are equivalent:

i) The space T is of metric type p for some $p > 1$.
ii) For all $\epsilon > 0$, T does not contain C_1^n's $(1+\epsilon)$-uniformly.
iii) For some $\epsilon > 0$, T does not contain C_1^n's $(1+\epsilon)$-uniformly.

Proof: Let us first check that i) $\Rightarrow$ ii). Assume that ii) fails. Then for some $\epsilon > 0$ there are subsets $T_n \subset T$ and bijections $f_n : C_1^n \to T_n$ such that

$$\|f_n\|_{\mathrm{Lip}} \|f_n^{-1}\|_{\mathrm{Lip}} \leq 1+\epsilon.$$

Let $a = \|f_n\|_{\mathrm{Lip}}$ and $b = \|f_n^{-1}\|_{\mathrm{Lip}}$ so that $ab \leq 1+\epsilon$. We have

$$d(f_n(\epsilon), f_n(-\epsilon)) \geq b^{-1} d_1(\epsilon, -\epsilon) = b^{-1} 2n$$

and

$$\Delta_i f(\epsilon) \leq 2a \quad \text{for all } \epsilon \text{ in } \{-1,+1\}^n.$$

If T satisfied 7.1 we would have

$$2n \leq bC2an^{1/p} \leq 2C(1+\epsilon)n^{1/p}$$

which is impossible when $n \to \infty$ and $p > 1$. This shows that (i) $\Rightarrow$ (ii). The implication (ii) $\Rightarrow$ (iii) is trivial. To show that (iii) $\Rightarrow$ (i), we introduce the number t_n which is the smallest constant C such that for all $f : \{-1,+1\}^n \to T$ we have

$$\int d(f(\epsilon), f(-\epsilon))^2 d\mu \leq C^2 \sum_1^n \int (\Delta_i f)^2 d\mu.$$

Clearly (7.2) implies $t_n \leq \sqrt{n}$. The following lemma will be important.

Lemma 7.2: $t_{nk} \leq t_n t_k$ for all integers n and k.

Proof: Consider $f : \{-1,+1\}^{nk} \to T$. For all $(\xi_1, \ldots, \xi_k)$ in $\{-1,+1\}^k$ let

$$F_\epsilon(\xi_1, \ldots \xi_k) = f(\epsilon_1\xi_1, \ldots, \epsilon_n\xi_1, \epsilon_{n+1}\xi_2, \ldots, \epsilon_{2n}\xi_2, \ldots, \epsilon_{n(k-1)+1}\xi_k, \ldots, \epsilon_{nk}\xi_k)$$

We have

(7.3)
$$\int d(F_\epsilon(\xi), F_\epsilon(-\xi))^2 d\mu(\xi) \leq t_k^2 \sum_1^k \int (\Delta_i F_\epsilon)^2 d\mu(\xi).$$

Observe that

$$\int d(f(\epsilon), f(-\epsilon))^2 d\mu(\epsilon) = \int d(F_\epsilon(\xi), F_\epsilon(-\xi))^2 d\mu(\xi) d\mu(\epsilon).$$

Moreover, if we consider $F_\epsilon(\xi)$ as a function of $\{\epsilon_j | (i-1)n < j \leq in\}$, we obtain (after integration)

$$\int \Delta_i F_\epsilon^2 d\mu(\xi) d\mu(\epsilon) \leq t_n^2 \sum_{(i-1)n<j\leq in} \int (\Delta_j f)^2 d\mu(\epsilon),$$

hence integrating (7.3) over ϵ we find

$$\int d(f(\epsilon), f(-\epsilon))^2 d\mu \leq t_n^2 t_k^2 \sum_{1\leq j\leq nk} \int (\Delta_j f)^2 d\mu$$

which implies lemma 7.2.

End of the proof of theorem 7.1: Note that $t_n \le t_{n+1}$ for all n. Therefore lemma 7.2 implies the following dichotomy. Either $t_n = \sqrt{n}$ for all n, or there is an $n_0 > 1$ such that $t_{n_0} < \sqrt{n_0}$. In the second case, let p be such that $t_{n_0} = (n_0)^{\frac{1}{p}-\frac{1}{2}}$. Note that $1 < p \le 2$. Let $\alpha = \frac{1}{p} - \frac{1}{2}$. By lemma 7.2 we have $t_{n_0}^k \le n_0^{k\alpha}$ hence $t_n \le n_0^{\alpha} n^{\alpha}$ for all integers n.

Therefore, the second case implies that T is of metric type p for some $p > 1$. Hence, to show (iii) $\Rightarrow$ (i), it suffices to show that (iii) contradicts the first case. In other words, it suffices to show that if $t_n = \sqrt{n}$ for all n then T contains C_1^n's uniformly. Assume that $t_n = \sqrt{n}$. Then for all $\delta > 0$ there is a function $f : \{-1,+1\}^n \to T$ and a number $a > 0$ such that

$$\Big(\sum_1^n \int (\Delta_i f)^2 d\mu\Big)^{1/2} \le a\sqrt{n}(1+\delta)^{1/2}$$

and

$$\Big(\int d(f(\epsilon), f(-\epsilon))^2 d\mu\Big)^{1/2} \ge an\ .$$

We will show that when δ is "very small", f is an "embedding" of C_1^n into T. We will use (7.2) Let

$$\phi_i(\epsilon) = a^{-1} d(f_{i-1}(\epsilon), f_i(\epsilon)).$$

By (7.2) we have

$$\|\sum \phi_i\|_2 \ge n \tag{7.4}$$

and (since ϕ_i and $a^{-1}\Delta_i f$ have the same distribution)

$$\|\Big(\sum \phi_i^2\Big)^{1/2}\|_2 \le \sqrt{n}(1+\delta)^{1/2}. \tag{7.5}$$

Observe that $\sum \phi_i \le \sqrt{n}(\sum \phi_i^2)^{1/2}$ pointwise. We claim that (7.4) and (7.5) imply the following pointwise inequalities

$$\sum_1^n \phi_i \ge n - \psi(\delta)$$

and

$$|\phi_i - 1| \le \psi(\delta)$$

where $\psi(\delta)$ is a certain function such that $\psi(\delta) \to 0$ when $\delta \to 0$.

(*Note:* Here n is fixed, or equivalently $\psi(\delta)$ depends on n.)

The idea underlying this statement is simple: if (7.4) and (7.5) hold with $\delta = 0$ then we must have $\phi_i = \phi_j$ for all i and j. Since ϕ_i does not depend on ϵ_i, this implies that each ϕ_i is constant, and the normalization forces $\phi_i \equiv 1$. To prove the claim, we observe

$$\frac{1}{2}\sum_{ij}(\phi_i - \phi_j)^2 = n\sum \phi_i^2 - \Big(\sum \phi_i\Big)^2. \tag{7.6}$$

Let $S_1 = (\sum \phi_i)^2$ and $S_2 = n\sum \phi_i^2$. Note $S_1 \le S_2$. By (7.4) and (7.5) we have (the integrals are with respect to $d\mu$)

$$\int S_2 \le \int S_1(1+\delta)$$

hence

$$\int S_2 - S_1 \le \delta \int S_1 \le \delta \int S_2 \le \delta(1+\delta)n^2.$$

By (7.6) this implies

$$\int (\phi_i - \phi_j)^2 \le \delta(1+\delta)n^2.$$

Since there are only 2^n points in $\{-1,+1\}^n$ this implies

$$\sup_\epsilon |\phi_i(\epsilon) - \phi_j(\epsilon)|^2 \le \delta(1+\delta)n^2 2^n,$$

hence

$$\|\phi_i - \phi_j\|_\infty \le \psi_1(\delta) \quad \text{for all } i,j = 1,2,\ldots,n, \tag{7.7}$$

with $\psi_1(\delta) = (\delta(1+\delta)n^2 2^n)^{1/2}$. Let $\phi = \frac{1}{n}\sum \phi_i$. Clearly (7.7) implies

$$\|\phi_i - \phi\|_\infty \le \psi_1(\delta) \quad \text{for all } i. \tag{7.8}$$

Let E_i be the conditional expectation with respect to $\{\epsilon_1,\ldots,\epsilon_i\}$. Since ϕ_i does not depend on ϵ_i we have

$$E_i\phi_i = E_{i-1}\phi_i \tag{7.9}$$

and by (7.8)

$$\|E_k\phi_i - E_k\phi\|_\infty \le \psi_1(\delta) \quad \text{for all } i,k.$$

Hence by (7.9)

$$\|E_i\phi - E_{i-1}\phi\|_\infty \le 2\psi_1(\delta)$$

which implies

$$\|\phi - \int \phi\|_\infty \le 2n\psi_1(\delta). \tag{7.10}$$

We know from (7.4) and (7.5) that

$$1 \le \|\phi\|_2 \le (1+\delta)^{1/2}.$$

By (7.10) we have

$$|\,\|\phi\|_2 - \int \phi| \le \|\phi - \int \phi\|_2 \le 2n\psi_1(\delta)$$

hence

$$|\int \phi - 1| \le \psi_2(\delta)$$

with

$$\psi_2(\delta) = |(1+\delta)^{1/2} - 1| + 2n\psi_1(\delta),$$

and by (7.10) again

$$\|\phi - 1\|_\infty \leq \psi_3(\delta) = 2n\psi_1(\delta) + \psi_2(\delta).$$

On the other hand, by (7.8) $\|\phi_i - 1\|_\infty \leq \psi_1(\delta) + \psi_3(\delta)$. This completes the proof of the above claim since $\psi(\delta) = n(\psi_1(\delta) + \psi_3(\delta))$ tends to zero with δ.

Now let $\chi(\epsilon) = \frac{1}{a}d(f(\epsilon), f(-\epsilon))$. We have $\|\chi\|_2 \geq n$ and $\chi \leq n + n\psi(\delta)$ pointwise. Let m be the maximum of χ. We have $n^2 \leq |\chi|_2^2 \leq \frac{1}{2^n}m^2 + (1 - \frac{1}{2^n})[n + n\psi(\delta)]^2$ hence $m^2 \geq n^2 - (2^n - 1)(2n^2\psi(\delta) + n^2\psi(\delta)^2)$. This implies the following pointwise inequality

$$\chi \geq n - \psi_4(\delta) \tag{7.11}$$

for some function $\psi_4(\delta)$ with $\psi_4(\delta) \to 0$ when $\delta \to 0$.

We can now finish the proof of theorem 7.1. Consider ϵ, ϵ' in $\{-1, +1\}^n$. Let $\Delta(\epsilon, \epsilon') = \text{card}\{i | \epsilon_i \neq \epsilon'_i\}$. Note that $d_1(\epsilon, \epsilon') = 2\Delta_1(\epsilon, \epsilon')$. The metric Δ is sometimes called the Hamming metric (it is more often considered on $\{0, 1\}^n$). Since $\phi_i \leq 1 + \psi(\delta)$ everywhere, the triangle inequality implies

$$\frac{1}{a}d(f(\epsilon, f(\epsilon')) \leq (1 + \psi(\delta))\Delta_1(\epsilon, \epsilon'). \tag{7.12}$$

On the other hand, by (7.11)

$$\begin{aligned}
\frac{1}{a}d(f(\epsilon), f(\epsilon')) &\geq \frac{1}{a}d(f(\epsilon), f(-\epsilon)) - \frac{1}{a}d(f(-\epsilon), f(\epsilon')) \\
&\geq n - \psi_4(\delta) - \frac{1}{a}d(f(-\epsilon), f(\epsilon')) \quad \text{hence by (7.12)} \\
&\geq n - \psi_4(\delta) - (1 + \psi(\delta))\Delta_1(-\epsilon, \epsilon') \\
&\geq \Delta_1(\epsilon, \epsilon') - \psi_4(\delta) - n\psi(\delta) \\
&\geq \Delta_1(\epsilon, \epsilon')(1 - \psi_4(\delta) - n\psi(\delta)).
\end{aligned}$$

We now conclude: if T_n is the image of the map $f : C_1^n \to T$, the last inequality implies

$$\|f^{-1}_{|T_n}\|_{\text{Lip}} \leq 2a^{-1}(1 - \psi_4(\delta) - n\psi(\delta))^{-1}$$

and by (7.12)

$$\|f\|_{\text{Lip}} \leq \frac{a}{2}(1 + \psi(\delta)).$$

This shows that

$$\|f^{-1}_{|T_n}\|_{\text{Lip}}\|f\|_{\text{Lip}} \leq 1 + \psi_5(\delta)$$

for some function $\psi_5(\delta)$ tending to zero with δ.

Since we can choose δ as small as we please for each fixed n, we conclude as announced that if $t_n = \sqrt{n}$ for all n then T contains C_1^n's $(1 + \epsilon)$-uniformly for all $\epsilon > 0$. This completes the proof that (iii) $\Rightarrow$ (i) in theorem 7.1.

Remark: The extension of theorem 7.1 with some number p_0 such that $1 < p_0 < 2$ replacing 1 is considered in [BMW] but with some restriction. We will not pursue this here.

We now turn to the natural problem to compare the metric type with the usual type in the case our metric space is actually a normed space. To do this, the following lemma will be very useful. It is analogous (but the $\log n$ factor makes it weaker than) corollary 2.4.

Lemma 7.3: Let B be a Banach space. Consider a function $f : \{-1,+1\}^n \to B$. Let

$$D_i f = \frac{1}{2}[f(\epsilon_1,\dots,\epsilon_i,\dots,\epsilon_n) - f(\epsilon_1,\dots,-\epsilon_i,\dots,\epsilon_n)].$$

We have then for any $p \geq 1$ and any $n > 1$

$$\Big(\int \|f - \int f d\mu\|^p d\mu\Big)^{1/p} \leq 2e \log n \Big(\int \|\sum y_i D_i f(x)\|^p d\mu(x) d\mu(y)\Big)^{1/p}.$$

We need to introduce more notation for the proof. For any $A \subset \mathbf{N}$ with $|A| < \infty$, let $w_A = \prod_{n \in A} \epsilon_n$. The functions $\{w_A | A \subset \mathbf{N},\ |A| < \infty\}$ form an orthonormal basis of $L_2(D,\mu)$. For $0 \leq \epsilon \leq 1$, we introduce the operator

$$T(\epsilon) : L_2(D,\mu) \to L_2(D,\mu)$$

defined by

$$T(\epsilon) w_A = \epsilon^{|A|} w_A \quad \forall\ A \subset \mathbf{N},\ |A| < \infty.$$

It is rather easy to check that $T(\epsilon)$ is a positive contraction on $L_p(\mu)$ for $1 \leq p \leq \infty$.

Clearly, $T(\epsilon)$ can be extended naturally to a contraction on $L_p(\mu; B)$ for $1 \leq p \leq \infty$. We still denote by $T(\epsilon)$ its extension on $L_p(\mu, B)$. Consider a function $f : \{-1,+1\}^n \to B$. We will use the following easy observation

$$\epsilon^n \|f\|_{L_p(B)} \leq \|f\|_{L_p(B)}. \tag{7.13}$$

Indeed this is immediate for $n = 1$ and can then be proved by induction on n by successive integrations. In the sequel, we will often write simply $\|f\|_p$ for the norm in $L_p(D,\mu; B)$.

Proof of lemma 7.3: We use a duality argument. We may as well assume that B is finite dimensional so that $L_p(B)^* = L_{p'}(B^*)$ with $\frac{1}{p} + \frac{1}{p'} = 1$. Let $0 \leq \epsilon \leq 1$. We use the letters x, y for elements of $\{-1,+1\}^n$. Consider a function $g : \{-1,+1\}^n \to B^*$. We can develop g over the Walsh system $\{w_A\}$ and write

$$g = \sum_{A \subset \{1,\dots,n\}} g_A w_A$$

with

$$g_A = \int w_A g d\mu.$$

We define g_ϵ on $\{-1,+1\}^n \times \{-1,+1\}^n$ as follows

$$g_\epsilon(x,y) = \sum_A g_A \prod_{i \in A} (\epsilon x_i + (1-\epsilon) y_i).$$

We observe the following crucial identity

$$g_\epsilon(x,y) = (1-\epsilon)\sum_1^n y_i T(\epsilon) D_i g(x) + \Phi(x,y) \tag{7.14}$$

where Φ satisfies $\int \Phi y_i d\mu(y) = 0$ for $i = 1,2,\ldots,n$. The identity (7.14) is easy to check when $g = w_A$ for some $A \subset \{1,2,\ldots,n\}$, by linearity it follows that (7.14) holds in general.

We will also use

$$\|g_\epsilon\|_{L^{p'}(d\mu\times d\mu, B^*)} \le \|g\|_{p'} \tag{7.15}$$

which follows from a simple convexity argument. {Note: we may view g as the restriction to $\{-1,+1\}^n$ of a polynomial function $g : \mathbf{R}^n \to B^*$ which is of degree at most one in each variable. Then $g_\epsilon(x,y)$ is nothing but $g(\epsilon x + (1-\epsilon)y)$.}

Let us define formally $T'(\epsilon)g = \sum |A|\epsilon^{|A|-1} g_A w_A$. Note that $T'(\epsilon)g$ is the derivative in ϵ of $T(\epsilon)g$.

Now let $f : \{-1,+1\}^n \to B$. Consider the integral

$$I_\epsilon = \int\int < g_\epsilon(x,y), \sum y_i D_i f(x) > d\mu(x) d\mu(y).$$

By (7.14) we have

$$I_\epsilon = (1-\epsilon)\sum_1^n \int < T(\epsilon) D_i g, D_i f > d\mu \tag{7.16}$$

This implies

$$I_\epsilon = (1-\epsilon)\int < g, T'(\epsilon) f > d\mu. \tag{7.17}$$

Indeed, this is easy to check using the following identities. Let D_i^* be the adjoint of D_i on $L_2(D,\mu)$. Then for any $f : \{-1,+1\}^n \to \mathbf{R}$, let $f_A =< f, w_A >$, we have by elementary computations

$$\Big(\sum_1^n D_i^* D_i\Big) f = \sum |A| f_A w_A$$

and

$$T(\epsilon) D_i = \frac{1}{\epsilon} D_i T(\epsilon).$$

Since $T(\epsilon)$ is self adjoint, find

$$\sum_1^n (T(\epsilon) D_i)^* D_i f = \sum_1^n |A|\epsilon^{|A|-1} f_A w_A,$$

from which (7.17) follows.

Let now g be in $L_{p'}(B^*)$ such that

$$\|g\|_{p'} \leq 1 \quad \text{and} \quad \int < g, T'(\epsilon)f > d\mu = \|T'(\epsilon)f\|_p.$$

We deduce from (7.17) and (7.15)

$$\|T'(\epsilon)f\|_p \leq (1-\epsilon)^{-1}|I_\epsilon|$$
$$\leq (1-\epsilon)^{-1}\|g\|_{p'} \; \|\sum y_i D_i f\|_{L_p(\mu\times\mu,B)}$$

Since $T(\epsilon) - T(0) = \int_0^\epsilon T'(u)du$, we find for all $\epsilon < 1$

$$\|(T(\epsilon) - T(0))f\|_p \leq \log(\frac{1}{1-\epsilon})\|\sum y_i D_i f\|_{L_p(\mu\times\mu,B)}$$

Note that $T(0)f = \int f d\mu$ and we may as well assume that this vanishes. Then we find for all $n > 1$

$$\|T(1-\frac{1}{n})f\|_p \leq \log n\|\sum y_i D_i f\|_p$$

and using (7.13) we find

$$(1-\frac{1}{n})^n\|f\|_p \leq \log n\|\sum y_i D_i f\|_p$$

from which lemma 7.3 follows immediately. As a consequence, we obtain a simple proof of the following result from [BMW].

Theorem 7.4: Let $1 \leq p < 2$ and let B be a Banach space.
(i) If B is of type p then B is of metric type p_1 for all $p_1 < p$.
(ii) Conversely, if B is of metric type p then it is of type p_1 for all $p_1 < p$.
(iii) If B contains C_p^n's uniformly, then B contains ℓ_p^n's uniformly. (The converse is obvious).

Proof: (i) This follows from Lemma 7.3. If B is of type p, there is a constant C such that for all $f = \{-1,+1\}^n \to B$ with $\int f d\mu = 0$ we have

$$\|f\|_2 \leq 2e\log nC\|\left(\sum\|D_i f\|^p\right)^{1/p}\|_2$$
$$\leq 2eC(\log n)n^{\frac{1}{p}-\frac{1}{2}}\|\left(\sum\|D_i f\|^2\right)^{1/2}\|_2.$$

The last inequality implies immediately that B is of metric type p_1 for all $p_1 < p$ since

$$\|f(\epsilon) - f(-\epsilon)\|_2 \leq 2\|f\|_2.$$

(ii) This part is very simple. Let $x_1, \ldots, x_n$ be elements of B, let $f(\epsilon) = \sum_1^n \epsilon_i x_i$, then $D_i f = x_i$, so that if B is of metric type p we find

$$\|\sum \epsilon_i x_i\|_2 \leq Cn^{\frac{1}{p}-\frac{1}{2}}\left(\sum\|x_i\|^2\right)^{1/2} \tag{7.18}$$

for some constant C. Clearly this implies that for $p_1 < p$ B cannot contain $\ell_{p_1}^n$'s uniformly, hence (cf. Corollary 4.8) B must be of type p_1. [A simple direct proof can also be given that (7.18) implies type p_1 for all $p_1 < p$].

(iii) It is easy to see that if B contains C_p^n's uniformly then B cannot be of metric type r for any $r > p$, hence (by part (i) above) it cannot be of type r for any $r > p$, which implies (cf. corollary 4.8) that it contains ℓ_p^n's uniformly.

Remark: It is rather striking that if we can uniformly embed $\{-1,+1\}^n$ in the ℓ_p^n metric into B, then we can embed linearly the entire space ℓ_p^n (uniformly) into B. This no longer holds for $p > 2$ (cf. [BMW]).

A notion of metric type was investigated under a different name by Enflo (cf [E]). Let us say that a metric space T is of Enflo-type p (in short E-type p) if there is a constant C such that for all $f : \{-1,+1\}^n \to T$ we have

$$\int d(f(\epsilon), f(-\epsilon))^p d\mu \leq C \sum_1^n \int \Delta_i(f)^p d\mu.$$

As we have remarked earlier every metric space is of Enflo type 1.

Clearly, if T contains C_q^n's uniformly for some $q < p$ then T cannot be of E-type p, therefore by the preceding theorem E-type p implies metric type p_1 for all $p_1 < p$. The converse direction is not quite clear. In the linear case however, we have the following result which answers (essentially) a question already raised in [E].

Theorem 7.5: Let B be a Banach space of type $p > 1$. Then B is of E-type p_1 for all $1 \leq p_1 < p$.

Proof: Let $1 < p_1 < p$. Actually we will show a slightly stronger result. For any $1 < r < \infty$, there is a constant β such that for all n and all $f : \{-1,+1\}^n \to B$ with $\int f d\mu = 0$ we have

$$\|f\|_r \leq \beta \| \left(\sum \|D_i f\|^{p_1}\right)^{1/p_1} \|_r.$$

Taking $r = p_1$ we obtain theorem 7.5.

We will use duality as for lemma 7.3. Fix a number $1 < q < \infty$. Since B is of type p, it is easy to see that there is a constant C such that for all $z_1, \ldots, z_n$ in B^* and for all Φ in $L_2(B^*)$ such that $\int \Phi y_i d\mu(y) = 0$ for $i = 1, \ldots, n$, we have

$$\left(\sum \|z_i\|^{p'}\right)^{1/p'} \leq c \| \sum y_i z_i + \Phi \|_{L^q(\mu(dy); B^*)}.$$

Using the identity (7.14) and (7.15) this implies for all g in $L_q(B^*)$ (depending on $(\epsilon_1, \ldots, \epsilon_n)$ only)

$$(7.19) \qquad (1-\epsilon) \| \left(\sum \|T(\epsilon) D_i g\|^{p'}\right)^{1/p'} \|_q \leq C \|g\|_q.$$

On the other hand, since $T(\epsilon)$ is positive we have clearly

$$\|T(\epsilon) D_i g\|_\infty \leq \|D_i g\|_\infty \leq \|g\|_\infty,$$

so that

$$(7.20) \qquad \| \sup_i \|T(\epsilon) D_i g\| \|_\infty \leq \|g\|_\infty.$$

Interpolating between (7.19) and (7.20) we find that if $0 < \theta < 1$ is defined by the relation $\frac{1}{p_1} = \frac{\theta}{p} + \frac{1-\theta}{1}$, i.e. $\theta = \frac{p'}{p_1'}$, and if r' satisfies $\frac{1}{r'} = \frac{\theta}{q}$

$$(7.21) \qquad \|\Big(\sum \|T(\epsilon)D_i g\|^{p_1'}\Big)^{1/p_1'}\|_{r'} \le C^\theta (1-\epsilon)^{-\theta}\|g\|_{r'}.$$

Now we use (7.16) and (7.17). We have for any $f : \{-1,+1\}^n \to B$

$$\int < g, T'(\epsilon) f > = \sum < T(\epsilon) D_i g, D_i f > .$$

Hence if $\|(\sum \|D_i f\|^{p_1})^{1/p_1}\|_r \le 1$ and if g is such that $\|T'(\epsilon)f\|_r = \int < g, T'(\epsilon)f >$ and $\|g\|_{r'} \le 1$, we have by (7.21)

$$\|T'(\epsilon)f\|_r \le \|\Big(\sum \|T(\epsilon)D_i g\|^{p_1'}\Big)^{1/p_1'}\|_{r'}$$
$$\le C^\theta (1-\epsilon)^{-\theta}.$$

Integrating and assuming $T(0)f = \int f d\mu = 0$, we find

$$\|f\|_r = \|T(1)f\|_r \le \int_0^1 C^\theta (1-\epsilon)^{-\theta} d\epsilon < \infty.$$

By homogeneity, this establishes the announced result since we may adjust q so that r is any number such that $1 < r < \infty$.

Remark: The paper [BMW] also includes a generalization of Dvoretzky's theorem for metric spaces. Let us quote the following refined version which will appear in a paper in preparation by Bourgain, Figiel and Milman. For each $\epsilon > 0$, there is a constant $C(\epsilon) > 0$ such that any finite metric space T with cardinality N contains a subset $S \subset T$ with cardinality $[C(\epsilon)\log N]$ such that there is a subset $\tilde{S} \subset \ell_2$ and a bijection $\phi : S \to \tilde{S}$ satisfying

$$\|\phi\|_{\mathrm{Lip}} \|\phi^{-1}\|_{\mathrm{Lip}} < 1 + \epsilon.$$

In other words, S is $(1+\epsilon)$-isomorphic (in the Lipschitz sense) to a subset of a Hilbert space. We refer to the above mentioned forthcoming paper for more details.

Chapter 8
Notes and References for Further Reading

In this chapter, we would like to survey briefly several other applications of probability theory to the Geometry of Banach spaces which we chose not to develop here.

First we should say that the *general principle* in the proof of theorem 1.2 goes back to Milman [Mi]. It reappears in various contexts besides [FLM]. This "concentration of measure phenomenon" (in Milman's terminology) can take many different forms, cf. [MS]. The martingale methods in this context go back to Maurey [M2] who used it to extract symmetric basic sequences with large cardinality from finite subsymmetric ones. Schechtman developed Maurey's ideas in [Sc2]. See also [Sc1]. The paper [AM1] contains several theorems of the same nature, in particular it gives estimates on the cardinality of unconditional basic sequences which one can find in a f.d. space as blocks of a given sequence. See also [AM2].

Some important progress was made in [El] concerning the cardinality of the ℓ_1^n-sequences which can be extracted from a given sequence. Precisely, John Elton proved the following: suppose $x_1, \ldots, x_n$ are in the unit ball of some Banach space B and satisfy

$$\mathbf{E}\Big\|\sum_{i=1}^{n} \epsilon_i x_i\Big\| \geq \delta n \quad \text{for some } \delta > 0\ ,$$

then for some positive numbers $\alpha = \alpha(\delta)$ and $\beta = \beta(\delta)$ (depending *only* on δ) we can find a subset $A \subset \{1, \ldots, n\}$ with $|A| \geq \alpha n$ such that

$$\forall\ (\alpha_i) \in \mathbf{R}^A \quad \Big\|\sum_{i \in A} \alpha_i x_i\Big\| \geq \beta \sum_{i \in A} |\alpha_i|\ .$$

In other words, the extracted subset $(x_i)_{i \in A}$ spans a subspace $1/\beta$-isomorphic to $\ell_1^{|A|}$.

The extension to the complex case is surprisingly a non trivial result which was obtained by Pajor [Pa1] together with several improvements of the behaviour of $\delta \to \alpha(\delta)$ and $\delta \to \beta(\delta)$, as well as a simplified proof of Elton's theorem. See [Pa2] for a more complete exposition. Elton's theorem suggests to estimate the cardinality of A as a function of $M = \mathbf{E}\|\sum_1^n \epsilon_i x_i\|$. In the particular case when B is an L_∞-space and x_i are $\{-1, 1\}$-valued functions, some surprisingly sharp estimates can be obtained; we refer the reader to [Mi2] and [P6] for more details.

The paper [P2] contains also an estimate of the dimension of the ℓ_1^n's subspaces in terms of the stable type 1 constant, which we chose not to include here. (By the way, there is a mistake in the proof of the case $p = 1$ of the main result of [P2], on line 8 from the bottom of page 208 in [P2], but it is not difficult to correct this.)

Similar questions can be raised concerning the dimension of the ℓ_∞^n subspaces of a Banach space. We refer to [AMi] for more details in this direction.

In a completely different direction, the very important work of Gluskin [G1], [G2] can be viewed as an application of probability to Banach space theory. Let us briefly review the main results that his approach has generated. Gluskin is the first one who used "random Banach spaces" in the following manner. Let n be an integer. Let $X_1, \ldots, X_{2n}$ be a sequence of i.i.d. random variables with values in $\mathbf{R}^n$ and uniformly distributed over the Euclidean sphere of $\mathbf{R}^n$. Let C_ω be the closed convex and symmetric hull of the set $\{e_1, \ldots, e_n, X_1(\omega), \ldots, X_{2n}(\omega)\}$. Let us denote by E_ω the space $\mathbf{R}^n$ equipped with the norm for which C_ω is the unit ball.

Gluskin [G1] used these spaces to study the extremes of the "Banach-Mazur distance" defined in the beginning of chapter 1.

By a classical result of F. John, it is known that any n-dimensional Banach space E satisfies $d(E, \ell_2^n) \leq \sqrt{n}$. Moreover, simple examples such as ℓ_∞^n and ℓ_1^n show that this cannot be improved in general. Now, if E and F are two n-dimensional spaces, we can majorize their "distance" in an obvious way

$$d(E,F) \leq d(E,\ell_2^n) d(\ell_2^n, F) \leq n \; . \tag{8.1}$$

For a long time, it remained an open question whether or not this bound could be improved. Prior to Gluskin's work, in all the known cases an estimate of the form $d(E,F) \leq$ Constant $\times \sqrt{n}$ had been found. However, Gluskin showed that there is a number $\delta > 0$ such that

$$\mathbf{P} \times \mathbf{P}\{(\omega, \omega') | d(E_\omega, E_{\omega'}) < \delta n\} \to 0$$

when $n \to \infty$. In particular, this implies the existence of two sequences of Banach spaces $\{E_n\}$ and $\{F_n\}$ which $\dim E_n = \dim F_n = n$ such that $d(E_n, F_n) \geq \delta n$ for all n. This shows that the "diameter"

$$D_n = \sup\{d(E,F) | \dim E = \dim F = n\}$$

of the compact set formed by the n-dimensional Banach spaces satisfies $D_n \geq \delta n$. In other words (8.1) cannot be improved, at least asymptotically. Not much seems to be known about δ. It is likely that $D_n/n \to 1$ when $n \to \infty$ but this is unknown (as well as for the analogous questions in the sequel). Gluskin's method was used later by Gluskin [G2] and Szarek [S1] to study the basis constant $b(E)$ of an n-dimensional Banach space E. This constant is defined as follows. Let $(x_i)_{i \leq n}$ be any linear basis of E. Let P_k be the partial sum projection defined by

$$P_k x_i = \begin{cases} x_i & \text{if } i \leq k \\ 0 & \text{if } i > k \; . \end{cases}$$

Let us denote by $\|P_k\|$ its norm as an operator from E into itself. The basis constant of $\{x_i\}$ is defined as $\sup \| P_k \|$ and that of E as

$$b(E) = \inf b(\{x_i\})$$

where the infimum runs over all possible bases $\{x_i\}$. Since $b(\ell_2^n) = 1$ and $d(E, \ell_2^n) \leq \sqrt{n}$, we clearly have

$$b(E) \leq \sqrt{n} \tag{8.2}$$

The question whether this bound can be improved remained opened until Szarek [S1] proved that there is a $\delta > 0$ such that for a suitable modification $\hat{E}_\omega$ of E_ω one has

$$\mathbf{P}\{b(\hat{E}_\omega) \leq \delta\sqrt{n}\} \to 0 \quad \text{when } n \to \infty .$$

Therefore, there is a sequence of Banach spaces E_n with $\dim E_n = n$ such that $b(E_n) \geq \delta\sqrt{n}$ for all n, so that (8.2) cannot be improved at least asymptotically.

Let us recall a classical fact (due to Kadec and Snobar): if F is a k-dimensional subspace of an arbitrary Banach space B, there is a projection $P : B \to F$ with norm $\|P\| \leq \sqrt{k}$. The preceding results show that this result cannot be (asymptotically) improved; indeed, Szarek constructed a sequence $\{E_n\}$ as above and such that $\|P\| \geq \delta\sqrt{n}$ for *all* projections $P : E_n \to E_n$ with rank $P = [n/2]$.

Independently and slightly before Szarek, Gluskin had obtained all these results with $(n/\log n)^{1/2}$ instead of $\sqrt{n}$ (cf. [G2]).

In [Ma], Mankiewicz showed that similar random spaces have with "large" probability a large asymmetry constant. The latter constant is defined as follows. Let G be a group of invertible operators on E with the property that only the multiples of the identity commute with every element of G. Let then

$$s(E) = \inf \sup_{T \in G} \|T\|$$

where the infimum runs over all possible such groups G. Obviously $s(\ell_2^n) = 1$ (consider the group of all isometries) so that for a general n-dimensional Banach space E we have

$$s(E) \leq \sqrt{n} . \tag{8.3}$$

Mankiewicz proved that for some $\delta > 0$ there is a sequence of normed spaces E_n with $\dim E_n = n$ and $s(E_n) \geq \delta\sqrt{n}$ for all n, so that (8.3) also cannot be improved (asymptotically). More recently, the Gluskin approach was used to study the complex Banach spaces (i.e. Banach spaces over $\mathbf{C}$). Clearly, any complex Banach space can be considered as a real Banach space, but the converse raises all kinds of questions both in the finite and infinite dimensional cases. For instance, can every infinite dimensional real Banach space be equipped with a complex Banach space structure? For this, a negative answer has been known for a long time since

there are examples which fail this, for instance the classical James space J such that $\dim J^{**}/J = 1$. We can ask a similar "quantitative" question in the finite dimensional case (but of course only for spaces with an even dimension over the reals). The corresponding question was settled by Mankiewicz. Let n be an integer. Any n dimensional complex Banach space defines a fortiori a $2n$-dimensional real Banach space. Let us denote by C_n the class of all the $2n$-dimensional real Banach spaces obtained in this way. These obviously have an underlying complex structure, but a general $2n$-dimensional real space E can be quite far from these. Indeed, let

$$\Delta_{\mathbf{C}}(E) = \inf\{d(E,F) | F \in C_n\} .$$

Since $d(E, \ell_2^{2n}) \leq \sqrt{2n}$, we have clearly

$$\Delta_{\mathbf{C}}(E) \leq \sqrt{2n}\ . \tag{8.4}$$

Mankiewicz showed that for some $\delta > 0$ there is a sequence E_n with $\dim E_n = 2n$ and such that $\Delta_{\mathbf{C}}(E_n) \geq \delta\sqrt{n}$ for all n. Hence (8.4) cannot be (asymptotically) improved.

Now let us consider an infinite dimensional complex Banach space B. Can several complex structures yield the same underlying real structure? In other words, if $\tilde{B}$ is another complex space and if B and $\tilde{B}$ are $\mathbf{R}$-isomorphic, must they be C-isomorphic? In particular, consider the "conjugate" space $\overline{B}$ which is defined as the same as B except that the complex multiplication $\odot$ is defined on $\overline{B}$ by the following $\forall\ \lambda \in \mathbf{C}\ \forall\ x \in \overline{B}\ \lambda_{\odot}x = \overline{\lambda}x$. Clearly, B and $\overline{B}$ admit the same underlying real spaces. Are B and $\overline{B}$ always C-isomorphic? This question was recently answered negatively by Bourgain [B3], using (rather surprisingly) a finite dimensional method.

Note that for any n-dimensional complex Banach space E, we have

$$\begin{aligned} d(E,\overline{E}) &\leq d(E,\ell_2^n)d(\ell_2^n,\overline{\ell_2^n})d(\overline{\ell_2^n},\overline{E}) \\ &\leq \sqrt{n} \times 1 \times \sqrt{n} = n\ , \end{aligned} \tag{8.5}$$

here, of course, the distance is meant *in the complex sense*.

Szarek [S2] improved an earlier result of Bourgain [B3] and showed that for some $\delta > 0$ and for a suitable modification of E_ω which we will denote by F_ω (these are now complex spaces of dimension n) we have

$$\mathbf{P}\{d(F_\omega, \overline{F}_\omega) \leq \delta n\} \to 0 \quad \text{when } n \to \infty\ .$$

Thus (8.5) cannot be improved (at least asymptotically) although E and $\overline{E}$ are obviously $\mathbf{R}$-isometric.

The recent papers of Szarek [S3] [S4] develop his ideas from [S2]; in particular he obtains in [S3] an example of a uniformly convex real Banach space which does not admit any complex structure and hence is not isomorphic to its square.

REFERENCES

[A1] A. DE ACOSTA. Inequalities for B-valued random vectors with applications to the strong law of large numbers. Ann. Probab. **9** (1981) 157-161.

[A2] ____________. Stable measures and semi-norms. Ann. Probab. **3** (1975) 365-375.

[AMi] N. ALON, V. D. MILMAN. Embedding of ℓ_∞^k in finite dimensional Banach spaces. Israel J. Math. 45 (1983) 265-280.

[AM1] D. AMIR, V. D. MILMAN. Unconditional and symmetric sets in n-dimensional normed spaces. Israel J. Math. **37** (1980) 3-20.

[AM2] ____________. A quantitative finite-dimensional Krivine theorem. Israel J. Math. 50 (1985) 1-12.

[AG] A. ARAUJO, E. GINÉ. The central limit theorem for real and Banach space valued random variables. Wiley (1980).

[BG] Y. BENYAMINI, Y, GORDON. Random factorizations of operators between Banach spaces. J. d'Analyse Jerusalem **39** (1981) 45-74.

[Be] A. BEURLING. On analytic extensions of semi-groups of operators. J. Funct. Anal. 6 (1970) 387-400.

[Bo] C. BORELL. The Brunn-Minkowski inequality in Gauss spaces. Invent. Math. **30** (1975) 207-216.

[B1] J. BOURGAIN. New Banach space properties of the disc algebra and H^∞. Acta Math. **152** (1984) 1-48.

[B2] ____________. On martingale transforms on spaces with unconditional and symmetric basis with an appendix on K-convexity constant. Math. Nachrichten (1984).

[B3] ____________. A complex Banach space such that X and $\overline{X}$ are not isomorphic. Proc. A.M.S. To appear.

[BMW] J. BOURGAIN, V. D. MILMAN, H. WOLFSON, On the type of metric spaces. Trans. AMS. To appear.

[BDK] J. BRETAGNOLLE, D. DACUNHA-CASTELLE, J. L. KRIVINE, Lois stables et espaces L^p. Ann. Inst. Henri Poincaré. sect. B. **2** (1966) 231-259.

[Bu] D. BURKHOLDER. Distribution function inequalities for martingales. Ann. Probab. **1** (1973) 19-42.

[D] A. DVORETZKY. Some results on convex bodies and Banach spaces. Proc. Symp. on Linear Spaces. Jerusalem 1961. p. 123-160.

[DR] A. DVORETZKY, C. A. ROGERS. Absolute and unconditional convergence in normed linear spaces. Proc. Nat. Acad. Sci. (USA) **36** (1950) 192-197.

[E] P. ENFLO. Uniform homeomorphisms between Banach spaces - Séminaire Maurey-Schwartz 75-76. Exposé no. 18. Ecole Polytechnique. Paris.

[Eh] A. EHRHARD. Inégalités isopérimétriques et intégrales de Dirichlet gaussiennes Annales E.N.S. **17** (1984) 317-332.

[FA] T. FACK. Type and cotype inequalities for non-commutative L^p spaces. (Preprint Université Paris 6).

[Fe1] X. FERNIQUE. Intégrabilité des vecteurs gaussiens. C. R. Acad. Sci. Paris A **270** (1970) 1698-1699.

[Fe2] ―――――――. Régularité des trajectoires des fonctions aléatoires gaussiennes. Springer Lecture Notes in Math. no. 480 (1975) 1-96.

[F1] T. FIGIEL. On the moduli of convexity and smoothness. Studia Math. **56** (1976) 121-155.

[F2] ―――――――. On a recent result of G. Pisier. Longhorn Notes. Univ. of Texas. Functional Analysis Seminar 82/83 p. 1-15.

[FLM] T. FIGIEL, J. LINDENSTRAUSS, V. D. MILMAN. The dimensions of almost spherical sections of convex bodies. Acta. Math. **139** (1977) 53-94.

[FT] T. FIGIEL, N. TOMCZAK-JAEGERMANN. Projections onto Hilbertian subspaces of Banach spaces. Israel J. Math. **33** (1979) 155-171.

[Ga] D. J. H. GARLING. Convexity smoothness and martingale inequalities. Israel J. Math. **29** (1978) 189-198.

[G1] E. D. GLUSKIN. The diameter of the Minkowski compactum is roughly equal to n. Functional Anal. Appl. **15** (1981) 72-73.

[G2] E. D. GLUSKIN. Finite dimensional analogues of spaces without a basis. Dokl. Akad. Nauk. SSSR. 261 (1981) 1046-1050 (Russian).

[Go1] Y. GORDON. On Dvoretzky's theorem and extensions of Slepian's lemma. Lecture no. II Israel Seminar on Geometrical Aspects of Functional Analysis 83-84. Tel Aviv University.

[Go2] Some inequalities for Gaussian processes and applications. To appear.

[HJ1] J. HOFFMANN-JØRGENSEN. Sums of independent Banach space valued random variables. Aarhus University. Preprint series 1972-73 no. 15.

[HJ2] ―――――――. Sums of independent Banach space valued random variables. Studia Math. 52 (1974) 159-186.

[IN] K. ITO, M. NISIO. On the convergence of sums of independent Banach space valued random variables. Osaka J. Math. **5** (1968) 35-48.

[J] R. C. JAMES. Non reflexive spaces of type 2. Israel J. Math. 30 (1978) 1-13.

[JS] W. B. JOHNSON, G. SCHECHTMAN. Embedding ℓ_p^m into ℓ_1^n. Acta Math. 149 (1982) 71-85.

[Ka] J. P. KAHANE. Some random series of functions. (1968) Heath Mathematical Monographs. Second Edition, Cambridge University Press (1985).

[Kat] T. KATÔ. A characterization of holomorphic semi-groups. Proc. Amer. Math. Soc. 25 (1970) 495-498.

[KT] H. KÖNIG, L. TZAFRIRI. Some estimates for type and cotype constants. Math. Ann. 256 (1981) 85-94.

[K] J. L. KRIVINE. Sons-espaces de dimension finie des espaces de Banach réticulés. Annals of Maths. 104 (1976) 1-29.

[Ku] J. KUELBS. Kolmogorov's law of the iterated logarithm for Banach space valued random variables. Illinois J. Math. 21 (1977) 784-800.

[Kw] S. KWAPIEŃ. Isomorphic characterizations of inner product spaces by orthogonal series with vector coefficients. Studia Math. **44** (1972) 583-595.

[LS] H. LANDAU, L. SHEPP. On the supremum of a Gaussian process. Sankhya, **A32** (1970) 369-378.

[LWZ] R. LEPAGE, M. WOODROOFE, J. ZINN. Convergence to a stable distribution via order statistics. Ann. Probab. **9** (1981) 624-632.

[L] D. LEWIS. Ellipsoids defined by Banach ideal norms. Mathematika **26** (1979) 18-29.

[Ma] P. MANKIEWICZ. Finite dimensional Banach spaces with symmetry constant of order $\sqrt{n}$. Studia Math. To appear.

[MaP] M. B. MARCUS, G. PISIER. Characterizations of almost surely continuous p-stable random Fourier series and strongly stationary processes. Acta Math. **152** (1984) 245-301.

[M1] B. MAUREY. Type et cotype dans les espaces munis de structure locale inconditionnelle. Séminaire Maurey-Schwartz 73-74. Exposé no. 24-25 Ecole Polytechnique, Paris.

[M2] ———. Construction de suites symétriques C.R.A.S. Sci. Paris. A. 288 (1979) 679-681.

[MP] B. MAUREY, G. PISIER. Séries de variables aléatoires vectorielles indépendantes et géométrie des espaces de Banach. Studia Math **58** (1976) 45-90.

[Me] P. A. MEYER. Note sur les processus d'Ornstein-Uhlenbeck. Séminaire de Probabilités XVI. p. 95-132. Lecture notes in Math no. 920. Springer 1982.

[Mi1] V. D. MILMAN. A new proof of the theorem of A. Dvoretzky on sections of convex bodies. Functional Anal. Appl. **5** (1971) 28-37.

[Mi2] ———. Some remarks about embedding of ℓ_1^k in finite dimensional spaces. Israel J. Math . 43 (1982), 129-138.

[MS] V. D. MILMAN, G. SCHECHTMAN. Asymptotic theory of finite dimensional normed spaces. Springer Lecture Notes. Vol. 1200.

[Pa1] A. PAJOR. Plongement de ℓ_1^n dans les espaces de Banach complexes. C. R. Acad. Sci. Paris A 296 (1983) 741-743.

[Pa2] ———. Thèse de 3e cycle. Université Paris 6. November 84.

[P1] G. PISIER, Les inégalités de Khintchine-Kahane d'aprés C. Borell. Exposé no. 7. Séminaire sur la géométrie des espaces de Banach - 1977-78. Ecole Polytechnique Palaiseau.

[P2] ———. On the dimension of the ℓ_p^n-subspaces of Banach spaces, for $1 < p < 2$. Trans. A.M.S. **276** (1983) 201-211.

[P3] ———. Martingales with values in uniformly convex spaces. Israel J. Math. **20** (1975) 326-350.

[P4] ———. Un exemple concernant la super-réflexivité. Annexe no. 2. Séminaire Maurey-Schwartz 1974-75. Ecole Polytechnique. Paris.

[P5] ———. Holomorphic semi-groups and the geometry of Banach spaces. Annals of Maths. 115 (1982) 375-392.

[P6] ———. Remarques sur les classes de Vapnik-Červonenkis. Ann. Inst. Henri Poincaré, Probabilités et Statistiques, 20 (1984) 287-298.

[R] J. ROSINSKI. Remarks on Banach spaces of stable type. Probability and Math. Statist. **1** (1980) 67-71.

[Sc1] G. SCHECHTMAN. Random embeddings of Euclidean spaces in sequence spaces. Israel J. Math. **40** (1981) 187-192.

[Sc2] ———. Lévy type inequality for a class of finite metric spaces. Martingale Theory in Harmonic Analysis and Banach spaces. Cleveland 1981. Springer Lecture Notes no. 939, p. 211-215.

[Sz] A. SZANKOWSKI. On Dvoretzky's theorem on almost spherical sections of convex bodies. Israel J. Math. **17** (1974) 325-338.

[S1] S. SZAREK. The finite dimensional basis problem with an appendix on nets of Grassman manifolds. Acta Math. **151** (1983) 153-179.

[S2] ———. On the existence and uniqueness of complex structure and spaces with few operators. Trans. A.M.S. to appear.

[S3] ———. A superreflexive Banach space which does not admit complex structure. To appear.

[S4] ———. A Banach space without a basis which has the bounded approximation property. To appear.

[TJ] N. TOMCZAK-JAEGERMANN. On the moduli of convexity and smoothness and the Rademacher averages of the trace classes S_p $(1 \leq p < \infty)$. Studia Math. **50** (1974) 163-182.

[T1] L. TZAFRIRI. On Banach spaces with unconditional basis. Israel J. Math. **17** (1974) 84-93.

[T2] ———. On the type and cotype of Banach spaces. Israel J. Math. 32 (1979) 32-38.

[Y] V. YURINSKII. Exponential bounds for large deviations. Theor. Probability Appl. **19** (1974) 154-155.

SOME REMARKS ON INTEGRAL OPERATORS AND EQUIMEASURABLE SETS

Walter Schachermayer

Linz, Austria

Abstract: We give a characterisation of equimeasurable sets in terms of the difference between the notions of almost everywhere convergence and convergence in measure. We apply this characterisation to obtain a direct proof of a criterion for integral representability of operators, due to A. V. Bukhvalov (obtained in 1974) by a criterion of the present author (obtained in 1979).

In the second part - following an idea due to A. Costé - we show that convolution with a suitably chosen singular measure defines a positive operator on L^2, which is of trace class p, for $p > 2$, but fails to be integral. This sharpens a result, due to D. H. Fremlin.

Introduction: This paper is stimulated by the question of characterisation of integral operators. This problem, which was raised by J. v. Neumann [v.N.] in 1935 was solved in 1974 by Bukhvalov [B2] using the theory of order-bounded operators (see theorem 2.3 below). In 1979, the present author - working on problems of [H-S] - obtained independently a different characterisation using methods from the theory of differentiation of vector measures [S1].

In 1981 A. Schep [S2] has indicated that the latter characterisation may be deduced from Bukhvalov's, while in 1982 L. Weis [W] showed (among many other interesting results) how to prove Bukhvalov's characterisation by means of arguments similar to those used in [S1].

In the present paper we show that the notion of equimeasurability, due to A. Grothendieck, on which our characterisation is based, is intimately linked with the difference between convergence in measure and convergence almost everywhere, on which Bukhvalov's criterion is based. This furnishes a better understanding why the two conditions are equivalent.

We start the paper by defining the concept of equimeasurability and pointing out its relevance to the question of representing an operator from $L^1(\mu)$ by a Bochner-derivative (proposition 2.2 below). We then state the characterisations of integral operators due to Bukhvalov and the present author (theorem 2.3 below) and analyse the structure of equimeasurable sets to obtain some equivalent characterisations, one of them involving the difference between convergence in measure and a.e. With the help of this characterisation we then deduce the criterion of Bukhvalov from ours by a very direct argument.

In the last part of the paper we deal with a different topic. In 1975 D. H. Fremlin [F] gave an example of a positive compact operator on L^2 which is not integral. This important and highly non-trivial result was achieved by means of a somewhat ad-hoc-construction.

On the other hand J. J. Uhl [U] noted - based on work of A. Costé - that "convolution with a biased coin" furnishes an example of a completely continuous non-representable operator from L^1 to L^1. Although Fremlin had applied his example to answer a question of Schaefer in the negative (i.e., there exists a positive compact (Grothendieck-)intergral operator from L^∞ to L^1 which is not nuclear - which comes up to the same as a positive completely continuous non-representable operator from L^1 to L^1), apparently the connection between these two examples has not been noticed.

Costê's argument is based on the classical result, due to Menchoff, that there is a Lebesgue-singular measure on the torus such that the Fourier-coefficients tend to zero.

Translating Costê's example into the language of integral operators on L^2 we obtain a considerable sharpening of Fremlin's result: There is a positive operator on $L^2(0,1)$ which is of trace-class p for every $p > 2$, but not integral. The operator is just the convolution with a suitably chosen "sequence of biased coins".

2. Equimeasurable sets

Let (X,Σ,μ) and (Y,T,ν) denote finite measure spaces (the σ-finite case reduces to this case by a simple change of density). For $1 \leq p \leq \infty$, let $L^p(\mu)$ and $L^p(\nu)$ denote the usual Lebesgue-spaces (over the reals) and let $L^0(\mu)$ be the F-space of (equivalence classes of) μ-measurable real-valued functions, equipped with the metric

$$d(f,g) = \inf\{\varepsilon : \mu(\omega : |f(\omega)-g(\omega)| \geq \varepsilon) < \varepsilon\}.$$

2.1. Definition (Grothendieck, [G]): A subset M of $L^0(\mu)$ is called equimeasurable if, for $\varepsilon > 0$, there is $X_\varepsilon \subset X$ with $\mu(X \setminus X_\varepsilon) < \varepsilon$ such that M restricted to X_ε is relatively norm-compact in $L^\infty(X_\varepsilon,\mu|_{X_\varepsilon})$.

The following proposition, which goes back to Grothendieck's memoir [G] shows the importance of this notion for the representability of operators.

2.2. Proposition ([S2], [S1]): An operator T from $L^1(\mu)$ to a Banach space E is representable by a Bochner-integrable function $F : X \to E$, i.e. for $f \in L^1(\mu)$

$$Tf = \int_X F(\omega) f(\omega) d\mu(\omega),$$

iff T^* maps the unitball of E^* into an equimeasurable subset of $L^\infty(\mu)$.

The proof of the theorem in fact essentially goes back to Dunford and Pettis (see e.g. [D-U], prop. III. 2, 21, p. 78). It reduces quickly to the correspondence between compact operators $T : L^1(\mu) \to E$ and relatively compact valued μ-measurable functions $F : X \to E$. Note in passing that for a weak-star dense subset C of the unitball of E^* we have that $T^*(C)$ is equimeasurable iff T^* (unitball (E^*)) is equimeasurable.

Despite its simplicity proposition 2.2 seems to us to be a key result in understanding questions about representability of operators by Bochner-derivatives or - what is closely related - by kernel functions.

We shall now state the two criteria for integral representation. For simplicity we restrict ourselves to the classical L^2-case. See, however, the subsequent remark 2.9.

<u>2.3. Theorem:</u> Let $T : L^2(\nu) \to L^2(\mu)$ be a linear map. T.f.a.e.

(i) T is an integral operator (for a definition see, e.g. [B1], [H-S] or [S1]).

(ii) For $(g_n)_{n=1}^\infty \in L^2(\nu)$ such that there is $g \in L^2_+(\nu)$ with $|g_n| \le g$ and $g_n \to 0$ in measure, the sequence $(Tg_n)_{n=1}^\infty$ converges to zero μ-a.e. (Bukhvalov, 1974).

(iii) T transforms order bounded subsets of $L^2(\nu)$ into equimeasurable sets (Schachermayer, 1979).

Of course, the preceding theorem is well-known: The equivalence of (i) and (ii) was shown in [B2], that of (i) and (iii) in [S1]. The question we are now dealing with is somewhat aesthetical: How direct a proof of (ii) $\Rightarrow$ (iii) and (iii) $\Rightarrow$ (ii) may be given? More precisely: What is the relation between equimeasurable sets and the distinction between convergence in measure and convergence almost everywhere?

The subsequent result shows that there is in fact an intimate relation for convex sets (absolute convexity is assumed just for the convenience of the formulation).

2.4. Proposition: Let M be an absolutely convex subset of $L^0(\mu)$. T.f.a.e.

(i) There is a function $\varphi \in L_+^\infty(\mu)$, $\varphi > 0$ μ-a.e., such that

$$\varphi.M = \{\varphi.f : f \in M\}$$

is relatively compact in $(L^\infty(\mu), \|.\|_\infty)$.

(ii) M is order-bounded and every sequence $(f_n)_{n=1}^\infty \subset M$ which converges to zero in measure converges to zero μ-a.e.

(iii) M is equimeasurable.

Proof:

(iii) $\Rightarrow$ (i): If M is equimeasurable we may find a μ-a.e. partition of X into disjoint sets $(A_k)_{k=1}^\infty$ in Σ such that M restricted to A_k is relatively $\|.\|_\infty$-compact. In particular there are constants $a_k \in \mathbb{R}_+$ such that M restricted to A_k is $\|.\|_\infty$-bounded by the constant a_k. It is obvious that

$$\varphi = \sum_{k=1}^\infty (k.a_k)^{-1} \chi_{A_k}$$

does the desired job.

(i) ⇒ (ii): The order-boundedness of M is obvious. If $(f_n)_{n=1}^{\infty} \subset M$ converges to zero in measure then $(\varphi f_n)_{n=1}^{\infty}$ does so too. As convergence in measure defines a coarser Hausdorff-topology on M than the $\|.\|_{\infty}$-topology, we infer from the relative $\|.\|_{\infty}$-compactness of M that $(\varphi.f_n)_{n=1}^{\infty}$ converges to zero uniformly. Hence $(f_n)_{n=1}^{\infty}$ converges to zero μ-a.e.

(ii) ⇒ (iii): This is the (relatively) non-trivial part of the proposition and we shall delay its proof to the subsequent proposition, which states a more general "local version", as the proof for the present case is not easier than the general one.

□

<u>2.5. Remark:</u> The conditions (i), (ii) and (iii) of the above proposition 2.4 correspond essentially to the conditions of theorem 2.3. The implication (iii) ⇒ (ii) of 2.4 was noted by A. Schep [S2] and from this he easily deduced the implication (iii) ⇒ (ii) of 2.3. But it is worth remarking that (ii) and (iii) of 2.4 are in fact equivalent (for absolutely convex sets). The reader should note that a subset $M \subset L^0(\mu)$ is equimeasurable iff its absolutely convex, closed hull has this property.

The reformulation 2.4 (i) of the concept of equimeasurability was stressed out to us by J. B. Cooper and it emphazises that - up to multiplication with a weight function - the equimeasurable sets are just the relatively compact subsets of $L^{\infty}(\mu)$. Representing $L^{\infty}(\mu)$ as a C(K)-space they correspond to the equicontinuous sets (by the Ascoli-Arzela theorem). Hence "equimeasurability" may be viewed as a kind of analogue to "equicontinuity", the former applying to measurable the latter to continuous functions.

Let us recall in this context that the original definition of

A. Grothendieck [G] was given in the "french style", i.e. μ is a Radon-measure, X_ε a compact subset of X and M, restricted to X_ε, an equicontinuous subset of $C(X_\varepsilon)$.

We now pass to the announced "local version" of proposition 2.4. It will be convenient to state it for bounded subsets of $L^\infty(\mu)$ (instead of order-bounded subsets of $L^0(\mu)$) and to assume that the measure space (X,Σ,μ) is separable, i.e. that $L^1(\mu)$ is a separable Banach space.

Both restrictions are inessential: Passing from order-bounded subsets of $L^0(\mu)$ to bounded subsets of $L^\infty(\mu)$ is a matter of multiplication with a weight function. Also the conditions of 2.6 are easily seen to be separably determined which allows the reduction of the case of a general (finite) measure space to the separable one. However, for a separable measure space (X,Σ,μ) the weak-star-topology is metrisable on bounded subsets of $L^\infty(\mu)$, which allows us to formulate the subsequent proposition more elegantly: Fix δ to be a metric on $L^\infty(\mu)$ inducing the weak-star-topology on its bounded subsets.

<u>2.6. Proposition:</u> Let M be an absolutely convex bounded subset of $L^\infty(\mu)$ and let $A \in \Sigma$. T.f.a.e.

(a) Let $F_n = \sup\{f \in M : d(f,0) \leq n^{-1}\}$ and $F = \lim F_n$. Then F equals zero μ-a.e. on A.

(a') Let $G_n = \sup\{f \in M : \delta(f,0) \leq n^{-1}\}$ and $G = \lim G_n$. Then G equals zero μ-a.e. on A.

(b) If $(f_n)_{n=1}^\infty \subset M$ converges to zero in measure, then $(f_n)_{n=1}^\infty$ converges to zero μ-a.e. on A.

(b') If $(f_n)_{n=1}^\infty \subset M$ weak-star-convergerges to zero, then $(f_n)_{n=1}^\infty$ converges to zero μ-a.e. on A.

(c) The restriction of M to A is equimeasurable.

Note that the suprema appearing in (a) and (a') (taken over a (possibly) uncountably set of functions) have to be interpreted in the lattice-sense.

<u>Proof:</u> (c) $\Rightarrow$ (b'): Let $(A_k)_{k=1}^{\infty}$ be a μ-a.e. partition of A such that M restricted to A_k is relatively $\|.\|_\infty$-compact. The operator

$$\chi_{A_k} : L^\infty(\mu) \to L^\infty(\mu)$$
$$f \to f.\chi_{A_k}$$

is weak-star-continuous. On $\chi_{A_k} M$ the σ^*-topology and the $\|.\|_\infty$-topology coincide, hence the operator χ_{A_k} is σ^*-$\|.\|_\infty$-continuous on M. So, if $(f_n)_{n=1}^{\infty} \subset M$ converges σ^* to zero, then it converges to zero (uniformly) on every A_k, hence μ-a.e. on A.

(a') $\Rightarrow$ (c): If G equals zero μ-a.e. on A then Egoroff's theorem tells us that there is a μ-a.e. partition $(A_k)_{k=1}^{\infty}$ of A such that G_n converges to zero uniformly on every A_k. This means - as above - that the restriction of the operator χ_{A_k} to M is σ^*-$\|.\|_\infty$-continuous at the point $0 \in M$. From the absolute convexity of M we conclude that the restriction of χ_{A_k} to M is σ^*-$\|.\|_\infty$ uniformly continuous. As M is relatively σ^*- compact we deduce the relative $\|.\|_\infty$-compactness of the restriction of M to A_k.

(b') $\Rightarrow$ (a'): If (a') fails then $d(\chi_A G, 0) = \alpha > 0$. As $G_n \geq G \geq 0$ we may find, for every $n \in \mathbb{N}$, a finite sequence $f_1^n, \ldots, f_{m_n}^n$ in M such that

$$\delta(f_i^n, 0) \leq n^{-1} \qquad i = 1, \ldots, m_n$$

while

$$\delta(\sup\{\chi_A \cdot f_i^n : 1 \leq i \leq m_n\}, 0) > \alpha/2.$$

The sequence $(((f_i^n)_{i=1}^{m_n})_{n=1}^{\infty}$ converges weak-star to zero, while it cannot converge to zero μ-a.e. on A as the lim sup of the sequence has a distance from zero (in measure) of at least $\alpha/2$.

(b) $\Rightarrow$ (a) uses the same argument as (b') $\Rightarrow$ (a') (replacing G by F and δ by d).

(b') $\Rightarrow$ (b) and (a') $\Rightarrow$ (a) are obvious.

(a) $\Rightarrow$ (a'): We shall in fact show that F equals G μ-a.e. As we shall use a pointwise reasoning it will be convenient to argue with (countably many) representants of the equivalence-classes of functions: For $n \in \mathbb{N}$, find sequences $(\tilde{g}_i^n)_{i=1}^{\infty}$ of representants of elements of M with $\delta(\tilde{g}_i^n, 0) \leq n^{-1}$ and such that the function

$$\tilde{G}_n(\omega) = \sup_i \tilde{g}_i^n(\omega)$$

is a representant of G_n. We may assume that the sequence $(\tilde{G}_n)_{n=1}^{\infty}$ is pointwise decreasing at every $\omega \in X$. Hence

$$\tilde{G}(\omega) = \lim_n \tilde{G}_n(\omega)$$

exists for every $\omega \in X$ and is a representant of G. Let C be the collection of all functions of the form

$$C = \{k^{-1}(\tilde{g}_{i_1}^{n_1} + \ldots + \tilde{g}_{i_k}^{n_k})\}$$

which is a countable set of representants of elements of M. Define

$$\tilde{H}_n(\omega) = \sup\{\tilde{f}(\omega) : \tilde{f} \in C, \|\tilde{f}\|_2 \leq n^{-1}\}$$

and

$$\tilde{H}(\omega) = \lim \tilde{H}_n(\omega).$$

We shall show that

$$\tilde{H}(\omega) \geq \tilde{G}(\omega) \qquad \forall \omega \in X.$$

Indeed, fixing $\omega_o \in X$, we may find for every $n \in \mathbb{N}$ an $i \in \mathbb{N}$ such that $\tilde{g}_i^n(\omega_o) \geq \tilde{G}(\omega_o)$ and $\delta(\tilde{g}_i^n, 0) \leq n^{-1}$. Given $\varepsilon > 0$, let $k \in \mathbb{N}$ s.t.

$$k \geq 2(\sup\{\|f\|_2 : f \in M\}/\varepsilon)^2$$

and, for $1 \leq l \leq k$, find inductively functions $\tilde{g}_{i_l}^{n_l}$, such that

$$\tilde{g}_{i_l}^{n_l}(\omega_o) \geq \tilde{G}(\omega_o)$$

and such that the $\tilde{g}_{i_l}^{n_l}$'s are mutually almost orthogonal, say

$$|(\tilde{g}_{i_l}^{n_l}, \tilde{g}_{i_m}^{n_m})| < \varepsilon^2/2 \qquad \forall l \neq m.$$

Then

$$\tilde{f} = k^{-1}(\tilde{g}_{i_i}^{n_1} + \ldots + \tilde{g}_{i_k}^{n_k})$$

is an element of C with

$$\tilde{f}(\omega_o) \geq \tilde{G}(\omega_o)$$

and we may estimate its $\|.\|_2$-norm as follows:

$$\|\tilde{f}\|_2^2 = (\tilde{f}, \tilde{f}) \leq k^{-1}[k.\sup\{\|f\|_2.f \in M\}^2 + k^2.\frac{\varepsilon^2}{2}] \leq \frac{\varepsilon^2}{2} + \frac{\varepsilon^2}{2}.$$

Thus we have found an $\tilde{f} \in C$ of $\|.\|_2$-norm less than ε, that majorizes G at the point ω_o. This readily shows that $\tilde{H}(\omega) \geq \tilde{G}(\omega)$ for all $\omega \in X$. If H denotes the equivalence class of the function $\tilde{H}$ then $F \geq H \geq G$; on the other hand the inequality $F \leq G$ is obvious. Hence $F = G$ (as equivalence classes of functions). □

<u>2.7. Remark:</u> It seems worth noting that the proof of (a) ⇒ (a') really shows that the functions F, G (as defined in the statement of 2.6) as well as the function

$$H = \lim H_n$$

for

$$H_n = \sup \{f : \|f\|_2 \leq n^{-1}\}$$

are identical.

This gives a connection to a result of Mokobodzki [M] from 1972 stating that $T : L^1(\nu) \to L^1(\mu)$ is representable by a Bochner-integrable function (equivalently: is an integral operator) iff T maps dominated $\sigma(L^1(\nu), L^\infty(\nu))$-convergent sequences to almost everywhere convergent sequences. This resembles (for the special case of L^1) Bukhvalov's criterion with convergence in measure replaced by weak convergence. Proposition 2.6 clarifies why both conditions are equivalent.

Let us also note that taking in proposition 2.6 $A = X$ the implication (b) $\Rightarrow$ (c) of 2.4 together with the remarks preceding 2.6 furnishes the missing proof of the (ii) $\Rightarrow$ (iii) of 2.4.

2.8.: Let us now show how Bukhvalov's criterion may be deduced directly from ours (i.e. (ii) $\Rightarrow$ (iii) of th. 2.3) with the help of the above proposition 2.6. Consider the direct sum of the two measure spaces (X,Σ,μ) and $(Y,\mathcal{T},\nu)$, i.e. $(X \cup Y, \Sigma \oplus \mathcal{T}, \mu \oplus \nu)$.

It is easily seen that 2.3 (ii) implies the order-continuity of T. Thus if 2.3 (ii) holds while 2.3 (iii) fails, we can find $\psi \in L^2_+(\nu)$, say $\psi \geq 1$, such that there is $\varphi \in L^\infty(\mu)$, $\varphi > 0$ μ-a.e. with

$$T([-\psi,\psi]) \subset [-\varphi^{-1},\varphi^{-1}]$$

but such that $T([-\psi,\psi])$ is not equimeasurable. If M_φ and $M_{\psi^{-1}}$ denote the multiplication operator with φ and ψ^{-1} resp. define

$$S : L^2(\nu) \to L^\infty(X \cup Y, \mu \oplus \nu)$$
$$g \to (M_\varphi Tg, M_{\psi^{-1}} g).$$

The set $S([-\psi,\psi])$ is an absolutely convex, bounded subset of $L^{\infty}(\mu\otimes\nu)$, such that the restriction to X fails to be equimeasurable and we infer from 2.8 (b) that there is a sequence $(g_n)_{n=1}^{\infty}$ in $[-\psi,\psi]$ such that $(M_{\varphi}Tg_n, M_{\psi^{-1}}g_n)_{n=1}^{\infty}$ tends to zero in measure but $(M_{\varphi}Tg_n)_{n=1}^{\infty}$ does not converge μ-a.e. Hence $(g_n)_{n=1}^{\infty}$ is a sequence in $L^2(\nu)$, $|g_n| \leq \psi$ for $n \in \mathbb{N}$, which converges to zero in measure while Tg_n does not converge to zero μ-a.e.; with this contradiction we are done.

2.9. Remark: We have stated theorem 2.3 for the case of L^2-spaces but the arguments carry over to operators from F to E, where E and F are general order ideal spaces (on the finite measure spaces (X,Σ,μ) and (Y,T,ν) resp.). Indeed, note first that $T : F \to E$ is integral iff T is integral as an operator from F to $L^0(\mu)$, hence the question of integral representability does not depend on the space E on the right hand side. In fact, it only depends on the collection of order-intervals of F. Precisely the same arguments as in the L^2-case work in the general case and again proposition 2.6 gives the link between Bukhvalov's and our criterion.

3. An example of a positive, compact operator on L^2, which is not integral

We now turn to a different question: We shall show that convolution with a suitably chosen "sequence of biased coins" furnishes an example of a positive, compact operator on L^2 which fails to be integral. The example is to a large extent just a translation of an example due to A. Costé ([C] and [D-U], p. 90). It seems more natural than D. H. Fremlin's construction [F] and gives a sharper result: The operator is not only compact, if is even of trace class p, $p > 2$. I would like to thank V. Losert, who pointed out to me the use of an infinite product to obtain the estimate relevant for the S_p-norm.

Let X be the compact group $\Delta = \{-1,1\}^{\mathbb{N}}$, equipped with normalized Haar-measure μ on the Borel-σ-algebra Σ. For $1/2 < \alpha < 1$ let $\lambda(a)$ be the measure on the two-point-set $\{-1,1\}$ given by

$$\lambda(\alpha)(\{1\}) = \alpha$$
$$\lambda(\alpha)(\{-1\}) = 1-\alpha.$$

Given a sequence $(\alpha_n)_{n=1}^{\infty}$ in $]1/2,1[$ define the probability measure $\lambda((\alpha_n)_{n=1}^{\infty})$ on Δ as the product of the $\lambda(\alpha_n)$, i.e.

$$\lambda((\alpha_n)_{n=1}^{\infty}) = \bigotimes_{n=1}^{\infty} \lambda(\alpha_n).$$

We have the following dichotomy result:

<u>3.1. Proposition (Kakutani [K], [U]):</u> We have $\lambda \perp \mu$ or $\lambda \ll \mu$ according as $\sum_{n=1}^{\infty} (2\alpha_n - 1)^2$ diverges or converges.

We now fix a sequence $(\alpha_n)_{n=1}^{\infty}$ in $]1/2,1[$ such that

$$\sum_{n=1}^{\infty} (2\alpha_n - 1)^2 = \infty$$

while

$$\sum_{n=1}^{\infty} (2\alpha_n - 1)^p < \infty \quad \text{for} \quad p > 2.$$

The proceeding proposition tells us that the probability measure $\lambda = \lambda((\alpha_n)_{n=1}^{\infty})$ is singular with respect to μ. Let

$$T_\lambda : L^2(\mu) \to L^2(\mu)$$
$$f \to f * \lambda$$

be the operator of convolution with λ. Clearly T is positive (in the lattice sense) since λ is positive. The fact that λ is singular with respect to μ corresponds to the fact that T_λ is not an integral operator. Indeed, viewing T_λ as an operator from $C(\Delta)$ to $C(\Delta)$ the restriction of the adjoint operator $T_\lambda^* : M(\Delta) \to M(\Delta)$ to $L^1(\mu)$ is represented by a μ-essentially uniquely weak-star-measurable function $F : \Delta \to M(\Delta)$ (see [D-S], p. 503). It is obvious from the definition of the convolution that this F is given by

$$F : \omega \to \lambda_\omega$$

where λ_ω denotes the translate of λ by $\omega \in \Omega$; hence F takes its values in $M(\Delta) \setminus L^1(\mu)$. So there can not exist a Halmos-function $\gamma : \Delta \to L^1(\mu)$ representing T_λ (for the definition see [S1]) as γ would have to equal F μ-a.e.

This shows that T is not an integral operator. Let

$$\varepsilon_n : \{-1,1\}^{\mathbb{N}} \to \{-1,1\}$$

be the projection onto the n'th coordinate and, for a finite subset $A \subset \mathbb{N}$, define the Walsh-function

$$w_A(\omega) = \prod_{n\in A} \varepsilon_n(\omega).$$

It is wellknown (e.g., [K2]) that the Walsh-functions are the characters of the group Δ and that T_λ is a diagonal operator with respect to the Walsh-basis. The corresponding eigenvalues are given by

$$\int w_A(\omega)\,d\lambda(\omega) = \int\Big(\prod_{n\in A} \varepsilon_n(\omega)\Big)d\lambda(\omega) = \prod_{n\in A}(1.\alpha_n + (-1).(1-\alpha_n)) = $$
$$= \prod_{n\in A}(2\alpha_n - 1).$$

The norm $\|T_\lambda\|_p$ of T_λ with respect to the trace class p, for $p > 2$, can therefore be estimated by

$$\|T_\lambda\|_p^p = \sum_{A\subset\mathbb{N}}\Big(\prod_{n\in A}(2\alpha_n-1)\Big)^p = \prod_{n=1}^{\infty}(1+(2\alpha_n-1)^p) =$$
$$= \exp\Big(\sum_{n=1}^{\infty} \ln\,(1+(2\alpha_n-1)^p)\Big) \leq$$
$$\leq \exp\Big(\sum_{n=1}^{\infty}(2\alpha_n-1)^p\Big) < \infty.$$

This shows that T_λ is of trace class p, for every $p > 2$, and finishes the presentation of the example.

<u>3.2. Remark:</u> To point out the flavour of the different criteria of theorem 2.3 we shall show how to use our or Bukhvalov's integral representability criterion to see that the above operator T_λ is not integral. The alert reader will notice that these arguments are just different aspects of the same issue.

a) We shall show that T_λ transforms the unit-ball of $C(\Delta)$ into a non-equi-measurable set. Indeed if T_λ (ball $C(\Delta)$) were equi-measurable then - by translation-invariance - it would be relatively $\|.\|_\infty$-compact, i.e. T_λ would induce a compact operator from $C(\Delta)$ to $C(\Delta)$. The adjoint $T_\lambda^* : M(\Delta) \to M(\Delta)$ would also be compact and, since $T_\lambda^*(L^1(\mu)) = T_\lambda(L^1(\mu)) \subset L^1(\mu)$, this would imply that T_λ^* maps $M(\Delta)$ into $L^1(\mu)$. But if δ_e denotes the Dirac-measure located at the unit-element e of the group Δ, then $T_\lambda^*(\delta_e) = \lambda$, which is in $M(\Delta) \setminus L^1(\mu)$; this furnishes the desired contradiction.

b) To apply Bukhvalov's criterion note that for every $n \in \mathbb{N}$ there is a compact set $K_n \subset \Delta$ such that $\lambda(K_n) > 1/2$ while $\mu(K_n) \leq n^{-1}$. Let f^n be a $[0,1]$-valued continuous function on Δ, which equals 1 on K_n and zero on a set of μ-measure greater than $1-2n^{-1}$. Note that $T_\lambda(f^n)$ is a continuous function on Δ, s.t.

$$T_\lambda(f^n)(e) = \int_\Delta f^n(\omega)d\lambda(\omega) > 1/2,$$

hence there is a neighbourhood of the unit-element e on which $T_\lambda f^n$ is greater than $1/2$. By the compactness of Δ we may find finitely many translates of $T_\lambda f^n$ such that the supremum is greater than $1/2$ on all of Δ. As T_λ commutes with the translation, there are finitely many translates $f_1^n,\dots,f_{m_n}^n$ of f^n such that, for every $\omega \in \Delta$,

$$\sup\{T_\lambda f_i^n(\omega) : 1 \leq i \leq m_n\} > 1/2.$$

The sequence $((f_i^n)_{i=1}^{m_n})_{n=1}^{\infty}$ is dominated by the constant function 1, converges to zero measure, while $((T_\lambda f_i^n)_{i=1}^{m_n})_{n=1}^{\infty}$ does not converge to zero at any point of Δ; this gives the desired contradiction to Bukhvalov's criterion.

References

[B1] A.V. Bukhvalov: Application of methods of the theory of order-bounded operators to the theory of operators in L^p-spaces, Russian Math. Surveys 38:6 (1983), p. 43 - 98.

[B2] A.V. Bukhvalov: On integral representation of linear operators, Zap. Nauchm. Sem. LOMI 47 (1974), 5 - 14. MR 53 no. 3767.

[C] A. Costé: An example of an operator in $L^1[0,1]$, unpublished communication.

[D-U] J. Diestel, J.J. Uhl: Vector Measures, Math. Surveys No. 15, A.M.S., Providence, RI (1977).

[D-S] N. Dunford, J.T. Schwartz: Linear Operators, Part I, Interscience, New York 1958.

[F] D. H. Fremlin: A positive compact operator, Manuscripta Math. 15 (1975), p. 323 - 327.

[G] A. Grothendieck: Produits tensoriels topologiques et espaces nucléaires, Mem. A.M.S. 16 (1955).

[H-S] P. R. Halmos, V. S. Sunder: Bounded Intergral Operators on L^2 Spaces, Springer (1978).

[K1] S. Kakutani: On equivalence of product measures, Ann. Math. 49 (1948).

[K2] Y. Katznelson: An Introduction to Harmonic Analysis, Wylie & Sons, New York 1968.

[M] G. Mokobodzki: Noyaux absolutement mesurables et opérateurs nucléaires. Sém. Goulaounic-Schwartz (1971 - 1972), exp. 6.

[v.N.] J. v. Neumann: Characterisierung des Spektrums eines Integral-Operators, Actualités Sci. et Ind., Paris 1935, no. 229.

[S1] W. Schachermayer: Integral Operators on L^p-spaces I, Indiana University Math. Journal 30, p. 123 - 140, (1981).

[S-W] W. Schachermayer, L. Weis: Almost compactness and decomposability of integral operators. Proc. A.M.S. 81 (1981), p. 595 - 599.

[S2] A.R. Schep: Compactness properties of an operator which imply that it is an integral operator, T.A.M.S. 265 (1981), p. 111-119.

[S3] Ch. Stegall: The Radon-Nikodym property in conjugate Banach spaces II, T.A.M.S. 264 (1981), p. 507 - 519.

[U] J. J. Uhl jr.: Kakutani's theorem on infinite product measures and operators on L^1, The Altgeld Book, p. III.1 - III.13, University of Illinois, Urbana (1975/76).

[W] L. Weis: Integral operators and changes of density, Indiana Univ. Math. J. 31 (1982), 83 - 96.

CYLINDER MEASURES, LOCAL BASES AND NUCLEARITY

Maurice Sion
The University of British Columbia
121 - 1984 Mathematics Road
Vancouver, B.C. Canada V6T 1Y4

0. Introduction

In this paper we are concerned with the problem of finding a measure on an infinite dimensional vector space X having given finite dimensional distributions. The general idea for tackling this problem is to view X as a limit of finite dimensional spaces S_j, as j runs over some index set J, consider measures μ_j on S_j which satisfy a (clearly necessary) consistency condition, and then try to construct a limit measure ν on X. A major source of difficulty is the lack of uniqueness of candidates for a limit space of the system $(S_j)_{j \in J}$. So, unless X is connected with the system of measures $(\mu_j)_{j \in J}$ as well, it is unlikely that it will be able to support a desired limit measure ν. Our goal is to indicate connections between X and the μ_j which guarantee the existence of such a ν.

The problem considered here is part of the more general one of finding a limit of an inverse (or projective) system of measure spaces. We shall discuss very briefly such systems first, not only because a cylinder measure is a special case of an inverse system of measures, but because our main result holds in a more general context than that of cylinder measures and topological vector spaces.

1. Inverse Systems and Limits

Consider an index set J directed by some partial order relation $<$, spaces S_j for $j \in J$ and surjective maps $p_{ij}: S_j \to S_i$ for $i < j$ satisfying the conditions that p_{jj} is the identity and

$$p_{ik} = p_{ij} \circ p_{jk} \qquad \text{for } i < j < k.$$

We refer to $(S_j, p_{ij})_{i,j \in J}$ as an inverse system of spaces.

1.1. Definition

$(X, p_i)_{i \in J}$ is a (inverse) <u>limit space</u> of $(S_j, p_{ij})_{i,j \in J}$ iff, for every i, $j \in J$, p_i: $X \to S_i$ is surjective and

$p_i = p_{ij} \circ p_j$ whenever $i < j$.

Now suppose that, for each $j \in J$, A_j is a σ-field of subsets of S_j and μ_j is a (countably additive) measure on A_j.

1.2. Definition

$(\mu_j)_{j \in J}$ is an inverse system of measures iff, for every i, j, α: $i < j$ and $\alpha \in A_i \Rightarrow p_{ij}^{-1}(\alpha) \in A_j$ and $\mu_j(p_{ij}^{-1}(\alpha)) = \mu_i(\alpha)$.

Finally, suppose that $(X, p_i)_{i \in J}$ is a limit space of $(S_j, p_{ij})_{i,j \in J}$, B is a σ-field of subsets of X and ν is a measure on B.

1.3. Definition

ν is a (inverse) limit measure of $(\mu_j)_{j \in J}$ iff, for every $j \in J$ and $\alpha \in A_j$,

$$p_j^{-1}(\alpha) \in B \text{ and } \nu(p_j^{-1}(\alpha)) = \mu_j(\alpha).$$

Example

The classical example of an inverse system and its possible limits is the following.

Let T be any (infinite) space,

$J = \{j: j \subset T \text{ and } j \text{ is finite}\}$

$S_j = R^j = \{y: y \text{ is a function on } j \text{ to } R\}$ for $j \in J$

$\pi_{ij}: y \in R^j \to y|_i \in R^i$ for $i \subset j \in J$

$\pi_i: x \in R^T \to x|_i \in R^i$ for $i \in J$.

Then J is directed by inclusion ($\subset$), the π_{ij} and π_i are the canonical projections, so $(S_j, \pi_{ij})_{i,j \in J}$ constitutes an inverse system of spaces and $(R^T, \pi_i)_{i \in J}$ is clearly a limit space of the system. However, it is just as clear that, for any $X \subset R^T$ with $\pi_i(X) = S_i$ for every $i \in J$, $(X, \pi_i)_{i \in J}$ will also be a limit space of the system.

Now if, for each $t \in T$, μ_t is a probability measure on $R(\simeq R^{\{t\}})$ and, for each $j \in J$,

$$\mu_j = \bigotimes_{t \in j} \mu_t$$

is the product measure on S_j then $(\mu_j)_{j \in J}$ constitutes an inverse system of measures. The fact that a limit measure ν exists on $\mathcal{R}^T$ is a consequence of theorems due to Daniell [2] and Kolmogorov [9]. When $\mathcal{R}$ is replaced by an arbitrary space, the existence of a limit measure is due to von Neumann (see [6]). When $T = \mathcal{N}$ and μ_t is a fixed Gaussian measure (say, of mean 0 and variance 1) on $\mathcal{R}$ for every t then, in view of the above, a limit measure will exist on $\mathcal{R}^{\mathcal{N}}$. However it will not exist on $\ell_2 \subset \mathcal{R}^{\mathcal{N}}$, nor for that matter on any ℓ_p, $1 \leq p \leq \infty$. This is a consequence of the easily checked fact that, for $k < \infty$ and $\varepsilon > 0$, if $N \in \mathcal{N}$ is large enough and $j = \{0,\cdots,N\}$ and $B_j(k)$ is the ball of radius k in S_j then $\mu_j(B_j(k)) < \varepsilon$.

2. Cylinder Measures

Throughout this section, let E be a vector space over $\mathcal{R}$. For any (linear) subspace F of E, let F* denote the algebraic dual of F, i.e.

$$F^* = \{y: \ y \text{ is a linear functional on } F\}.$$

We set

$$J = \{j: \ j \text{ is a finite dimensional subspace of } E\}$$

$$S_j = j^* \qquad \text{for } j \in J$$

$$\pi_{ij}: \ y \in j^* \to y|_i \in i^* \qquad \text{for } i, j \in J \text{ and } i \subset j$$

$$\pi_i: \ x \in E^* \to x|_i \in i^* \qquad \text{for } i \in J.$$

Then, clearly $(S_j, \pi_{ij})_{i,j \in J}$ constitutes an inverse system of spaces (with J directed by $\subset$) and, for any $X \subset E^*$ with $\pi_i(X) = S_i$ for $i \in J$, $(X, \pi_i)_{i \in J}$ is a limit space of the system.

2.1. Definition

A cylinder measure over E is a system of measures μ_j on σ-fields A_j of subsets of S_j for $j \in J$, such that $(\mu_j)_{j \in J}$ constitutes an inverse system.

Since for each $j \in J$ the space S_j is finite dimensional, S_j has a unique locally convex topology induced by any isomorphism of S_j with a Euclidean space. All topological notions on S_j will refer to this topology.

2.2. Assumptions

We suppose from now on that:

(1) for each $j \in J$, A_j is the Borel family generated by the closed subsets of S_j and μ_j is a probability measure on A_j;

(2) $(\mu_j)_{j \in J}$ is a cylinder measure;

(3) X is a subspace of E* and $\pi_i(X) = S_i$ for $i \in J$.

Thus, $(X,\pi_i)_{i \in J}$ is a limit space of $(S_j,\pi_{ij})_{i,j \in J}$. We want to study conditions under which the system $(\mu_j)_{j \in J}$ has a limit measure ν on X. Our first step is to construct a canonical measure μ^* on E*.

2.3. Definitions Let

$$C = \{\pi_j^{-1}(\alpha);\ j \in J \text{ and } \alpha \in A_j\},$$

$$\tau:\quad \pi_j^{-1}(\alpha) \longmapsto \mu_j(\alpha),$$

μ^* = The Caratheodory outer measure on E* generated by τ,

M = The family of μ^*-measurable sets in the sense of Caratheodory

Note: (1) C is the family of cylinders in E* having as a base a Borel subset of a finite dimensional space.

(2) τ is well defined on C because the μ_j constitute an inverse system.

(3) μ^* is defined for any $A \subset E^*$ by:

$$\mu^*(A) = \inf\{\sum_{C \in F} \tau(C);\ F \text{ a countable subfamily of } C \text{ which covers } A\}.$$

(4) By definition:

$$A \in M \iff \mu^*(T) = \mu^*(T \cap A) + \mu^*(T \setminus A) \text{ for every } T \subset E^*.$$

From elementary measure theory (see e.g. [15]) we know that M is a σ-field and that μ^* is countably additive on M. Since C is clearly an algebra and τ is finitely additive on C, we also know that $C \subset M$. The crux of the problem is to check that τ is in fact countably additive on C so μ^* is an extension of τ. The main results are the following.

2.4. Theorem μ^* is a limit measure of $(\mu_j)_{j \in J}$ on E*.

2.5. Lemma The system $(\mu_j)_{j \in J}$ has a limit measure on X $\iff \mu^*(X) = 1$.

Proofs The proof of 2.4 is an immediate consequence of any one of well known results, for example from the classical Kolmogorov theorem [9]. (Take any Hamel basis T for E and identify E^* with R^T). Also, a direct proof in this case is fairly straightforward. For this theorem, the key point is that E^* possesses the following property.

For any countable nested $J_0 \subset J$:

$$y_j \in S_j \text{ and } \pi_{ij}(y_j) = y_i \text{ for every } i,j \in J_0 \text{ with } i \subset j \Rightarrow$$

there exists an $x \in E^*$ such that $\pi_j(x) = \pi_j(y)$ for every $j \in J_0$.

The proof of 2.5 is elementary and straight forward (see [7]). We should point out here that, in general, X is not a measurable subset of E^*, i.e. $X \notin \mathcal{M}$, so the use of an outer measure defined on the family of all subsets of E^* is more than a notational convenience. To ensure that $X \in \mathcal{M}$, one frequently imposes certain countability conditions on the system J, by requiring E to be separable for example. By making use of lemma 2.5 we shall avoid requiring such conditions.

Our aim now is to find relations between X and the given μ_j themselves which guarantee that $\mu^*(X) = 1$. At this level of generality, the fundamental criterion is the following.

2.6. Theorem. Suppose that, for any countable nested $J_0 \subset J$ and $\varepsilon > 0$, we can find $B_j(\varepsilon) \in \mathcal{A}_j$ for $j \in J_0$ satisfying the following conditions:

(1) $\mu_j(S_j \setminus B_j) \leq \varepsilon$ for $j \in J_0$;

(2) $i \subset j \Rightarrow \pi_{ij}(B_j) \subset B_i$;

(3) $y \in E^*$ and $\pi_j(y) \in B_j$ for every $j \in J_0 \Rightarrow$ there exists an $x \in X$ such that $\pi_j(x) = \pi_j(y)$ for every $j \in J_0$.

Then $\mu^*(X) = 1$ so $(\mu_j)_{j \in J}$ has a limit measure on X.

Proof. We shall show that any countable cover of X by cylinders in $\mathcal{C}$ covers almost all of E^*. Let $J_0 \subset J$ be countable, $\alpha_j \in \mathcal{A}_j$ for $j \in J_0$, and

$$X' = \bigcup_{j \in J_0} \pi_j^{-1}(\alpha_j)$$

be such that $X \subset X'$. We suppose without loss of generality that J_0 is nested. To see that $\mu^*(E^* \setminus X') = 0$, given any $\varepsilon > 0$ choose sets $B_j(\varepsilon)$

as in the hypothesis of the theorem. We have

$$(E^*\backslash X') \subset \bigcup_{j\in J_0} \pi_j^{-1}(S_j\backslash B_j(\varepsilon))$$

for, if $y \in E^*$ and $\pi_j(y) \notin (S_j\backslash B_j(\varepsilon))$ so $\pi_j(y) \in B_j(\varepsilon)$ for every $j \in J_0$, then by condition (3) there exists an $x \in X$ with $\pi_j(x) = \pi_j(y)$ for every $j \in J_0$ and, since $x \in X \subset X'$, $\pi_i(x) \in \alpha_i$ for some $i \in J_0$ so $\pi_i(y) \in \alpha_i$ and $y \in \pi_i^{-1}(\alpha_i) \subset X'$. By condition (2):

$$i \subset j \Rightarrow \pi_i^{-1}(S_i\backslash B_i(\varepsilon)) \subset \pi_j^{-1}(S_j\backslash B_j(\varepsilon)).$$

Hence, by condition (1):

$$\mu^*(E^*\backslash X') \le \lim_{j\in J_0} \mu^*(S_j\backslash B_j(\varepsilon)) = \lim_{j\in J_0} \mu_j(S_j\backslash B_j(\varepsilon)) \le \varepsilon.$$

Letting $\varepsilon \to 0$ yields the desired result.

By Bochner's theorem, we know that the μ_j are determined by their one-dimensional distributions so we concentrate our attention on these. More precisely, we consider the following.

2.7. Definition. For $\varepsilon > 0$ and $u \in E$:

$$\lambda(\varepsilon,u) = \inf\{r > 0: \quad \mu_i(\{y \in S_i: \quad y(u) > r\}) < \varepsilon\}$$

where i = space spanned by u.

Note that $0 \le \lambda(\varepsilon,u) < \infty$ since μ_i is a bounded Borel measure on a space isomorphic to $\mathbb{R}$. Also, from the definition of inverse system it follows immediately that

$$u \in j \in J \Rightarrow \mu_j(\{y \in S_j: \quad y(u) > \lambda(\varepsilon,u)\}) \le \varepsilon$$

$$\text{and} \qquad \mu^*(\{y \in E^*: \quad y(u) > \lambda(\varepsilon,u)\}) \le \varepsilon.$$

The desired connection between X and the μ_j to guarantee the existence of a limit measure on X will be described in the next sections in terms of the numbers $\lambda(\varepsilon,u)$.

3. Local Bases

We continue with the notation and assumptions introduced in section 2. Since each $j \in J$ is a finite dimensional space we can treat it (and its dual S_j) as a Euclidean space by introducing a basis for it. Normally, this is done by choosing a Hamel basis for the infinite dimensional space E (or a dense subspace of E in the topological case) and then taking the subspaces generated by a finite number of basis elements. For our purposes and the applications we have in mind, this is too restrictive. For each $j \in J$, we shall consider an isomorphism with Euclidean space without a priori concern for E, E* or X.

3.1. <u>Definitions</u>. For each $j \in J$ let

n_j = dimension of j (= dimension of S_j)

$f_j\colon \mathcal{R}^{n_j} \to j$ be an isomorphism (1-1 and linear)

$f_j^*\colon j^* = S_j \to (\mathcal{R}^{n_j})^* = \mathcal{R}^{n_j}$ be the adjoint of f_j.

We refer to $(f_j)_{j\in J}$ as a system of local bases for E. Note that each f_j is associated with a unique basis $\{e_1,\cdots,e_{n_j}\}$ for j such that

$$f_j\colon \; s\in \mathcal{R}^{n_j} \longmapsto \sum_{k=1}^{n_j} s_k e_k \qquad \text{and}$$

$$f_j^*\colon \; y\in S_j \longmapsto (y(e_1),\cdots,y(e_{n_j}))\in \mathcal{R}^{n_j}.$$

3.2. <u>Notation</u>. For $j\in J$, s, $t\in \mathcal{R}^{n_j}$ and $\varepsilon > 0$:

$$t\cdot s = \sum_{k=1}^{n_j} s_k t_k \;;$$

$$\|s\|_j = \sqrt{s\cdot s}$$

$$\mathcal{S}_j = \{s\in \mathcal{R}^{n_j}\colon \; \|s\|_j = 1\} \;;$$

ξ_j is the uniform probability measure on $\mathcal{S}_j$;

$$\bar\lambda(\varepsilon,f_j) = \int \lambda(\varepsilon,\, f_j(s))d\,\xi_j\,(s) \;;$$

$$\beta_j(\varepsilon,f_j) = \{y\in S_j\colon \; \|f_j^*(y)\|_j \le \frac{\sqrt{n_j}}{\varepsilon}\;\bar\lambda(\varepsilon,f_j)\}.$$

The key result in this section is the following.

3.3. <u>Theorem</u>. There is an absolute constant $C < \infty$ such that, for every $j \in J$,

$\mu_j(S_j \setminus \beta_j(\varepsilon, f_j)) \leq 2C\varepsilon$.

The proof is based on the following technical lemma.

Lemma. There is an absolute constant C such that $0 < C < \infty$ and, for any $j \in J$ and $t \in R^{n_j}$ with $t \neq 0$,

$$\frac{1}{C} < \xi_j(\{s \in S_j: \frac{t \cdot s}{\|t\|_j} > \frac{1}{\sqrt{n_j}}\}).$$

Proof. see lemma A2 in the appendix.

For the proof of theorem 3.3, we fix j and drop the subscript j from our notation. Let

$$y \cdot s = y(f(s)) = f^*(y) \cdot s \qquad \text{for } y \in S ;$$

$$r = \frac{1}{\varepsilon} \bar{\lambda}(\varepsilon, f) ;$$

$$B = S \setminus \beta(\varepsilon, f) = \{y \in S: \|f^*(y)\| > \frac{\sqrt{n}}{\varepsilon} \bar{\lambda}(\varepsilon, f)\};$$

$$L = \{(y, s): y \in S, s \in S \text{ and } y \cdot s > r\}.$$

Then, by the above lemma and the Fubini theorem, we have

$$\begin{aligned}
\frac{1}{C} \mu(B) &\leq \int_B \xi(\{s \in S: \frac{y \cdot s}{\|f^*(y)\|} > \frac{1}{\sqrt{n}}\})\, d\mu(y) \\
&\leq \int_B \xi(\{s \in S: y \cdot s > r\})\, d\mu(y) \\
&\leq \int \xi(\{s \in S: y \cdot s > r\})\, d\mu(y) \\
&= \mu \otimes \xi(L) \\
&= \int \mu(\{y \in S: y \cdot s > r\})\, d\xi(s).
\end{aligned}$$

Now, let

$$A = \{s \in S: r > \lambda(\varepsilon, f(s))\} \qquad \text{and}$$

$$A' = \{s \in S: r \leq \lambda(\varepsilon, f(s))\}.$$

Then

$$s \in A \Rightarrow \mu(\{y \in S: y \cdot s > r\}) \leq \mu(\{y \in S: y \cdot s > \lambda(\varepsilon, f(s))\}) \leq \varepsilon$$

and

$$s \in A' \Rightarrow 1 \leq \frac{1}{r} \lambda(\varepsilon, f(s)) \qquad \text{so}$$

$$\xi(A') = \int 1_{A'}(s) d\xi(s) \leq \frac{1}{r} \int \lambda(\varepsilon, f(s)) d\xi(s) = \frac{1}{r} \bar{\lambda}(\varepsilon, f) = \varepsilon.$$

Thus,

$$\int_A \mu(\{y \in S: y \cdot s > r\})\, d\xi(s) \leq \varepsilon \cdot \xi(A) \leq \varepsilon$$

$$\int_{A'} \mu(\{y \in S: y \cdot s > r\})\, d\xi(s) \leq \int_{A'} 1 \cdot d\xi(s) \leq \varepsilon$$

so

$$\frac{1}{C} \mu(B) \leq 2\varepsilon.$$

4. Cylinder Measures over Locally Convex Spaces

We shall now consider the case when E is a locally convex topological vector space so the topology is induced by a family τ of semi norms. In referring to topological notions we shall confuse τ with the topology induced by τ whenever convenient.

4.0. Notation.

$E'(\tau) = \{x \in E^*: \ x \text{ is } \tau\text{-continuous}\}$

$\mathcal{U}(\tau) = \{V: \ V \text{ is a convex, symmetric } \tau\text{-neighborhood of } 0 \text{ in } E\}$

$V^0 = \{x \in E^*: \ |x(u)| \leq 1 \ \text{ for every } u \in V\}$.

We shall drop reference to τ in the notation when there is no danger of confusion.

We continue with the notation and general assumptions of section 2 so $(\mu_j)_{j \in J}$ is a probability cylinder measure over E. The question now is to determine conditions under which a limit measure will exist on the topological dual E'.

As an immediate application of theorem 2.6 we have the following useful general lemma.

4.1. Lemma. Suppose that for every countable nested $J_0 \subset J$ and $\varepsilon > 0$ there exists a $W \in \mathcal{U}(\tau)$ such that

$$\mu_j(S_j \setminus \pi_j(W^0)) < \varepsilon \qquad \text{for every } j \in J_0.$$

Then $(\mu_j)_{j \in J}$ has a limit measure on $E'(\tau)$.

Proof. Apply theorem 2.6 with $B_j(\varepsilon) = \pi_j(W^0)$ after noting that condition (2) is trivially satisfied and condition (3) is a consequence of the Hahn-Banach theorem, for if $y \in E^*$ and $\pi_j(y) \in \pi_j(W^0)$ for $j \in J_0$ then

$$|y(u)| \leq 1 \quad \text{for every} \quad u \in W \cap \bigcup_{j \in J_0} j$$

hence there is an $x \in W^0 \subset E'$ such that $\pi_j(x) = \pi_j(y)$ for every $j \in J_0$.

Combining 4.1 with theorem 3.3, we get the following.

4.2. Theorem. Suppose that, for any countable nested $J_0 \subset J$ and $\varepsilon > 0$, there exist a system of local bases $(f_j)_{j\in J_0}$ and a $W \in \mathcal{U}(\tau)$ such that

$$\beta_j(\varepsilon, f_j) \subset \pi_j(W^0) \qquad \text{for every } j \in J_0.$$

Then $(\mu_j)_{j\in J}$ has a limit measure on $E'(\tau)$.

In view of the above, we are interested in relating the numbers $\lambda(\varepsilon,u)$ introduced in definition 2.7 with the topology on E. This link is provided by the following definition.

4.3. Definition.

$(\mu_j)_{j\in J}$ is τ-continuous iff for every $\varepsilon > 0$ there exists a $V \in \mathcal{U}(\tau)$ such that

$$u \in V \Rightarrow \mu_i(\{y \in S_i: \ |y(u)| > 1\}) < \varepsilon$$

where i = space spanned by u.

Note: The above amounts to saying that for every $\varepsilon > 0$ there exists a semi norm $p \in \tau$ and $r < \infty$ such that $\lambda(\varepsilon,u) \le r \cdot p(u)$ for every $u \in E$.

Two major results concerning continuous cylinder measures are the Minlos theorem, which states that a continuous cylinder measure over a nuclear space E has a limit measure on E', and the Sazonov extension of Bochner's theorem to a Hilbert space E which states that, for any given positive definite function F on E, if F satisfies a continuity condition then there exists a measure on E(=E') whose Fourier transform is F. We want to show that the concepts involved in the statements of these theorems are subsumed by the point of view of local bases introduced in section 3 and that the theorems are direct consequences of theorem 4.2. In the process, we shall remove separability conditions from the usual hypotheses and extend both the concepts and the results to more general settings.

5. Nuclearity

The notion of a nuclear space was introduced by Grothendieck [4,5] and plays an important role in the study of certain function spaces and distributions. In [14], Sazonov considered a topology on a Hilbert space to discuss a continuity condition on the Fourier transform of a measure. Subsequently, Kolmogorov in [10] introduced an I-topology which contained as special cases the continuity conditions used by Minlos and Sazonov. In presenting these concepts, we follow the formulations due to Ito in [8] rather than those found in most textbooks because they are both simpler and better suited for our purposes here. We drop however the condition of separability. Variations and extensions of these notions will be considered later.

5.0. Definitions.

(1) A Hilbertian semi norm on E is a semi norm induced by a semi inner product. Orthogonality in a Hilbertian semi norm refers to orthogonality in the associated semi inner product.

(2) For any two Hilbertian semi norms p,q on E:
$p < q$ (in the Hilbert-Schmidt sense) iff

$$\sup\{ \sum_{k=1}^{n} (p(e_k))^2; \quad n \in \mathbb{N} \text{ and } \{e_1, \cdots, e_n\} \text{ orthonormal in } q\} < \infty.$$

(3) E is nuclear iff E is a topological vector space whose topology is induced by a family of Hilbertian semi norms directed by $<$.

(4) For any family τ of Hilbertian semi norms on E:
$I(\tau) = \{p\colon$ p is a Hilbertian semi norm and, for some $q \in \tau$, $p < q\}$.

Remarks. In [8], Ito points out the following (assuming separability):

(1) The above definition of nuclear space is equivalent to the usual one found in most textbooks involving extensions of the identity map to Hilbert Schmidt operators on the completions of E in each of the semi norms.

(2) The family τ is directed by $<$ iff the topology induced by τ is the same as that induced by $I(\tau)$. Hence, for a nuclear space, the I-topology coincides with the original one.

(3) When E is a Hilbert space, so that τ consists of a single Hilbert norm, the I-topology is in fact the Sazonov topology.

He also states the following theorem (under separability conditions) which includes as special cases the Minlos theorem about nuclear spaces and the Sazonov extension of Bochner's theorem to Hilbert spaces.

5.1. Theorem. Let τ be a family of Hilbertian semi norms on E. If the cylinder measure $(\mu_j)_{j \in J}$ is $I(\tau)$-continuous then it has a limit measure on $E'(\tau)$.

Proof. Given $\varepsilon > 0$, by the $I(\tau)$-continuity of the cylinder measure, there exists a $p \in I(\tau)$ and $r < \infty$ such that $\lambda(\varepsilon,u) \leq rp(u)$ for $u \in E$. Choose $q \in \tau$ such that $p < q$ and let

$$M^2 = \sup\{\sum_{k=1}^{n} (p(e_k))^2; \; n \in N \text{ and } \{e_1,\cdots.e_n\} \text{ orthonormal in } q\}$$

so $M < \infty$, and

$$W = \{u \in E: \; q(u) < \frac{\varepsilon}{rM}\}.$$

For any $j \in J$, by the spectral theorem on finite dimensional spaces, choose a basis $\{e_1,\cdots,e_{n_j}\}$ for j which is orthogonal in both p and q and with $q(e_k) = 1$ for $k = 1,\cdots,n_j$. Let

$$f_j: \; s \in R^{n_j} \longrightarrow \sum_{k=1}^{n_j} s_k e_k.$$

Then

$$\lambda^2(\varepsilon,f_j(s)) \leq r^2 p^2(f_j(s)) = r^2 \sum_{k=1}^{n_j} s_k^2 \, p^2(e_k)$$

so, by the Cauchy-Schwarz inequality and lemma A1 in the appendix,

$$\bar{\lambda}^2(\varepsilon,f_j) \leq \int \lambda^2(\varepsilon,f_j(s))d\,\xi_j(s) \leq r^2 \sum_{k=1}^{n_j} (\int s_k^2 \, d\,\xi_j(s)\; p^2(e_k))$$

$$= \frac{r^2}{n_j} \sum_{k=1}^{n_j} p^2(e_k) \leq \frac{r^2}{n_j} M^2.$$

Now, for $y \in S_j$, $f_j^*(y) = (y(e_1)),\cdots,y(e_{n_j}))$ so

$$\| f_j^*(y) \|_j \leq \frac{\sqrt{n_j}}{\varepsilon} \bar{\lambda}(\varepsilon,f_j) \Rightarrow \text{ for any } s \in R^{n_j}:$$

$$|y(f_j(s))| = |f_j^*(y)\cdot s| \leq \frac{\sqrt{n_j}}{\varepsilon} \bar{\lambda}(\varepsilon,f_j) \, \|s\|_j \leq \frac{rM}{\varepsilon} q(f_j(s))$$

$$\Rightarrow y \in \pi_j(W^0).$$

Thus, $\beta_j(\varepsilon,f_j) \subset \pi_j(W^0)$ for every $j \in J_0$ and theorem 4.2 yields the desired result.

We now turn our attention to variations of the above notions in a more general setting and take fuller advantage of theorem 4.2 to obtain a stronger result.

5.2. Definitions.

For any $J_0 \subset J$ and semi norms p, q on $\bigcup_{j \in J_0} j$:

$$d_j(p,q) = \inf \int p(f(s))\, d\xi_j(s) \tag{1}$$

as f runs over all isomorphisms: $R^{n_j} \to j$ such that

$\|s\|_j \le q(f(s))$ for every $s \in R^{n_j}$.

$$M(J_0,p,q) = \sup_{j \in J_0} \sqrt{n_j}\, d_j(p,q). \tag{2}$$

$$p \ll q \quad \text{iff} \quad M(J_0,p,q) < \infty. \tag{3}$$

5.3. Definition. For any family τ of semi norms on E and $J_0 \subset J$:
$I(J_0,\tau) = \{p: \ p$ is a semi norm on $\bigcup_{j \in J_0} j$ and, for some $q \in \tau$, $p \ll q\}$.

5.4. Theorem. Let τ be any family of semi norms on E. If, for every countable nested $J_0 \subset J$, the subsystem $(\mu_j)_{j \in J_0}$ is $I(J_0,\tau)$-continuous on $\bigcup_{j \in J_0} j$ then the cylinder measure $(\mu_j)_{j \in J}$ has a limit measure on $E'(\tau)$.

Proof. We shall show that the hypothesis of theorem 4.2 is satisfied. Given any countable nested $J_0 \subset J$ and $\varepsilon > 0$, let $E_0 = \bigcup_{j \in J_0} j$ and choose a $p \in I(J_0,\tau)$ and $r < \infty$ such that $\lambda(\varepsilon,u) \le rp(u)$ for $u \in E_0$. Then choose $q \in \tau$ with $p \ll q$. For each $j \in J_0$, consider an isomorphism f_j: $R^{n_j} \to j$ such that

(a) $\|s\|_j \le q(f_j(s))$ for $s \in R^{n_j}$ and

(b) $\int p(f_j(s))\, d\xi_j(s) < d_j(p,q) + \dfrac{1}{\sqrt{n_j}}$.

Then, for any $y \in S_j$:

$\|f_j^*(y)\|_j \leq \sqrt{n_j}\,\bar{\lambda}(\varepsilon,f_j) \Rightarrow$ for any $s \in R^{n_j}$,

$$|y(f_j(s))| = |f_j^*(y)\cdot s| \leq \sqrt{n_j}\,\bar{\lambda}(\varepsilon,f_j)\,\|s\|_j \leq r(\sqrt{n_j}\,d_j(p,q) + 1)\|s\|_j$$

$$\leq r(M(J_0,\tau) + 1)\,q(f_j(s)).$$

Hence, letting

$$W = \{u \in E:\ q(u) < \frac{\varepsilon}{r(M(J_0,\tau) + 1)}\},$$

we see that $\beta_j(\varepsilon,f_j) \subset \pi_j(W^0)$ for $j \in J_0$ so theorem 4.2 applies and yields the desired result.

Actually, the above results and those in the preceding sections can be reformulated in a very general setting without any reference to E or E*.

Let J be any index set directed by a partial order relation $<$, $(R^{n_j}, p_{ij})_{i,j \in J}$ be an inverse system of Euclidean spaces, with $1 \leq n_j$, and $(\mu_j)_{j\in J}$ be an inverse system of probability Borel measures. Using the notation in R^{n_j} introduced in 3.2, let

$$\lambda_j(\varepsilon,s) = \inf\{r > 0:\ \mu_j(\{y \in R^{n_j}:\ y\cdot s > r\}) < \varepsilon\}$$

and $\bar{\lambda}_j(\varepsilon) = \int \lambda_j(\varepsilon,s)d\xi_j(s)$.

The fundamental result corresponding to theorem 4.2 is then the following.

5.5. Theorem.

Suppose that $(X,p_i)_{i\in J}$ is a limit space of $(R^{n_j},p_{ij})_{i,j\in J}$ and that for any countable ordered $J_0 \subset J$ and $\varepsilon > 0$ there exist $B_j(\varepsilon) \subset R^{n_j}$ for $j \in J_0$ such that:

$$\{y \in R^{n_j}:\ \|y\|_j \leq \frac{\sqrt{n_j}}{\varepsilon}\bar{\lambda}_j(\varepsilon)\} \subset B_j(\varepsilon); \tag{1}$$

$$i < j \Rightarrow p_{ij}(B_j(\varepsilon)) \subset B_i(\varepsilon); \tag{2}$$

$y_j \in B_j(\varepsilon)$ and $p_{ij}(y_j) = y_i$ for $i, j \in J_0$ with $i < j \Rightarrow$ (3)

there exists an $x \in X$ with $p_j(x) = y_j$ for $j \in J_0$.
Then $(\mu_j)_{j \in J}$ has a limit measure on X.

Proof. Let $S_j = R^{n_j}$. By theorem 3.3 we have $\mu_j(S_j \setminus B_j(\varepsilon)) \leq 2C\varepsilon$. We can then apply the same argument as in the proof of theorem 2.6 with E^* replaced by $S^* = \prod_{j \in J} S_j$, μ^* by a corresponding outer measure on S^*, and X by its image $X^* \subset S^*$ under the map: $x \in X \longmapsto (p_j(x))_{j \in J}$.

6. Appendix

A. Results in R^n.

We use the notation introduced in 3.2 (but with the subscript j omitted) so, for $n \in N$, S_n in the unit sphere in R^n and ξ_n is the uniform probability measure on S_n.

A1. Lemma.

$$\int s_k^2 \, d\xi_n(s) = \frac{1}{n} \qquad \text{for } k = 1, \cdots, n.$$

Proof. By symmetry, the above integral has the same value c for all $k = 1, \cdots, n$ and, for any $s \in S$, $\sum_{k=1}^{n} s_k^2 = 1$. Hence

$$nc = \sum_{k=1}^{n} \int s_k^2 \, d\xi_n(s) = \int \sum_{k=1}^{n} s_k^2 \, d\xi_n(s) = \int 1 \cdot d\xi_n(s) = 1.$$

A2. Lemma.

There is an absolute constant C such that $0 < C < \infty$ and, for $n \in N$, $n \geq 1$ and any $a \in S_n$

$$\frac{1}{C} < \xi_n(\{s \in S_n : \; a \cdot s \geq \frac{1}{\sqrt{n}}\}).$$

Proof. Let $2 \leq n \in N$ and, for $k = 2, \cdots, n$, let σ_k be the $(k-1)$-dimensional measure on R^k and

$$S_k(r) = \{s \in R^k : \; \|s\| = r\}.$$

Then

$$\sigma_2(S_2(r)) = 2\pi r$$

and for $2 < k$:

$$\sigma_k(S_k(r)) = \int_0^\pi \sigma_{k-1}(S_{k-1}(r \sin \theta))\, r d\theta,$$

so that by induction we have:

$$\sigma_k(S_k(r)) = \alpha_k r^{k-1} \qquad \text{for some } \alpha_k \geq 0.$$

Let $a \in S_n$ and

$$A_n = \{s \in S_n: \quad a \cdot s \geq \frac{1}{\sqrt{n}}\}.$$

Since $S_n = S_n(1)$ and $\xi_n = \dfrac{\sigma_n}{\sigma_n(S_n)}$, we have

$$\xi_n(A_n) = \frac{\int_0^{\cos^{-1}(\frac{1}{\sqrt{n}})} \alpha_{n-1}(\sin \theta)^{n-2}\, d\theta}{\int_0^\pi \alpha_{n-1}(\sin \theta)^{n-2}\, d\theta}$$

Letting $t = \cos \theta$ and $u = \sqrt{n}\, t$, we get

$$\xi_n(A_n) = \frac{\int_1^{\sqrt{n}} (1 - \frac{u^2}{n})^{\frac{n-3}{2}}\, du}{2 \int_0^{\sqrt{n}} (1 - \frac{u^2}{n})^{\frac{n-3}{2}}\, du}$$

As $n \to \infty$ the above tends to the limit

$$\frac{\int_1^\infty e^{-\frac{u^2}{2}}\, du}{2 \int_0^\infty e^{-\frac{u^2}{2}}\, du} > 0.$$

By symmetry, the value of $\xi_n(A_n)$ is independent of the choice of $a \in S_n$, hence

$$\inf_{n \geq 1} \xi_n(A_n) > 0.$$

B. Fourier Transforms

For a vector space E and a probability cylinder measure $(\mu_j)_{j \in J}$ over E, the Fourier transform is the function

$$\hat{\mu}: \quad u \in E \longmapsto \int e^{ix(u)} \, d\mu_j(x) \in \mathbb{C}$$

where j = space spanned by u. Actually, the value of the integral is the same for all $j \in J$ with $u \in j$ so the Fourier transform of any limit measure of $(\mu_j)_{j\in J}$ is the same as $\hat{\mu}$. One easily checks that $\hat{\mu}$ is positive definite, $\hat{\mu}(0) = 1$ and, for any fixed $j \in J$, $\hat{\mu}$ is continuous at 0 on j.

Given a function $F: E \to \mathbb{C}$ with the above properties and $X \subset E^*$, a major problem is to determine whether there is a measure on X whose Fourier transform is F. By Bochner's theorem [1], for each $j \in J$, there is a unique probability measure μ_j on j^* with $\hat{\mu}_j = F|_j$ and it is a simple matter to check that $(\mu_j)_{j\in J}$ is a cylinder measure. Thus, there is always a unique cylinder measure over E whose Fourier transform is F. The problem therefore reduces to determining if this cylinder measure has a limit measure on X.

When E is a topological vector space, it is natural to impose the extra condition that F be continuous at 0 on E (rather just on each $j \in J$) and to let X = E', the topological dual of E. In view of lemma B1 below, this is equivalent to requiring that the associated cylinder measure be continuous in the sense of definition 4.3. Thus, the theorems of section 5 are directly applicable to this question. The well known result in this context, Sazonov's theorem stated below, is a special case of the results in section 5.

B1. <u>Lemma</u>. When E is a locally convex space: $\hat{\mu}$ is continuous at 0 on E iff $(\mu_j)_{j\in J}$ is continuous.

<u>Proof</u>. (a) Suppose μ is continuous at 0 on E. Given $\varepsilon > 0$, choose a convex neighborhood U of 0 such that

$$u \in U \Rightarrow |\hat{\mu}(u) - 1| < \varepsilon \quad \Rightarrow \quad 1 - \varepsilon < \operatorname{Re} \hat{\mu}(u).$$

For $u \in U$, let j = space spanned by u and $\mu = \mu_j$. Then

$$0 \le t \le 1 \quad \Rightarrow \quad tu \in U \quad \Rightarrow \quad 1 - \varepsilon < \operatorname{Re} \hat{\mu}(tu) = \int \cos(t \cdot x(u)) \, d\mu(x).$$

Hence

$$1 - \varepsilon \le \int_0^1 dt \int \cos(t \cdot x(u)) d\mu(x) = \int d\mu(x) \int_0^1 \cos(t \cdot x(u)) dt$$

$$= \int \frac{\sin x(u)}{x(u)} \, d\mu(x).$$

Let $A = \{x: \ |x(u)| \le \frac{1}{\varepsilon}\}$ and $B = \{x: \ |x(u)| > \frac{1}{\varepsilon}\}$. Then

$$1 - \varepsilon \le \int \left|\frac{\sin x(u)}{x(u)}\right| d\mu(x) \le \int_A 1 d\mu(x) + \int_B \varepsilon d\mu(x)$$

$$\le \mu(A) + \varepsilon = 1 - \mu(B) + \varepsilon$$

so $\mu(B) \le 2\varepsilon$. Thus, if we let $V = \varepsilon U$ we get $v \in V \Rightarrow$ for some $u \in U$, $v = \varepsilon u$ and

$$\mu(\{x: |x(v)| > 1\}) = \mu(\{x: |x(u)| > \frac{1}{\varepsilon}\}) \le 2\varepsilon.$$

(b) Next, suppose $(\mu_j)_{j \in J}$ is continuous. Given $0 < \varepsilon < 1$, choose a convex neighborhood U of 0 such that for any $u \in U$, letting j = space spanned by u and $\mu = \mu_j$ and $B = \{x: |x(u)| > 1\}$, we have $\mu(B) < \varepsilon$. Then for $v = \varepsilon u$ we have

$$\int [1 - \cos x(v)] d\mu(x) = \int 2 \sin^2 \frac{x(v)}{2} d\mu(x)$$

$$\le 2 \int_{B^c} \frac{|x(v)|^2}{4} d\mu(x) + 2 \int_B 1 \, d\mu(x)$$

$$\le \frac{\varepsilon^2}{2} + 2\varepsilon < 3\varepsilon.$$

Similarly,

$$\left|\int \sin x(v) d\mu(x)\right| < 2\varepsilon.$$

Thus, for $v \in \varepsilon U$ we have:

$$|\hat{\mu}(v) - 1| = \left|\int [e^{ix(v)} - 1] d\mu(x)\right| < 5\varepsilon.$$

B2. <u>Theorem</u> (Sazonov).

Let E be a separable Hilbert space with inner product $\langle \cdot, \cdot \rangle$ so $E = E'$ and let $F: E \to \mathbb{C}$ be continuous, positive definite and $F(0) = 1$. Then there is a probability measure ν on E with $\hat{\nu} = F$ iff, for every $\varepsilon > 0$, there is a symmetric, positive Hilbert-Schmidt operator A such that

$$\langle Au, u \rangle \le 1 \Rightarrow |\text{Re } F(u) - 1| < \varepsilon.$$

<u>Note</u>: (1) If $p_A(u) = \langle Au, u \rangle$, we see that p_A is a semi norm.

(2) If τ is the set consisting of the single norm induced by $\langle \cdot, \cdot \rangle$ then

$$I(\tau) = \{p_A; \text{ A is a symmetric, positive Hilbert-Schmidt operator}\}$$

so the condition on F amounts to saying that Re F is $I(\tau)$-continuous at 0.

(3) Continuity of Re F at 0 implies continuity of F at 0 in this situation.

C. Brownian Motion

It may be interesting to look at Brownian motion in the context of local bases. Let

$$T = (0;\infty)$$

$$J = \{j:\ \phi \neq j \subset T \text{ and } j \text{ is finite}\}$$

$$\pi_{ij}: y \in R^j \longrightarrow y|_i \in R^i \qquad \text{for } i \subset j$$

$$\pi_i: y \in R^T \longrightarrow y|_i \in R^i$$

so J is directed by inclusion, $(R^j, \pi_{ij})_{i,j \in J}$ is an inverse system of spaces and, for any $X \in R^T$ with $\pi_i(X) = R^i$ for every $i \in J$, $(X, \pi_i)_{i \in J}$ is a limit space of the system. A stochastic process on R is then an inverse system of probability measures $(\mu_j)_{j \in J}$. Such a system always has a limit measure μ^* on R^T (by Kolmogorov's theorem) and much of the theory is concerned with finding $\mu^*(X)$ for given $X \in R^T$, mainly with showing that $\mu^*(X) = 1$ or $\mu^*(X) = 0$.

The description of Brownian motion as a Gaussian process with independent increments leads very naturally to consideration of the following system of local bases. For each $j \in J$, let n_j be the number of elements in j and let $j = \{t_1, \cdots, t_{n_j}\}$ with $t_1 < \cdots < t_{n_j}$. Then consider the isomorphism

$$g_j: y \in R^j \longrightarrow (y(t_1),\ y(t_2) - y(t_1),\ \cdots,\ y(t_{n_j}) - y(t_{n_j - 1})) \in R^{n_j}$$

and set

$$p_{ij} = g_i \circ \pi_{ij} \circ g_i^{-1} : \quad R^{n_j} \to R^{n_i} \qquad \text{for } i \subset j$$

$$p_i = g_i \circ \pi_i : \quad R^T \to R^{n_i}$$

so $(R^{n_j}, p_{ij})_{i,j \in J}$ is also an inverse system of spaces and $(X, p_i)_{i,j \in J}$ is a limit space of the system whenever $X \subset R^T$ and $\pi_i(X) = R^i$ for $i \in J$. For each $k = 1, \cdots, n_j$ let γ_k be Gaussian measure on R with mean 0 and variance $\sigma_k^2 = t_k - t_{k-1}$, i.e.

$$\gamma_k(\alpha) = \frac{1}{\sqrt{2\pi(t_k - t_{k-1})}} \int_\alpha \exp \frac{-u^2}{2(t_k - t_{k-1})}\, du$$

where $t_0 = 0$. Let ν_j be the product measure

$$\nu_j = \bigotimes_{k=1}^{n_j} \gamma_k$$

and μ_j be the image of ν_j onto R^j, i.e.

$$\mu_j(\alpha) = \nu_j(g_j(\alpha)) \qquad \text{for } \alpha \subset R^j.$$

One checks by straight forward computation that $(\nu_j)_{j\in J}$ is an inverse system. Clearly, the limit measures μ^* and ν^* on R^T are the same so the systems $(\mu_j)_{j\in J}$ and $(\nu_j)_{j\in J}$ are viewed as equivalent and one can shift from one to the other at will. In practice, most of the computations as well as much of the intuitive understanding of Brownian motion involve using the ν_j. For example, the variance σ_s^2 of the one-dimensional distribution of ν_j in any given direction $s \in S_j$ is:

$$\sigma_s^2 = \int |u\cdot s|^2 \, d\nu_j(u) = \sum_{k=1}^{n_j} (s_k^2 \int u_k^2 \, d\nu_j(u)) = \sum_{k=1}^{n_j} s_k^2 \sigma_k^2$$

so we can get a rough estimate of the numbers $\lambda(\varepsilon,s)$ for $\varepsilon > 0$ using Chebychev's inequality:

$$\lambda^2(\varepsilon,s) \le \frac{1}{\varepsilon} \sigma_s^2 = \frac{1}{\varepsilon} \sum_{k=1}^{n_j} s_k^2 \sigma_k^2.$$

Then, by the Cauchy-Schwarz inequality and lemma A1 in the appendix:

$$\bar{\lambda}^2(\varepsilon) \le \int \lambda^2(\varepsilon,s) d\xi_j(s) \le \frac{1}{\varepsilon\, n_j} \sum_{k=1}^{n_j} \sigma_k^2 = \frac{1}{\varepsilon\, n_j} t_{n_j}$$

so far any $y \in R^T$:

$$\|p_j(y)\|_j > \frac{\sqrt{n_j}}{\varepsilon} \bar{\lambda}(\varepsilon) \Rightarrow \sum_{k=1}^{n_j} |y(t_k) - y(t_{k-1})|^2 > \frac{1}{\varepsilon^3} t_{n_j}.$$

Hence

$$\mu_j(\{y \in R^j: \sum_{k=1}^{n_j} |y(t_k) - y(t_{k-1})|^2 > \frac{1}{\varepsilon^3} t_{n_j}\}) < 2C\varepsilon.$$

The isomorphisms g_j also play a very natural part in the discussion of more general processes such as martingales or quasi-martingales. For example, by simply requiring that the probability measure ν_j satisfy the conditions

$$\int u_k u_{k'} \, d\nu_j(u) = 0 \qquad \text{for } k \neq k'$$

and $\sup_j \int \|u\|_j^2 \, d\nu_j(u) < \infty$

one develops by well known arguments much of the theory of square integrable martingales.

References

1. Bochner, S. Monotone Funktionen, Stieltjessche Integrale und harmonische Analyse. Math Annalen 108 (1933) pp. 378-410.

2. Daniell, P.J. Integrals on an infinite number of dimensions. Ann. of Math. (2), 20 (1919), pp 281-288.

3. Gelfand, I. and Vilenkin, N. Generalized functions, Vol. 4, Springer, Berlin 1964.

4. Grothendieck, A. Sur une notion de produit tensoriel topologique d'espaces vectoriels topologiques et une classe remarquable d'espaces vectoriels lies a cette notion. C.R. Acad. Sci. Paris, 233 (1951), pp. 1556-1558.

5. ———, Produits tensoriels topologiques et espaces nucléaires. Memoir AMS, 16 (1955).

6. Hewitt, E. and Stromberg, K. Real and Abstract analysis. Springer-Verlag 1969.

7. Millington, H. and Sion, M. Inverse systems of group-valued measures. Pac. J. of Math, 44 (1973), pp. 637-650.

8. Ito, K. Foundations of stochastic differential equations in infinite dimensional spaces. SIAM Lecture Notes 1984.

9. Kolmogorov, A.N. Grundbegriff der Wahrscheinlichkeit. Berlin 1933, English translation, Chelsea, N.Y. 1956.

10. ———, A note on the papers of R.A. Minlos and V. Sazonov. Theory of Prob. and Appl., 4 (1959), pp. 221-223.

11. Mallory, D.J. and Sion, M. Limits of inverse systems of measures. Ann. Inst. Fourier, Univ. de Grenoble, 21 (1971), pp. 25-57.

12. Minlos, R.A. Continuation of a generalized random process to a completely additive measure. Doklady Akad. Nauk SSSR(N.S) 119 (1958) pp. 439-442. (MR 20 (1959) #5522).

13. ———, Generalized random processes and their extension in measure. Trudy Moskov. Mat. Obsc. 8 (1959) pp. 497-518. AMS Translation (S) 3 pp. 291-313 (MR 21 (1960) #7563).

14. Sazonov, V. On characteristic functionals. Theory of Prob. and Appl. 3 (1958), pp. 201-205. (MR 20 (1959) #4882)

15. Sion, M. Introduction to the methods of real analysis. Holt, Rinehart and Winston, New York 1968.

LIST OF C.I.M.E. SEMINARS

		Publisher
1954 -	1. Analisi funzionale	C.I.M.E.
	2. Quadratura delle superficie e questioni connesse	"
	3. Equazioni differenziali non lineari	"
1955 -	4. Teorema di Riemann-Roch e questioni connesse	"
	5. Teoria dei numeri	"
	6. Topologia	"
	7. Teorie non linearizzate in elasticità, idrodinamica, aerodinamica	"
	8. Geometria proiettivo-differenziale	"
1956 -	9. Equazioni alle derivate parziali a caratteristiche reali	"
	10. Propagazione delle onde elettromagnetiche	"
	11. Teoria della funzioni di più variabili complesse e delle funzioni automorfe	"
1957 -	12. Geometria aritmetica e algebrica (2 vol.)	"
	13. Integrali singolari e questioni connesse	"
	14. Teoria della turbolenza (2 vol.)	"
1958 -	15. Vedute e problemi attuali in relatività generale	"
	16. Problemi di geometria differenziale in grande	"
	17. Il principio di minimo e le sue applicazioni alle equazioni funzionali	"
1959 -	18. Induzione e statistica	"
	19. Teoria algebrica dei meccanismi automatici (2 vol.)	"
	20. Gruppi, anelli di Lie e teoria della coomologia	"
1960 -	21. Sistemi dinamici e teoremi ergodici	"
	22. Forme differenziali e loro integrali	"
1961 -	23. Geometria del calcolo delle variazioni (2 vol.)	"
	24. Teoria delle distribuzioni	"
	25. Onde superficiali	"
1962 -	26. Topologia differenziale	"
	27. Autovalori e autosoluzioni	"
	28. Magnetofluidodinamica	"
1963 -	29. Equazioni differenziali astratte	"
	30. Funzioni e varietà complesse	"
	31. Proprietà di media e teoremi di confronto in Fisica Matematica	"

1964 -	32. Relatività generale	C.I.M.E.
	33. Dinamica dei gas rarefatti	"
	34. Alcune questioni di analisi numerica	"
	35. Equazioni differenziali non lineari	"
1965 -	36. Non-linear continuum theories	"
	37. Some aspects of ring theory	"
	38. Mathematical optimization in economics	"
1966 -	39. Calculus of variations	Ed. Cremonese, Firenze
	40. Economia matematica	"
	41. Classi caratteristiche e questioni connesse	"
	42. Some aspects of diffusion theory	"
1967 -	43. Modern questions of celestial mechanics	"
	44. Numerical analysis of partial differential equations	"
	45. Geometry of homogeneous bounded domains	"
1968 -	46. Controllability and observability	"
	47. Pseudo-differential operators	"
	48. Aspects of mathematical logic	"
1969 -	49. Potential theory	"
	50. Non-linear continuum theories in mechanics and physics and their applications	"
	51. Questions of algebraic varieties	"
1970 -	52. Relativistic fluid dynamics	"
	53. Theory of group representations and Fourier analysis	"
	54. Functional equations and inequalities	"
	55. Problems in non-linear analysis	"
1971 -	56. Stereodynamics	"
	57. Constructive aspects of functional analysis (2 vol.)	"
	58. Categories and commutative algebra	"
1972 -	59. Non-linear mechanics	"
	60. Finite geometric structures and their applications	"
	61. Geometric measure theory and minimal surfaces	"
1973 -	62. Complex analysis	"
	63. New variational techniques in mathematical physics	"
	64. Spectral analysis	"

1974 -	65. Stability problems		Ed. Cremonese, Firenze
	66. Singularities of analytic spaces		"
	67. Eigenvalues of non linear problems		"
1975 -	68. Theoretical computer sciences		"
	69. Model theory and applications		"
	70. Differential operators and manifolds		"
1976 -	71. Statistical Mechanics		Ed. Liguori, Napoli
	72. Hyperbolicity		"
	73. Differential topology		"
1977 -	74. Materials with memory		"
	75. Pseudodifferential operators with applications		"
	76. Algebraic surfaces		"
1978 -	77. Stochastic differential equations		"
	78. Dynamical systems		Ed. Liguori, Napoli and Birkhäuser Verlag
1979 -	79. Recursion theory and computational complexity		Ed. Liguori, Napoli
	80. Mathematics of biology		"
1980 -	81. Wave propagation		"
	82. Harmonic analysis and group representations		"
	83. Matroid theory and its applications		"
1981 -	84. Kinetic Theories and the Boltzmann Equation	(LNM 1048)	Springer-Verlag
	85. Algebraic Threefolds	(LNM 947)	"
	86. Nonlinear Filtering and Stochastic Control	(LNM 972)	"
1982 -	87. Invariant Theory	(LNM 996)	"
	88. Thermodynamics and Constitutive Equations	(LN Physics 228)	"
	89. Fluid Dynamics	(LNM 1047)	"
1983 -	90. Complete Intersections	(LNM 1092)	"
	91. Bifurcation Theory and Applications	(LNM 1057)	"
	92. Numerical Methods in Fluid Dynamics	(LNM 1127)	"
1984	93. Harmonic Mappings and Minimal Immersions	(LNM 1161)	"
	94. Schrödinger Operators	(LNM 1159)	"
	95. Buildings and the Geometry of Diagrams	(LNM 1181)	"
1985 -	96. Probability and Analysis	(LNM 1206)	"
	97. Some Problems in Nonlinear Diffusion	to appear	"
	98. Theory of Moduli	to appear	"

Note: Volumes 1 to 38 are out of print. A few copies of volumes 23,28,31,32,33,34,36,38 are available on request from C.I.M.E.

Vol. 1062: J. Jost, Harmonic Maps Between Surfaces. X, 133 pages. 1984.

Vol. 1063: Orienting Polymers. Proceedings, 1983. Edited by J.L. Ericksen. VII, 166 pages. 1984.

Vol. 1064: Probability Measures on Groups VII. Proceedings, 1983. Edited by H. Heyer. X, 588 pages. 1984.

Vol. 1065: A. Cuyt, Padé Approximants for Operators: Theory and Applications. IX, 138 pages. 1984.

Vol. 1066: Numerical Analysis. Proceedings, 1983. Edited by D.F. Griffiths. XI, 275 pages. 1984.

Vol. 1067: Yasuo Okuyama, Absolute Summability of Fourier Series and Orthogonal Series. VI, 118 pages. 1984.

Vol. 1068: Number Theory, Noordwijkerhout 1983. Proceedings. Edited by H. Jager. V, 296 pages. 1984.

Vol. 1069: M. Kreck, Bordism of Diffeomorphisms and Related Topics. III, 144 pages. 1984.

Vol. 1070: Interpolation Spaces and Allied Topics in Analysis. Proceedings, 1983. Edited by M. Cwikel and J. Peetre. III, 239 pages. 1984.

Vol. 1071: Padé Approximation and its Applications, Bad Honnef 1983. Prodeedings. Edited by H. Werner and H.J. Bünger. VI, 264 pages. 1984.

Vol. 1072: F. Rothe, Global Solutions of Reaction-Diffusion Systems. V, 216 pages. 1984.

Vol. 1073: Graph Theory, Singapore 1983. Proceedings. Edited by K.M. Koh and H.P. Yap. XIII, 335 pages. 1984.

Vol. 1074: E.W. Stredulinsky, Weighted Inequalities and Degenerate Elliptic Partial Differential Equations. III, 143 pages. 1984.

Vol. 1075: H. Majima, Asymptotic Analysis for Integrable Connections with Irregular Singular Points. IX, 159 pages. 1984.

Vol. 1076: Infinite-Dimensional Systems. Proceedings, 1983. Edited by F. Kappel and W. Schappacher. VII, 278 pages. 1984.

Vol. 1077: Lie Group Representations III. Proceedings, 1982–1983. Edited by R. Herb, R. Johnson, R. Lipsman, J. Rosenberg. XI, 454 pages. 1984.

Vol. 1078: A.J.E.M. Janssen, P. van der Steen, Integration Theory. V, 224 pages. 1984.

Vol. 1079: W. Ruppert. Compact Semitopological Semigroups: An Intrinsic Theory. V, 260 pages. 1984

Vol. 1080: Probability Theory on Vector Spaces III. Proceedings, 1983. Edited by D. Szynal and A. Weron. V, 373 pages. 1984.

Vol. 1081: D. Benson, Modular Representation Theory: New Trends and Methods. XI, 231 pages. 1984.

Vol. 1082: C.-G. Schmidt, Arithmetik Abelscher Varietäten mit komplexer Multiplikation. X, 96 Seiten. 1984.

Vol. 1083: D. Bump, Automorphic Forms on GL (3,IR). XI, 184 pages. 1984.

Vol. 1084: D. Kletzing, Structure and Representations of Q-Groups. VI, 290 pages. 1984.

Vol. 1085: G.K. Immink, Asymptotics of Analytic Difference Equations. V, 134 pages. 1984.

Vol. 1086: Sensitivity of Functionals with Applications to Engineering Sciences. Proceedings, 1983. Edited by V. Komkov. V, 130 pages. 1984

Vol. 1087: W. Narkiewicz, Uniform Distribution of Sequences of Integers in Residue Classes. VIII, 125 pages. 1984.

Vol. 1088: A.V. Kakosyan, L.B. Klebanov, J.A. Melamed, Characterization of Distributions by the Method of Intensively Monotone Operators. X, 175 pages. 1984.

Vol. 1089: Measure Theory, Oberwolfach 1983. Proceedings. Edited by D. Kölzow and D. Maharam-Stone. XIII, 327 pages. 1984.

Vol. 1090: Differential Geometry of Submanifolds. Proceedings, 1984. Edited by K. Kenmotsu. VI, 132 pages. 1984.

Vol. 1091: Multifunctions and Integrands. Proceedings, 1983. Edited by G. Salinetti. V, 234 pages. 1984.

Vol. 1092: Complete Intersections. Seminar, 1983. Edited by S. Greco and R. Strano. VII, 299 pages. 1984.

Vol. 1093: A. Prestel, Lectures on Formally Real Fields. XI, 125 pages. 1984.

Vol. 1094: Analyse Complexe. Proceedings, 1983. Edité par E. Amar, R. Gay et Nguyen Thanh Van. IX, 184 pages. 1984.

Vol. 1095: Stochastic Analysis and Applications. Proceedings, 1983. Edited by A. Truman and D. Williams. V, 199 pages. 1984.

Vol. 1096: Théorie du Potentiel. Proceedings, 1983. Edité par G. Mokobodzki et D. Pinchon. IX, 601 pages. 1984.

Vol. 1097: R.M. Dudley, H. Kunita, F. Ledrappier, École d'Éte de Probabilités de Saint-Flour XII – 1982. Edité par P.L. Hennequin. X, 396 pages. 1984.

Vol. 1098: Groups – Korea 1983. Proceedings. Edited by A.C. Kim and B.H. Neumann. VII, 183 pages. 1984.

Vol. 1099: C.M. Ringel, Tame Algebras and Integral Quadratic Forms. XIII, 376 pages. 1984.

Vol. 1100: V. Ivrii, Precise Spectral Asymptotics for Elliptic Operators Acting in Fiberings over Manifolds with Boundary. V, 237 pages. 1984.

Vol. 1101: V. Cossart, J. Giraud, U. Orbanz, Resolution of Surface Singularities. Seminar. VII, 132 pages. 1984.

Vol. 1102: A. Verona, Stratified Mappings – Structure and Triangulability. IX, 160 pages. 1984.

Vol. 1103: Models and Sets. Proceedings, Logic Colloquium, 1983, Part I. Edited by G.H. Müller and M.M. Richter. VIII, 484 pages. 1984.

Vol. 1104: Computation and Proof Theory. Proceedings, Logic Colloquium, 1983, Part II. Edited by M.M. Richter, E. Börger, W. Oberschelp, B. Schinzel and W. Thomas. VIII, 475 pages. 1984.

Vol. 1105: Rational Approximation and Interpolation. Proceedings, 1983. Edited by P.R. Graves-Morris, E.B. Saff and R.S. Varga. XII, 528 pages. 1984.

Vol. 1106: C.T. Chong, Techniques of Admissible Recursion Theory. IX, 214 pages. 1984.

Vol. 1107: Nonlinear Analysis and Optimization. Proceedings, 1982. Edited by C. Vinti. V, 224 pages. 1984.

Vol. 1108: Global Analysis – Studies and Applications I. Edited by Yu.G. Borisovich and Yu.E. Gliklikh. V, 301 pages. 1984.

Vol. 1109: Stochastic Aspects of Classical and Quantum Systems. Proceedings, 1983. Edited by S. Albeverio, P. Combe and M. Sirugue-Collin. IX, 227 pages. 1985.

Vol. 1110: R. Jajte, Strong Limit Theorems in Non-Commutative Probability. VI, 152 pages. 1985.

Vol. 1111: Arbeitstagung Bonn 1984. Proceedings. Edited by F. Hirzebruch, J. Schwermer and S. Suter. V, 481 pages. 1985.

Vol. 1112: Products of Conjugacy Classes in Groups. Edited by Z. Arad and M. Herzog. V, 244 pages. 1985.

Vol. 1113: P. Antosik, C. Swartz, Matrix Methods in Analysis. IV, 114 pages. 1985.

Vol. 1114: Zahlentheoretische Analysis. Seminar. Herausgegeben von E. Hlawka. V, 157 Seiten. 1985.

Vol. 1115: J. Moulin Ollagnier, Ergodic Theory and Statistical Mechanics. VI, 147 pages. 1985.

Vol. 1116: S. Stolz, Hochzusammenhängende Mannigfaltigkeiten und ihre Ränder. XXIII, 134 Seiten. 1985.

Vol. 1117: D.J. Aldous, J.A. Ibragimov, J. Jacod, Ecole d'Été de Probabilités de Saint-Flour XIII – 1983. Édité par P.L. Hennequin. IX, 409 pages. 1985.

Vol. 1118: Grossissements de filtrations: exemples et applications. Seminaire, 1982/83. Edité par Th. Jeulin et M. Yor. V, 315 pages. 1985.

Vol. 1119: Recent Mathematical Methods in Dynamic Programming. Proceedings, 1984. Edited by I. Capuzzo Dolcetta, W.H. Fleming and T. Zolezzi. VI, 202 pages. 1985.

Vol. 1120: K. Jarosz, Perturbations of Banach Algebras. V, 118 pages. 1985.

Vol. 1121: Singularities and Constructive Methods for Their Treatment. Proceedings, 1983. Edited by P. Grisvard, W. Wendland and J.R. Whiteman. IX, 346 pages. 1985.

Vol. 1122: Number Theory. Proceedings, 1984. Edited by K. Alladi. VII, 217 pages. 1985.

Vol. 1123: Séminaire de Probabilités XIX 1983/84. Proceedings. Edité par J. Azéma et M. Yor. IV, 504 pages. 1985.

Vol. 1124: Algebraic Geometry, Sitges (Barcelona) 1983. Proceedings. Edited by E. Casas-Alvero, G.E. Welters and S. Xambó-Descamps. XI, 416 pages. 1985.

Vol. 1125: Dynamical Systems and Bifurcations. Proceedings, 1984. Edited by B.L.J. Braaksma, H.W. Broer and F. Takens. V, 129 pages. 1985.

Vol. 1126: Algebraic and Geometric Topology. Proceedings, 1983. Edited by A. Ranicki, N. Levitt and F. Quinn. V, 523 pages. 1985.

Vol. 1127: Numerical Methods in Fluid Dynamics. Seminar. Edited by F. Brezzi, VII, 333 pages. 1985.

Vol. 1128: J. Elschner, Singular Ordinary Differential Operators and Pseudodifferential Equations. 200 pages. 1985.

Vol. 1129: Numerical Analysis, Lancaster 1984. Proceedings. Edited by P.R. Turner. XIV, 179 pages. 1985.

Vol. 1130: Methods in Mathematical Logic. Proceedings, 1983. Edited by C.A. Di Prisco. VII, 407 pages. 1985.

Vol. 1131: K. Sundaresan, S. Swaminathan, Geometry and Nonlinear Analysis in Banach Spaces. III, 116 pages. 1985.

Vol. 1132: Operator Algebras and their Connections with Topology and Ergodic Theory. Proceedings, 1983. Edited by H. Araki, C.C. Moore, Ş. Strătilă and C. Voiculescu. VI, 594 pages. 1985.

Vol. 1133: K.C. Kiwiel, Methods of Descent for Nondifferentiable Optimization. VI, 362 pages. 1985.

Vol. 1134: G.P. Galdi, S. Rionero, Weighted Energy Methods in Fluid Dynamics and Elasticity. VII, 126 pages. 1985.

Vol. 1135: Number Theory, New York 1983–84. Seminar. Edited by D.V. Chudnovsky, G.V. Chudnovsky, H. Cohn and M.B. Nathanson. V, 283 pages. 1985.

Vol. 1136: Quantum Probability and Applications II. Proceedings, 1984. Edited by L. Accardi and W. von Waldenfels. VI, 534 pages. 1985.

Vol. 1137: Xiao G., Surfaces fibrées en courbes de genre deux. IX, 103 pages. 1985.

Vol. 1138: A. Ocneanu, Actions of Discrete Amenable Groups on von Neumann Algebras. V, 115 pages. 1985.

Vol. 1139: Differential Geometric Methods in Mathematical Physics. Proceedings, 1983. Edited by H. D. Doebner and J. D. Hennig. VI, 337 pages. 1985.

Vol. 1140: S. Donkin, Rational Representations of Algebraic Groups. VII, 254 pages. 1985.

Vol. 1141: Recursion Theory Week. Proceedings, 1984. Edited by H.-D. Ebbinghaus, G.H. Müller and G.E. Sacks. IX, 418 pages. 1985.

Vol. 1142: Orders and their Applications. Proceedings, 1984. Edited by I. Reiner and K. W. Roggenkamp. X, 306 pages. 1985.

Vol. 1143: A. Krieg, Modular Forms on Half-Spaces of Quaternions. XIII, 203 pages. 1985.

Vol. 1144: Knot Theory and Manifolds. Proceedings, 1983. Edited by D. Rolfsen. V, 163 pages. 1985.

Vol. 1145: G. Winkler, Choquet Order and Simplices. VI, 143 pages. 1985.

Vol. 1146: Séminaire d'Algèbre Paul Dubreil et Marie-Paule Malliavin. Proceedings, 1983–1984. Edité par M.-P. Malliavin. IV, 420 pages. 1985.

Vol. 1147: M. Wschebor, Surfaces Aléatoires. VII, 111 pages. 1985.

Vol. 1148: Mark A. Kon, Probability Distributions in Quantum Statistical Mechanics. V, 121 pages. 1985.

Vol. 1149: Universal Algebra and Lattice Theory. Proceedings, 1984. Edited by S. D. Comer. VI, 282 pages. 1985.

Vol. 1150: B. Kawohl, Rearrangements and Convexity of Level Sets in PDE. V, 136 pages. 1985.

Vol 1151: Ordinary and Partial Differential Equations. Proceedings, 1984. Edited by B.D. Sleeman and R.J. Jarvis. XIV, 357 pages. 1985.

Vol. 1152: H. Widom, Asymptotic Expansions for Pseudodifferential Operators on Bounded Domains. V, 150 pages. 1985.

Vol. 1153: Probability in Banach Spaces V. Proceedings, 1984. Edited by A. Beck, R. Dudley, M. Hahn, J. Kuelbs and M. Marcus. VI, 457 pages. 1985.

Vol. 1154: D.S. Naidu, A.K. Rao, Singular Pertubation Analysis of Discrete Control Systems. IX, 195 pages. 1985.

Vol. 1155: Stability Problems for Stochastic Models. Proceedings, 1984. Edited by V.V. Kalashnikov and V.M. Zolotarev. VI, 447 pages. 1985.

Vol. 1156: Global Differential Geometry and Global Analysis 1984. Proceedings, 1984. Edited by D. Ferus, R.B. Gardner, S. Helgason and U. Simon. V, 339 pages. 1985.

Vol. 1157: H. Levine, Classifying Immersions into $\mathbb{R}^4$ over Stable Maps of 3-Manifolds into $\mathbb{R}^2$. V, 163 pages. 1985.

Vol. 1158: Stochastic Processes – Mathematics and Physics. Proceedings, 1984. Edited by S. Albeverio, Ph. Blanchard and L. Streit. VI, 230 pages. 1986.

Vol. 1159: Schrödinger Operators, Como 1984. Seminar. Edited by S. Graffi. VIII, 272 pages. 1986.

Vol. 1160: J.-C. van der Meer, The Hamiltonian Hopf Bifurcation. VI, 115 pages. 1985.

Vol. 1161: Harmonic Mappings and Minimal Immersions, Montecatini 1984. Seminar. Edited by E. Giusti. VII, 285 pages. 1985.

Vol. 1162: S.J.L. van Eijndhoven, J. de Graaf, Trajectory Spaces, Generalized Functions and Unbounded Operators. IV, 272 pages. 1985.

Vol. 1163: Iteration Theory and its Functional Equations. Proceedings, 1984. Edited by R. Liedl, L. Reich and Gy. Targonski. VIII, 231 pages. 1985.

Vol. 1164: M. Meschiari, J.H. Rawnsley, S. Salamon, Geometry Seminar "Luigi Bianchi" II – 1984. Edited by E. Vesentini. VI, 224 pages. 1985.

Vol. 1165: Seminar on Deformations. Proceedings, 1982/84. Edited by J. Ławrynowicz. IX, 331 pages. 1985.

Vol. 1166: Banach Spaces. Proceedings, 1984. Edited by N. Kalton and E. Saab. VI, 199 pages. 1985.

Vol. 1167: Geometry and Topology. Proceedings, 1985. Edited by J. Alexander and J. Harer. VI, 292 pages. 1985.

Vol. 1168: S.S. Agaian, Hadamard Matrices and their Applications. III, 227 pages. 1985.

Vol. 1169: W.A. Light, E.W. Cheney, Approximation Theory in Tensor Product Spaces. VII, 157 pages. 1985.

Vol. 1170: B.S. Thomson, Real Functions. VII, 229 pages. 1985.

Vol. 1171: Polynômes Orthogonaux et Applications. Proceedings, 1984. Edité par C. Brezinski, A. Draux, A.P. Magnus, P. Maroni et A. Ronveaux. XXXVII, 584 pages. 1985.

Vol. 1172: Algebraic Topology, Göttingen 1984. Proceedings. Edited by L. Smith. VI, 209 pages. 1985.

Vol. 1173: H. Delfs, M. Knebusch, Locally Semialgebraic Spaces. XVI, 329 pages. 1985.

Vol. 1174: Categories in Continuum Physics, Buffalo 1982. Seminar. Edited by F.W. Lawvere and S.H. Schanuel. V, 126 pages. 1986.

Vol. 1175: K. Mathiak, Valuations of Skew Fields and Projective Hjelmslev Spaces. VII, 116pages. 1986.

Vol. 1176: R.R. Bruner, J.P. May, J.E. McClure, M. Steinberger, H_∞ Ring Spectra and their Applications. VII, 388 pages. 1986.

Vol. 1177: Representation Theory I. Finite Dimensional Algebras. Proceedings, 1984. Edited by V. Dlab, P. Gabriel and G. Michler. XV, 340 pages. 1986.

Vol. 1178: Representation Theory II. Groups and Orders. Proceedings, 1984. Edited by V. Dlab, P. Gabriel and G. Michler. XV, 370 pages. 1986.

Vol. 1179: Shi J.-Y. The Kazhdan-Lusztig Cells in Certain Affine Weyl Groups. X, 307 pages. 1986.

Vol. 1180: R. Carmona, H. Kesten, J.B. Walsh, École d'Été de Probabilités de Saint-Flour XIV – 1984. Édité par P.L. Hennequin. X, 438 pages. 1986.

Vol. 1181: Buildings and the Geometry of Diagrams, Como 1984. Seminar. Edited by L. Rosati. VII, 277 pages. 1986.

Vol. 1182: S. Shelah, Around Classification Theory of Models. VII, 279 pages. 1986.

Vol. 1183: Algebra, Algebraic Topology and their Interactions. Proceedings, 1983. Edited by J.-E. Roos. XI, 396 pages. 1986.

Vol. 1184: W. Arendt, A. Grabosch, G. Greiner, U. Groh, H.P. Lotz, U. Moustakas, R. Nagel, F. Neubrander, U. Schlotterbeck, One-parameter Semigroups of Positive Operators. Edited by R. Nagel. X, 460 pages. 1986.

Vol. 1185: Group Theory, Beijing 1984. Proceedings. Edited by Tuan H.F. V, 403 pages. 1986.

Vol. 1186: Lyapunov Exponents, Proceedings, 1984. Edited by L. Arnold and V. Wihstutz. VI, 374 pages. 1986.

Vol. 1187: Y. Diers, Categories of Boolean Sheaves of Simple Algebras. VI, 168 pages. 1986.

Vol. 1188: Fonctions de Plusieurs Variables Complexes V. Séminaire, 1979–85. Edité par François Norguet. VI, 306 pages. 1986.

Vol. 1189: J. Lukeš, J. Malý, L. Zajíček, Fine Topology Methods in Real Analysis and Potential Theory. X, 472 pages. 1986.

Vol. 1190: Optimization and Related Fields. Proceedings, 1984. Edited by R. Conti, E. De Giorgi and F. Giannessi. VIII, 419 pages. 1986.

Vol. 1191: A.R. Its, V.Yu. Novokshenov, The Isomonodromic Deformation Method in the Theory of Painlevé Equations. IV, 313 pages. 1986.

Vol. 1192: Equadiff 6, Proceedings, Brno 1985. Edited by J. Vosmansky and M. Zlámal. XXIII, 404 pages. 1986.

Vol. 1193: Geometrical and Statistical Aspects of Probability in Banach Spaces, Proceedings, 1985. Edited by X. Fernique, B. Heinkel, M.B. Marcus and P.A. Meyer. IV, 128 pages. 1986.

Vol. 1194: Complex Analysis and Algebraic Geometry, Proceedings, 1985. Edited by H. Grauert. VI, 235 pages. 1986.

Vol.1195: J.M. Barbosa, A.G. Colares, Minimal Surfaces in $\mathbb{R}^3$. X, 124 pages. 1986.

Vol. 1196: E. Casas-Alvero, S. Xambó-Descamps, The Enumerative Theory of Conics after Halphen. IX, 130 pages. 1986.

Vol. 1197: Ring Theory. Proceedings, 1985. Edited by F.M.J. van Oystaeyen. V, 231 pages. 1986.

Vol. 1198: Séminaire d'Analyse, P. Lelong – P. Dolbeault – H. Skoda, Seminar 1983/84. X, 260 pages. 1986.

Vol. 1199: Analytic Theory of Continued Fractions II. Proceedings, 1985. Edited by W.J. Thron. VI, 299 pages. 1986.

Vol. 1200: V.D. Milman, G. Schechtman, Asymptotic Theory of Finite Dimensional Normed Spaces. With an Appendix by M. Gromov. VIII, 156 pages. 1986.

Vol. 1201: Curvature and Topology of Riemannian Manifolds. Proceedings 1985. Edited by K. Shiohama, T. Sakai and T. Sunada. VII, 336 pages. 1986.

Vol. 1202: A. Dür, Möbius Functions, Incidence Algebras and Power Series Representations. XI, 134 pages. 1986.

Vol. 1203: Stochastic Processes and Their Applications, Proceedings, 1985. Edited by K. Itô and T. Hida. VI, 222 pages. 1986.

Vol. 1204: Séminaire de Probabilités XX, 1984/85, Proceedings. Edité par J. Azéma et M. Yor. V, 639 pages. 1986.

Vol. 1205: B.Z. Moroz, Analytic Arithmetic in Algebraic Number Fields. VII, 177 pages. 1986.

Vol. 1206: Probability and Analysis. Varenna (Como) 1985. Seminar. Edited by G. Letta and M. Pratelli. VIII, 280 pages. 1986.